Design of Comparative Experiments

This book should be on the shelf of every practising statistician who designs experiments.

Good design considers units and treatments first, and *then* allocates treatments to units. It does not choose from a menu of named designs. This approach requires a notation for units that does not depend on the treatments applied. Most structure on the set of observational units, or on the set of treatments, can be defined by factors. This book develops a coherent framework for thinking about factors and their relationships, including the use of Hasse diagrams. These are used to elucidate structure, calculate degrees of freedom and allocate treatment subspaces to appropriate strata. Based on a one-term course the author has taught since 1989, the book is ideal for advanced undergraduate and beginning graduate courses. Examples, exercises and discussion questions are drawn from a wide range of real applications: from drug development, to agriculture, to manufacturing.

R. A. BAILEY has been Professor of Statistics at Queen Mary, University of London since 1994. She is a fellow of the Institute of Mathematical Statistics and a past president of the International Biometric Society, British Region. This book reflects her extensive experience teaching design of experiments and advising on its application. Her book *Association Schemes* was published by Cambridge University Press in 2004.

CAMBRIDGE SERIES IN STATISTICAL AND PROBABILISTIC MATHEMATICS

Editorial Board

R. Gill (Department of Mathematics, Utrecht University)
B. D. Ripley (Department of Statistics, University of Oxford)
S. Ross (Department of Industrial and Systems Engineering, University of Southern California)
B. W. Silverman (St. Peter's College, Oxford)
M. Stein (Department of Statistics, University of Chicago)

This series of high-quality upper-division textbooks and expository monographs covers all aspects of stochastic applicable mathematics. The topics range from pure and applied statistics to probability theory, operations research, optimization, and mathematical programming. The books contain clear presentations of new developments in the field and also of the state of the art in classical methods. While emphasizing rigorous treatment of theoretical methods, the books also contain applications and discussions of new techniques made possible by advances in computational practice.

Already published

1. *Bootstrap Methods and Their Application*, by A. C. Davison and D. V. Hinkley
2. *Markov Chains*, by J. Norris
3. *Asymptotic Statistics*, by A. W. van der Vaart
4. *Wavelet Methods for Time Series Analysis*, by Donald B. Percival and Andrew T. Walden
5. *Bayesian Methods*, by Thomas Leonard and John S. J. Hsu
6. *Empirical Processes in M-Estimation*, by Sara van de Geer
7. *Numerical Methods of Statistics*, by John F. Monahan
8. *A User's Guide to Measure Theoretic Probability*, by David Pollard
9. *The Estimation and Tracking of Frequency*, by B. G. Quinn and E. J. Hannan
10. *Data Analysis and Graphics using R*, by John Maindonald and John Braun
11. *Statistical Models*, by A. C. Davison
12. *Semiparametric Regression*, by D. Ruppert, M. P. Wand, R. J. Carroll
13. *Exercises in Probability*, by Loic Chaumont and Marc Yor
14. *Statistical Analysis of Stochastic Processes in Time*, by J. K. Lindsey
15. *Measure Theory and Filtering*, by Lakhdar Aggoun and Robert Elliott
16. *Essentials of Statistical Inference*, by G. A. Young and R. L. Smith
17. *Elements of Distribution Theory*, by Thomas A. Severini
18. *Statistical Mechanics of Disordered Systems*, by Anton Bovier
20. *Random Graph Dynamics*, by Rick Durrett
21. *Networks*, by Peter Whittle
22. *Saddlepoint Approximations with Applications*, by Ronald W. Butler
23. *Applied Asymptotics*, by A. R. Brazzale, A. C. Davison and N. Reid
24. *Random Networks for Communication*, by Massimo Franceschetti and Ronald Meester

Design of Comparative Experiments

R. A. Bailey
Queen Mary, University of London

CAMBRIDGE UNIVERSITY PRESS
Cambridge, New York, Melbourne, Madrid, Cape Town, Singapore,
São Paulo, Delhi, Dubai, Tokyo, Mexico City

Cambridge University Press
The Edinburgh Building, Cambridge CB2 8RU, UK

Published in the United States of America by Cambridge University Press, New York

www.cambridge.org
Information on this title: www.cambridge.org/9780521865067

© R. A. Bailey 2008

This publication is in copyright. Subject to statutory exception
and to the provisions of relevant collective licensing agreements,
no reproduction of any part may take place without the written
permission of Cambridge University Press.

First published 2008

A catalogue record for this publication is available from the British Library

ISBN 978-0-521-86506-7 Hardback
ISBN 978-0-521-68357-9 Paperback

Cambridge University Press has no responsibility for the persistence or
accuracy of URLs for external or third-party internet websites referred to in
this publication, and does not guarantee that any content on such websites is,
or will remain, accurate or appropriate. Information regarding prices, travel
timetables, and other factual information given in this work is correct at
the time of first printing but Cambridge University Press does not guarantee
the accuracy of such information thereafter.

Contents

Preface			*page* **xi**
1	**Forward look**		**1**
1.1	Stages in a statistically designed experiment		1
	1.1.1	Consultation	1
	1.1.2	Statistical design	2
	1.1.3	Data collection	2
	1.1.4	Data scrutiny	3
	1.1.5	Analysis	4
	1.1.6	Interpretation	5
1.2	The ideal and the reality		5
	1.2.1	Purpose of the experiment	5
	1.2.2	Replication	5
	1.2.3	Local control	6
	1.2.4	Constraints	6
	1.2.5	Choice	7
1.3	An example		7
1.4	Defining terms		8
1.5	Linear model		14
1.6	Summary		15
	Questions for discussion		16
2	**Unstructured experiments**		**19**
2.1	Completely randomized designs		19
2.2	Why and how to randomize		20
2.3	The treatment subspace		21
2.4	Orthogonal projection		23
2.5	Linear model		24
2.6	Estimation		24
2.7	Comparison with matrix notation		26
2.8	Sums of squares		26
2.9	Variance		28
2.10	Replication: equal or unequal?		30

v

	2.11	Allowing for the overall mean	30
	2.12	Hypothesis testing	33
	2.13	Sufficient replication for power	35
	2.14	A more general model	38
		Questions for discussion	41
3	**Simple treatment structure**		**43**
	3.1	Replication of control treatments	43
	3.2	Comparing new treatments in the presence of a control	44
	3.3	Other treatment groupings	47
		Questions for discussion	52
4	**Blocking**		**53**
	4.1	Types of block	53
		4.1.1 Natural discrete divisions	53
		4.1.2 Continuous gradients	55
		4.1.3 Choice of blocking for trial management	55
		4.1.4 How and when to block	56
	4.2	Orthogonal block designs	57
	4.3	Construction and randomization	59
	4.4	Models for block designs	59
	4.5	Analysis when blocks have fixed effects	61
	4.6	Analysis when blocks have random effects	67
	4.7	Why use blocks?	68
	4.8	Loss of power with blocking	69
		Questions for discussion	71
5	**Factorial treatment structure**		**75**
	5.1	Treatment factors and their subspaces	75
	5.2	Interaction	77
	5.3	Principles of expectation models	84
	5.4	Decomposing the treatment subspace	87
	5.5	Analysis	90
	5.6	Three treatment factors	92
	5.7	Factorial experiments	97
	5.8	Construction and randomization of factorial designs	98
	5.9	Factorial treatments plus control	99
		Questions for discussion	99
6	**Row–column designs**		**105**
	6.1	Double blocking	105
	6.2	Latin squares	106
	6.3	Construction and randomization	108
	6.4	Orthogonal subspaces	110
	6.5	Fixed row and column effects: model and analysis	110

Contents   vii

    6.6   Random row and column effects: model and analysis   112
    Questions for discussion   116

7 Experiments on people and animals   117
    7.1   Introduction   117
    7.2   Historical controls   118
    7.3   Cross-over trials   118
    7.4   Matched pairs, matched threes, and so on   119
    7.5   Completely randomized designs   120
    7.6   Body parts as experimental units   120
    7.7   Sequential allocation to an unknown number of patients   121
    7.8   Safeguards against bias   122
    7.9   Ethical issues   124
    7.10 Analysis by intention to treat   126
    Questions for discussion   127

8 Small units inside large units   131
    8.1   Experimental units bigger than observational units   131
        8.1.1   The context   131
        8.1.2   Construction and randomization   132
        8.1.3   Model and strata   132
        8.1.4   Analysis   132
        8.1.5   Hypothesis testing   135
        8.1.6   Decreasing variance   137
    8.2   Treatment factors in different strata   138
    8.3   Split-plot designs   146
        8.3.1   Blocking the large units   146
        8.3.2   Construction and randomization   147
        8.3.3   Model and strata   148
        8.3.4   Analysis   149
        8.3.5   Evaluation   152
    8.4   The split-plot principle   152
    Questions for discussion   154

9 More about Latin squares   157
    9.1   Uses of Latin squares   157
        9.1.1   One treatment factor in a square   157
        9.1.2   More general row–column designs   158
        9.1.3   Two treatment factors in a block design   159
        9.1.4   Three treatment factors in an unblocked design   161
    9.2   Graeco-Latin squares   162
    9.3   Uses of Graeco-Latin squares   166
        9.3.1   Superimposed design in a square   166
        9.3.2   Two treatment factors in a square   166
        9.3.3   Three treatment factors in a block design   166

	9.3.4	Four treatment factors in an unblocked design	167
	Questions for discussion		167

10 The calculus of factors — 169

- 10.1 Introduction — 169
- 10.2 Relations on factors — 169
  - 10.2.1 Factors and their classes — 169
  - 10.2.2 Aliasing — 170
  - 10.2.3 One factor finer than another — 171
  - 10.2.4 Two special factors — 171
- 10.3 Operations on factors — 171
  - 10.3.1 The infimum of two factors — 171
  - 10.3.2 The supremum of two factors — 172
  - 10.3.3 Uniform factors — 175
- 10.4 Hasse diagrams — 175
- 10.5 Subspaces defined by factors — 178
  - 10.5.1 One subspace per factor — 178
  - 10.5.2 Fitted values and crude sums of squares — 178
  - 10.5.3 Relations between subspaces — 178
- 10.6 Orthogonal factors — 178
  - 10.6.1 Definition of orthogonality — 178
  - 10.6.2 Projection matrices commute — 179
  - 10.6.3 Proportional meeting — 180
  - 10.6.4 How replication can affect orthogonality — 181
  - 10.6.5 A chain of factors — 181
- 10.7 Orthogonal decomposition — 182
  - 10.7.1 A second subspace for each factor — 182
  - 10.7.2 Effects and sums of squares — 184
- 10.8 Calculations on the Hasse diagram — 185
  - 10.8.1 Degrees of freedom — 185
  - 10.8.2 Sums of squares — 187
- 10.9 Orthogonal treatment structures — 189
  - 10.9.1 Conditions on treatment factors — 189
  - 10.9.2 Collections of expectation models — 190
- 10.10 Orthogonal plot structures — 193
  - 10.10.1 Conditions on plot factors — 193
  - 10.10.2 Variance and covariance — 194
  - 10.10.3 Matrix formulation — 195
  - 10.10.4 Strata — 196
- 10.11 Randomization — 196
- 10.12 Orthogonal designs — 197
  - 10.12.1 Desirable properties — 197
  - 10.12.2 General definition — 198
  - 10.12.3 Locating treatment subspaces — 198
  - 10.12.4 Analysis of variance — 200

10.13	Further examples	202
	Questions for discussion	215

11 Incomplete-block designs — 219
11.1	Introduction	219
11.2	Balance	219
11.3	Lattice designs	221
11.4	Randomization	223
11.5	Analysis of balanced incomplete-block designs	226
11.6	Efficiency	229
11.7	Analysis of lattice designs	230
11.8	Optimality	233
11.9	Supplemented balance	234
11.10	Row–column designs with incomplete columns	235
	Questions for discussion	238

12 Factorial designs in incomplete blocks — 241
12.1	Confounding	241
12.2	Decomposing interactions	242
12.3	Constructing designs with specified confounding	245
12.4	Confounding more than one character	249
12.5	Pseudofactors for mixed numbers of levels	251
12.6	Analysis of single-replicate designs	253
12.7	Several replicates	257
	Questions for discussion	258

13 Fractional factorial designs — 259
13.1	Fractional replicates	259
13.2	Choice of defining contrasts	260
13.3	Weight	262
13.4	Resolution	265
13.5	Analysis of fractional replicates	266
	Questions for discussion	270

14 Backward look — 271
14.1	Randomization	271
	14.1.1 Random sampling	271
	14.1.2 Random permutations of the plots	272
	14.1.3 Random choice of plan	273
	14.1.4 Randomizing treatment labels	273
	14.1.5 Randomizing instances of each treatment	275
	14.1.6 Random allocation to position	275
	14.1.7 Restricted randomization	278
14.2	Factors such as time, sex, age and breed	279
14.3	Writing a protocol	282

		14.3.1	What is the purpose of the experiment?	282
		14.3.2	What are the treatments?	282
		14.3.3	Methods	283
		14.3.4	What are the experimental units?	283
		14.3.5	What are the observational units?	283
		14.3.6	What measurements are to be recorded?	283
		14.3.7	What is the design?	283
		14.3.8	Justification for the design	284
		14.3.9	Randomization used	284
		14.3.10	Plan	284
		14.3.11	Proposed statistical analysis	284
	14.4	The eight stages		285
	14.5	A story		286
	Questions for discussion			290

Exercises 291

Sources of examples, questions and exercises 313

Further reading 319

References 321

Index 327

Preface

This textbook on the design of experiments is intended for students in their final year of a BSc in Mathematics or Statistics in the British system or for an MSc for students with a different background. It is based on lectures that I have given in the University of London and elsewhere since 1989. I would like it to become the book on design which every working statistician has on his or her shelves.

I assume a basic background in statistics: estimation, variance, hypothesis testing, linear models. I also assume the necessary linear algebra on which these rest, including orthogonal projections and eigenspaces of symmetric matrices. However, people's exposure to these topics varies, as does the notation they use, so I summarize what is needed at various points in Chapter 2. Skim that chapter to see if you need to brush up your knowledge of the background.

My philosophy is that you should not choose an experimental design from a list of named designs. Rather, you should think about all aspects of the current experiment, and then decide how to put them together appropriately. Think about the observational units, and what structure they have before treatments are applied. Think about the number and nature of the treatments. Only then should you begin to think about the design in the sense of which treatment is allocated to which experimental unit.

To do this requires a notation for observational units that does not depend on the treatments applied. The cost is a little more notation; the gain is a lot more clarity. Writing Y_{24} for the response on the fourth unit with treatment 2 goes with a mindset that ignores randomization, that manages the experiment by treatment, and that does not see the need for blindness. I label observational units by lower-case Greek letters: thus $Y(\omega)$ is the response on observational unit ω and $T(\omega)$ is the treatment on that unit. This notation merely mimics good practice in data recording, which has a row for each observational unit: three of the columns will be the one which names the units, the one which shows the treatments applied, and the one showing the responses. In this book, randomization, blindness and management by plot structure are recurring themes.

Most structure on the set of observational units, or on the set of treatments, can be defined by factors. I have developed a method for thinking about factors and their relationships, including the use of Hasse diagrams, which covers all orthogonal designs. The method uses the *infimum* $F \wedge G$ of two factors (which almost everybody else, except Tjur [113], writes as $F.G$) and the dual concept, the *supremum* $F \vee G$, which almost nobody else (again apart from Tjur) sees the need for, until degrees of freedom mysteriously go wrong. Everyone that I have taught this method to has reacted enthusiastically and adopted it. However, you need to have some idea of simple structure before you can appreciate the generality of this approach, which is therefore delayed until Chapter 10.

The Hasse diagrams, and the insistence on naming observational units, are two features of this book that do not appear in most other books on the design of experiments. The third difference, which is relatively minor, is my notation for models in factorial designs. Expressions such as $\mu + \alpha_i + \beta_j + (\alpha\beta)_{ij}$ are compact ways of parametrizing several models at once, but they do encourage the fitting of inappropriate models (what Nelder has called *the neglect of marginality* in [83]). I take the view, explained in Chapter 5, that when we analyse data we first choose which model to fit and then estimate the parameters of that model; we do not need to know how to parametrize any of the models that we did not fit. Also in Chapter 5 I spell out three principles of modelling. The first two (Sum and Intersection) are often implicit, but their neglect can lead to contradictions. The third is Orthogonality: not everyone will agree with this (see Mead [77], for example), but I believe that we should aim for orthogonality between treatment factors wherever possible.

Another relatively minor difference in my approach is that my analysis-of-variance tables always include the grand mean. This is partly to make all the calculations easier, especially when using the Hasse diagram. A more important reason is to demonstrate that fitting a larger model after a smaller one (such as a complete two-way table after an additive model) is in principle no different from fitting treatment effects after removing the grand mean.

Unlike some topics in mathematics, Design of Experiments can set out its stall early. Thus Chapter 1 introduces most of the issues, and points the way forward to where in the book they are covered in more detail. Read this chapter to see if this book is for you.

Chapter 2 covers the simplest experiments: there is no structure on either the observational units or the treatments. This gives an opportunity to discuss randomization, replication and analysis of variance without extra complications, as well as to revise prerequisite knowledge.

Structure on the observational units is developed in Chapters 4 (simple blocking), 6 (row–column designs) and 8 (observational units smaller than experimental units). Structure on the treatments is developed in parallel, in two independent chapters. Chapter 5 deals with factorial treatments (crossed factors) while Chapter 3 covers control treatments and other ways of recursively splitting up the treatments (nested factors). Chapter 3 can be omitted in a short course, but there are some areas of application where Chapter 3 is more relevant than Chapter 5. The 'mixed' case of factorial treatments plus a control is covered in some detail in Chapters 1, 5 and 10; this occurs surprisingly often in practice, and is frequently misunderstood.

Chapter 8 deals with the situation when one or more of the treatment factors must be applied to something larger than observational units. This topic is often misunderstood in practice, as a glance at too many journals of experimental science shows. Every working statistician should be aware of the danger of false replication.

Chapters 7 and 9 are somewhat light relief from the main development, and could be omitted without making later chapters inaccessible. Chapter 7 applies the ideas so far to experiments on people; it also describes issues peculiar to such experiments. The reader who is concerned exclusively with such experiments is advised to continue with one of the more specialized texts, such as those recommended in the Further Reading. Chapter 9 takes a single combinatorial object—the Latin square—and uses it in several ways to design different types of experiment. This demonstrates that there is no such thing as a 'Latin-square design', or, perhaps, that the phrase has many interpretations.

Chapter 10 is my favourite. It pulls all the preceding material together into a single general approach. Because it is so general, the proofs are more abstract than those in the earlier chapters, and you may want to omit them at the first reading.

Chapters 11–13 introduce three more advanced topics that a statistician needs to be aware of: incomplete-block designs, confounded factorial designs, and fractional factorial designs. Anyone who needs to use these techniques frequently should probably follow this with some more advanced reading on these topics: some suggestions are made in Further Reading.

Finally, Chapter 14 is a rerun of Chapter 1 in the light of what has been covered in the rest of the book. Confronted with an experiment to design, how should we think about it and what should we do?

Each chapter is followed by questions for discussion. Because so many aspects of designing an experiment have no single 'right' answer, I have used these discussion questions with my students rather than requiring written homework. Each student is required to lead the discussion at least once. Apart from the initial difficulty of persuading students that this is not a terrifying ordeal, this technique has worked remarkably well. Other students join in; they share ideas and offer helpful criticism. At the end, I comment on both the presenting student's work and the general discussion, commending what is good, correcting any outright mistakes, and pointing out any important features that they have all missed. Every year a new set of students finds new nuances in these questions.

Some instructors may want to supplement the discussion questions with written homeworks. The Exercises at the end are provided for this purpose. They are less closely linked to the individual chapters than the questions for discussion.

Acknowledgements

I should like to thank the following people. R. E. Waller taught me the basics at the Air Pollution Research Unit: careful data recording, meticulous verification at every stage, and the excitement of extracting patterns from data. H. D. Patterson taught me so much about designing experiments while I was working as a post-doctoral researcher under his guidance. D. J. Finney was bold enough to let me teach the course on Design of Experiments in the Statistics MSc at the University of Edinburgh within twelve months of my meeting the material. J. A. Nelder appointed me as a statistician at Rothamsted Experimental Station even though I had no formal statistical qualification. D. A. Preece introduced me to such practical matters as data sniffing. D. R. Cox has been gently nagging me to write this book for over twenty years.

Thanks, too, to all the scientists whose interesting experiments I have worked on and to all those statisticians who continue to bring me their interesting design problems. Many of these are named at the back of the book.

Finally, thanks to all those students, friends and colleagues, in the widest sense of the word, who have read drafts of part of the material and made helpful suggestions. Of course, all opinions and any remaining errors are my own.

R. A. Bailey
December 2007

Chapter 1
Forward look

1.1 Stages in a statistically designed experiment

There are several stages in designing an experiment and carrying it out.

1.1.1 Consultation

The scientist, or other investigator, comes to the statistician to ask advice on the design of the experiment. Sometimes an appointment is made; sometimes the approach is by telephone or email with the expectation of an instant answer. A fortunate statistician will already have a good working relationship with the scientist. In some cases the scientist and statistician will both view their joint work as a collaboration.

Ideally the consultation happens in plenty of time before the experiment. The statistician will have to ask questions to find out about the experiment, and the answers may not be immediately available. Then the statistician needs time to think, and to compare different possible designs. In complicated cases the statistician may need to consult other statisticians more specialized in some aspect of design.

Unfortunately, the statistician is sometimes consulted only the day before the experiment starts. What should you do then? If it is obvious that the scientist has contacted you just so that he can write 'Yes' on a form in response to the question 'Have you consulted a statistician?' then he is not worth spending time on. More commonly the scientist genuinely has no idea that statistical design takes time. In that case, ask enough questions to find out the main features of the experiment, and give a simple design that seems to answer the purpose. Impress on the scientist that this design may not be the best possible, and that you can do better if given more notice. Try to find out more about this sort of experiment so that you are better prepared the next time that this person, or one of her colleagues, comes to you.

Usually the scientist does not come with statistically precise requirements. You have to elicit this information by careful questioning. About 90% of the statistician's input at this stage is asking questions. These have to be phrased in terms that a non-statistician can understand. Equally, you must not be shy about asking the scientist to explain technical terms from his field if they seem relevant.

If the scientist does have a preconceived idea of a 'design', it may be chosen from an artificially short list, based on lack of knowledge of what is available. Too many books and courses give a list of three or four designs and manage to suggest that there are no others. Your job may be to persuade the scientist that a better design is available, even if it did not figure in the textbook from which she learnt statistics.

Example 1.1 (Ladybirds) A famous company (which I shall not name) had designed an experiment to compare a new pesticide which they had developed, a standard pesticide, and 'no treatment'. They wanted to convince the regulatory authority (the Ministry of Agriculture, Fisheries and Foods) that their new pesticide was effective but did not harm ladybirds. I investigated the data from the experiment, and noticed that they had divided a field into three areas, applied one pesticide (or nothing) to each area, and made measurements on three samples from each area. I asked the people who had designed it what the design was. They said that it was completely randomized (see Chapter 2). I said that I could see that it was not completely randomized, because all the samples for each pesticide came from the same area of the field. They replied that it must be completely randomized because there were no blocks (see Chapter 4) and it was not a Latin square (see Chapter 6). In defence of their argument they quoted a respectable textbook which gives only these three designs.

1.1.2 Statistical design

Most of this book is about statistical design. The only purpose in mentioning it here is to show how it fits into the process of experimentation.

1.1.3 Data collection

In collaboration with the scientist, design a form for collecting the data. This should either be on squared paper, with squares large enough to write on conveniently, or use the modern electronic equivalent, a spreadsheet or a hand-held data-logger. There should be a row for each observational unit (see Section 1.4) and a column for each variable that is to be recorded. It is better if these variables are decided before the experiment is started, but always leave space to include extra information whose relevance is not known until later.

Emphasize to the scientist that all relevant data should be recorded as soon as possible. They should never be copied into a 'neater' format; human beings almost always make errors when copying data. Nor should they be invented later.

Example 1.2 (Calf feeding) In a calf-feeding trial each calf was weighed several times, once at birth and thereafter on the nearest Tuesday to certain anniversaries, such as the nearest Tuesday to its eight-week birthday. The data included all these dates, which proved to be mutually inconsistent: some were not Tuesdays and some were the wrong length of time apart. When I queried this I was told that only the birthdate was reliable: all the other dates had been written down at the end of the experiment by a temporary worker who was doing her best to follow the 'nearest Tuesday' rule after the event. This labour was utterly pointless. If the dates had been recorded when the calves were weighed they would have provided evidence of how closely the 'nearest Tuesday' rule had been followed; deducing the dates after the event could more accurately and simply have been done by the computer as part of the data analysis.

Sometimes a scientist wants to take the data from his field notebooks and reorganize them into a more logical order for the statistician's benefit. Discourage this practice. Not only does

1.1. Stages in a statistically designed experiment

Plot 8	6
	0
	7
	3
	6
	0
	4
	5
	6
	4
Average	4.1

Plot 23	0
	0
	0
	0
	0
	0
	0
	28
	0
	0
Average	28

Fig. 1.1. Data sheets with intermediate calculations in Example 1.3

it introduce copying errors; reordering the data loses valuable information such as which plots were next to each other or what was the time sequence in which measurements were made: see Example 1.5.

For similar reasons, encourage the scientist to present you with the raw data, without making intermediate calculations. The data will be going into a computer in any case, so intermediate calculations do not produce any savings and may well produce errors. The only benefit brought by intermediate calculations is a rough check that certain numbers are the correct order of magnitude.

Example 1.3 (Leafstripe) In an experiment on leafstripe disease in barley, one measurement was apparently the percentage of disease on each plot. A preliminary graph of the data showed one outlier far away from the rest of the data. I asked to see the data for the outlying plot, and was given a collection of pieces of paper like those shown in Figure 1.1. It transpired that the agronomist had taken a random sample of ten quadrats in each plot, had inspected 100 tillers (sideshoots) in each quadrat to see how many were infected, and averaged the ten numbers. Only the average was recorded in the 'official' data. For the outlying plot the agronomist rightly thought that he did not need a calculator to add nine zeros to one nonzero number, but he did forget to divide the total by 10. Once I had corrected the average value for this plot, it fell into line with the rest of the data.

Also try to persuade the scientist that data collection is too important to be delegated to junior staff, especially temporary ones. An experiment cannot be better than its data, but a surprising number of good scientists will put much effort into their science while believing that the data can take care of themselves. Unless they really feel part of the team, junior or temporary staff simply do not have the same motivation to record the data carefully, even if they are conscientious. See also Example 1.2.

1.1.4 Data scrutiny

After the experiment is done, the data sheets or data files should be sent to the statistician for analysis. Look over these as soon as possible for obvious anomalies, outliers or evidence of bad practice. Can that number really be a calf's birthweight? Experienced statisticians become remarkably good at 'data sniffing'—looking over a sheet of figures and finding the one or two anomalies. That is how the errors in Example 1.2 were found. Simple tables and graphs can also show up errors: in Example 1.3 the outlier was revealed by a graph of yield in tonnes per hectare against percentage of infected tillers.

Examine the final digits in the data. If the number of significant figures changes at one point, this may indicate a change in the person recording the data or the machine being used. Occasionally it indicates a change such as from weighing in pounds to weighing in kilograms and dividing by 2.205. Any such change is likely to coincide with a change in conditions which is more serious than the appearance of the data. These checks are easier to conduct on paper data than on electronic data, because most spreadsheets give no facility for distinguishing between 29 and 29.00.

Example 1.4 (Kiwi fruit) At an agricultural research station in New Zealand, an instrument called a penetrometer was used to measure the hardness of kiwi fruit. After a preliminary analysis of the data, the statistician examined a graph of residuals and realized that there was something wrong with the data. He looked again at the data sheet, and noticed that two different handwritings had been used. He re-analysed the data, letting the data in one handwriting be an unknown constant multiple of those in the other. The fitted value of the constant was 2.2, indicating that one person had recorded in pounds, the other in kilograms.

Query dubious data while it is still fresh in the scientist's memory. That way there is a chance that either the data can be corrected or other explanatory information recorded.

Example 1.5 (Rain at harvest) In an experiment whose response was the yield of wheat on each plot, the numbers recorded on the last 12 plots out of a total of 72 were noticeably lower than the others. I asked if there was any reason for this, and was told that it had started to rain during the harvest, with the rain starting when the harvester was about 12 plots from the end. We were therefore able to include an extra variable 'rain', whose values were 60 zeros followed by 1, 2, ..., 12. Including 'rain' as a covariate in the analysis removed a lot of otherwise unexplained variation.

Example 1.6 (Eucalypts) In a forestry progeny trial in Asia, different families of eucalypts were grown in five-tree plots. After 36 months, a forestry worker measured the diameter of each tree at breast height. In the preliminary examination of the data, the statistician calculated the variance of the five responses in each plot, and found that every plot had exactly the same variance! Challenged on this, the forestry worker admitted that he had measured every tree in the first plot, but thereafter measured just tree 1 in each plot. For trees 2–5 he had added the constant c to the measurements from plot 1, where c was the difference between the diameter at breast height of tree 1 in this plot and the diameter at breast height of tree 1 in plot 1.

In this case, the statistician's preliminary scrutiny showed that the data were largely bogus.

1.1.5 Analysis

This means calculations with the data. It should be planned at the design stage, because you cannot decide if a design is good until you know how the data will be analysed. Also, this planning enables the experimenter to be sure that she is collecting the relevant data. If necessary, the analysis may be modified in the light of unforeseen circumstances: see Example 1.5.

For a simple design the statistician should, in principle, be able to analyse the data by hand, with a calculator. In practice, it is more sensible to use a reliable statistical computing package. A good package should ask the user to distinguish plot structure from treatment structure, as in Section 1.4; it should be able to handle all the structures given in Section 1.4; and it should automatically calculate the correct variance ratios for experiments like those in Chapters 8 and 10. It is a good idea to do the planned analysis on dummy data *before* the real data arrive, to avoid any unnecessary delay.

Many other statistics books are concerned almost exclusively with analysis. In this book we cover only enough of it to help with the process of designing experiments.

1.1.6 Interpretation

The data analysis will produce such things as analysis-of-variance tables, lists of means and standard errors, *P*-values and so on. None of these may mean very much to the scientist. It is the statistician's job to interpret the results of the analysis in terms which the scientist can understand, and which are pertinent to his original question.

1.2 The ideal and the reality

Here I discuss a few of the tensions between what the statistician thinks is desirable and what the experimenter wants.

1.2.1 Purpose of the experiment

Why is the experiment being done? If the answer is 'to use an empty greenhouse' or 'to publish another paper', do not put much statistical effort into it. A more legitimate answer is 'to find out about the differences between so-and-so', but even this is too vague for the statistician to be really helpful.

Ideally, the aims of the experiment should be phrased in terms of specific questions. The aim may be to estimate something: for example, 'How much better is Drug *A* than Drug *B*?' This question needs refining: how much of each drug? how administered? to whom? and how will 'better' be measured? For estimation questions we should aim to obtain unbiased estimators with low variance.

On the other hand, the aim may be to test a hypothesis, for example that there is no effective difference between organic and inorganic sources of nitrogen fertilizer. Again the question needs refining: how much fertilizer? applied to what crop? in what sorts of circumstances? is the effect on the yield or the taste or the colour? For hypothesis testing we want high power of detecting differences that are big enough to matter in the science involved.

1.2.2 Replication

This is the word for the number of times that each treatment is tested.

The well-known formula for the variance of the mean of n numbers is σ^2/n, on the assumption that the numbers are a random sample from a population with variance σ^2. Increasing the replication usually decreases the variance, because it increases the value of n.

On the other hand, increased replication may raise the variance. Typically, a larger number of experimental units are more variable than a small number, so increasing the replication may increase the value of σ^2. Sometimes this increase outweighs the increase in n.

Increased replication usually raises power. This is because it usually raises the number of residual degrees of freedom, and certain important families of distribution (such as t) have slimmer tails when they have more degrees of freedom.

The one thing that is almost certain about increased replication is that it increases costs, which the experimenter usually wants to keep down.

1.2.3 Local control

This means dividing the set of experimental units into blocks of alike units: see Chapter 4. It is also called *blocking*.

If it is done well, blocking lowers the variance, by removing some sources of variability from treatment contrasts. If each block is representative rather than homogeneous then blocking has the opposite effect.

Blocking can increase the variance if it forces the design to be non-orthogonal: see Chapter 11.

Because blocking almost always decreases the variance, it usually raises power. However, it decreases the number of residual degrees of freedom, so it can reduce power if numbers are small: see Example 4.15.

Blocking increases the complexity of the design. In turn this not only increases the complexity of the analysis and interpretation but gives more scope for mistakes in procedure during the experiment.

1.2.4 Constraints

The most obvious constraint is cost. Everybody will be pleased if the same results can be achieved for less money. If you can design a smaller, cheaper experiment than the scientist proposes, this is fine if it produces good estimators. On the other hand, it may be impossible to draw clear conclusions from an experiment that is too small, so then the entire cost is wasted. Part of your duty is to warn when you believe that the whole experiment will be wasted.

The availability of the test materials may provide a constraint. For example, in testing new varieties of wheat there may be limited quantities of seed of some or all of the new varieties.

Availability of the experimental units provides a different sort of constraint. There may be competition with other experimenters to use land or bench space. If results are needed by a certain deadline then time limits the number of experimental units. In a clinical trial it is unethical to use far too many patients because this unnecessarily increases the number of patients who do not get the best treatment. On the other hand, it is also unethical to use so few patients that no clear conclusions can be drawn, for then all the patients have been used in vain. Similar remarks apply to experiments on animals in which the animals have to be sacrificed.

If there are natural 'blocks' or divisions among the experimental units these may force constraints on the way that the experiment can be carried out. For example, it may be impossible to have all vaccinations administered by the same nurse.

There are often other constraints imposed by the management of the experiment. For example, temporary apple-pickers like to work with their friends: it may be unrealistic to expect them each to pick from separate rows of trees.

1.2.5 Choice

Given all the constraints, there are still two fundamentally important choices that have to be made and where the statistician can provide advice.

Which treatments are to be tested? The scientist usually has a clear idea, but questions can still be helpful. Why did he decide on these particular quantities? Why these combinations and not others? Should he consider changing two factors at a time? (see Chapter 5). Does the inclusion of less interesting treatments (such as the boss's favourite) mean that the replication for *all* treatments will be too low?

There is a strong belief in scientific circles that all new treatments should be compared with 'no treatment', which is often called *control*. You should always ask if a control is needed. Scientific orthodoxy says yes, but there are experiments where a control can be harmful. If there is already an effective therapy for a disease then it is unethical to run an experiment comparing a new therapy to 'do nothing'; in this case the treatments should be the new therapy and the one currently in use. In a trial of several pesticides in one field, if there is a 'do nothing' treatment on some plots then the pest may multiply on those plots and then spread to the others. A 'do nothing' treatment is also not useful if this would never be used in practice.

Sometimes it is already known that the 'do nothing' treatment has a very different effect from all the other treatments. Then the experiment may do nothing more than confirm this, as in Examples 3.2 and 6.3. In such cases, it is better to omit the 'do nothing' treatment so that more resources can be devoted to finding out whether there is any difference between the other treatments.

Which experimental units should be used? For example, is it better to use portions of representative farmers' fields or a well-controlled experimental farm? The latter is better if the effect to be detected is likely to be small, or if one of the treatments is sufficiently unknown that it might have disastrous economic or environmental consequences. The former is better for a large confirmatory experiment, before recommending varieties or treatments for use on a wide scale. Similarly, is it better to use 36 heifers from the same herd or 36 bought at the market specifically for this experiment? University students are a convenient source of experimental units for psychologists, but how far can results valid for such students be extrapolated to the general population?

1.3 An example

An example will help to fix ideas.

Example 1.7 (Rye-grass) An experiment was conducted to compare three different cultivars of rye-grass in combination with four quantities of nitrogen fertilizer. Two responses were measured: one was the total weight of dry matter harvested from each plot, and the other was the percentage of water-soluble carbohydrate in the crop.

The three cultivars of rye-grass were called Cropper, Melle and Melba. The four amounts

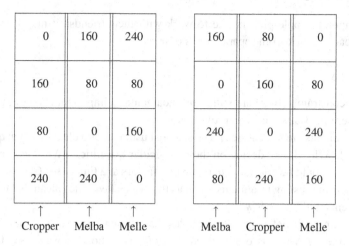

Fig. 1.2. Layout of the rye-grass experiment in Example 1.7

of fertilizer were 0 kg/ha, 80 kg/ha, 160 kg/ha and 240 kg/ha.

The experimental area consisted of two fields, each divided into three strips of land. Each strip consisted of four plots.

Cultivars were sown on whole strips because it is not practicable to sow them in small areas unless sowing is done by hand. In contrast, it is perfectly feasible to apply fertilizers to smaller areas of land, such as the plots. The layout for the experiment is shown in Figure 1.2.

Notice the pattern. Each amount of nitrogen is applied to one plot per strip, and each cultivar is applied to one strip per field. This pattern is the *combinatorial design*.

Notice the lack of pattern. There is no systematic order in the allocation of cultivars to strips in each field, nor any systematic order in the allocation of amounts of nitrogen to plots in each strip. This lack of pattern is the *randomization*.

1.4 Defining terms

Definition An *experimental unit* is the smallest unit to which a treatment can be applied.

Definition A *treatment* is the entire description of what can be applied to an experimental unit.

Although the previous two definitions appear to be circular, they work well enough in practice.

Definition An *observational unit* is the smallest unit on which a response will be measured.

Example 1.6 revisited (Eucalypts) The experimental units were the plots. The observational units should have been the trees.

Example 1.8 (Wheat varieties) The experiment compares different varieties of wheat grown in plots in a field. Here the experimental units are the plots and the treatments are the varieties. We cannot tell what the observational unit is without more information. Probably a plot is the observational unit, but it might be an individual plant. It might even be the whole field.

1.4. Defining terms

	whole class	small groups
one hour once per week	√	√
12 minutes every day	√	√

Fig. 1.3. Four treatments in Example 1.10

Example 1.7 revisited (Rye-grass) Here the treatments are the combinations of cultivars with amounts of fertilizer, so there are twelve treatments. The experimental unit is the plot. The observational unit is probably the plot but might be a plant or a strip.

Example 1.2 revisited (Calf feeding) Here the treatments were different compositions of feed for calves. The calves were not fed individually. They were housed in pens, with ten calves per pen. Each pen was allocated to a certain type of feed. Batches of this type of feed were put into the pen; calves were free to eat as much of this as they liked. Calves were weighed individually.

The experimental units were the pens but the observational units were the calves.

Example 1.9 (Asthma) Several patients take part in an experiment to compare drugs intended to alleviate the symptoms of chronic asthma. For each patient, the drugs are changed each month. From time to time each patient comes into the clinic, where the peak flow rate in their lungs is measured.

Here the treatments are the drugs. An experimental unit is a patient-month combination, so if 30 patients are used for 6 months then there are 180 experimental units. The observational unit is a visit by a patient to the clinic; we do not know how this relates to the patient-months without further information.

Example 1.10 (Mental arithmetic) After calculators became widespread, there was concern that children in primary schools were no longer becoming proficient in mental arithmetic. One suggested remedy was whole-class sessions, where the teacher would call out a question such as '5 + 7?' and children would put up their hands to offer to give the correct answer. An alternative suggestion was to do this in small groups of about four children, to encourage those who were shy of responding in front of the whole class. Another question was: is it better to have these sessions for one hour once a week or for 10–12 minutes every day?

The treatments are the four combinations of group size and timing shown in Figure 1.3. Each treatment can be applied only to a whole class, so the experimental units are classes. However, to measure the effectiveness of the treatments, each child must take an individual test of mental arithmetic after some set time. Thus the observational units are the children.

Example 1.11 (Detergents) In a consumer experiment, ten housewives test new detergents. Each housewife tests one detergent per washload for each of four washloads. She assesses the cleanliness of each washload on a given 5-point scale. Here the 40 washloads are the experimental units and the observational units; the detergents are the treatments.

Example 1.12 (Tomatoes) Different varieties of tomato are grown in pots, with different composts and different amounts of water. Each plant is supported on a vertical stick until it is 1.5 metres high, then all further new growth is wound around a horizontal rail. Groups of five adjacent plants are wound around the same rail. When the tomatoes are ripe they are harvested and the weight of saleable tomatoes per rail is recorded.

Now the treatment is the variety–compost–water combination. The pots are the experimental units but the rails are the observational units.

These examples show that there are four possible relationships between experimental units and observational units.

(i) The experimental units and the observational units are the same. This is the most usual situation. It occurs in Example 1.11; in Examples 1.7 and 1.8 if there is one measurement per plot; in Example 1.9 if there is one measurement of peak flow rate in lungs per patient per month.

(ii) Each experimental unit consists of several observational units. This is usually forced by practical considerations, as in Examples 1.2 and 1.10. Examples 1.7 and 1.8 are of this type if the observational unit is a plant. So is Example 1.9 if the observational unit is a patient-week. This situation is fine so long as the data are analysed properly: see Chapter 8.

(iii) Each observational unit consists of several experimental units. This would occur in Example 1.9 if each patient had their drugs changed monthly but their peak flow rate measured only every three months. It would also occur in Examples 1.7 and 1.8 if the observational unit were the strip or field respectively. In these cases the measurements cannot be linked to individual treatments so there is no point in conducting such an experiment.

Example 1.12 also appears to be of this form. Because the experiment would be useless if different pots in the same group (allocated to the same rail) had different treatments, in effect it is the group of pots that is the experimental unit, not the individual pot.

In fact, there are some unusual experiments where the response on the observational unit can be considered to be the sum of the (unknown) responses on the experimental units contained within it. However, these are beyond the scope of this book.

(iv) Experimental units and observational units have a partial overlap, but neither is contained in the other. This case is even sillier than the preceding one.

It is useful to write down the experimental units and the observational units in the experimental protocol. This should draw attention to cases (iii) and (iv) before it is too late to change the protocol.

Definition In cases (i) and (ii) an observational unit will often be called a *plot* for brevity.

This usage is justified by the large amount of early work on experimental design that took place in agricultural research. However, it can be a little disconcerting if the plot is actually a person or half a leaf. It is a useful shorthand in this book, but is not recommended for your conversations with scientists.

Notation In this book, general plots are denoted by lower-case Greek letters, such as α, β, γ, ω. The whole set of plots is denoted by Ω, and the number of plots by N.

1.4. Defining terms

	spray mid-season		no mid-season spray	
	spray late	no late spray	spray late	no late spray
spray early	√	√	√	√
no early spray	√	√	√	√

Fig. 1.4. Factorial treatment combinations in Example 1.15

Example 1.13 (Pullets) Two feeds for pullets are to be compared for their effect on the weight of their eggs. Ten pullets are selected for the experiment, are isolated and are individually fed, with five pullets receiving each of the two feeds. After the feeding regime has been in place for one month, the eggs laid thereafter by each pullet are individually weighed.

The individual feeding implies that the pullets are the experimental units, but what are the observational units? If the eggs are the observational units then we have two difficulties: we do not know the number of observational units in advance and the numbers will vary from one pullet to another. Both of these difficulties can be overcome by declaring that only the first so many eggs laid (or, more practically, collected) will be weighed. On the other hand, if the feeds affect the number of eggs laid as well as their weight then it might be more sensible to measure the total weight of eggs laid by each pullet; in this case the pullets are the observational units.

Example 1.14 (Simple fungicide) In a fungicide trial the treatments are doses of fungicide: full spray, half spray and 'no treatment'. The experimenter might say that there are two treatments and a control; in our vocabulary there are three treatments.

Example 1.15 (Fungicide factorial) In another fungicide trial on winter wheat the fungicide could be sprayed early, mid-season or late, or at any combination of those times. The treatments consisted of all combinations of 'spray' and 'no-spray' at each date. See Figure 1.4. Thus there were eight treatments; the experimenter told me that there were seven, because he did not consider 'never spray' to be a treatment.

Example 1.16 (Fungicide factorial plus control) In an important variant of Example 1.14, the spray is applied only once, but this can be early, mid-season or late. Thus the treatments are combinations of amount of fungicide with time of application. How many treatments are there? It is quite common to see the treatments in this example laid out schematically as in Figure 1.5(a), which suggests that there are nine treatments, being all combinations of amount of fungicide with time of application. I have seen data from such experiments analysed as if there were nine treatments. However, if there is no fungicide then it does not make sense to distinguish between time of application: the time of application should be regarded as 'not applicable'. This gives the seven treatments shown in Figure 1.5(b).

Example 1.17 (Oilseed rape) An experiment on methods of controlling the disease sclerotina in oilseed rape compared four new chemicals, coded A, B, C, D, with both 'no treatment' and the current (expensive) standard chemical X. Each of the new chemicals could be applied either early or late; the standard X was applied at both times. Thus there were two control treatments, and the treatments had the structure shown in Figure 1.6.

	early	mid-season	late
full spray	✓	✓	✓
half spray	✓	✓	✓
no spray	✓	✓	✓

(a) Inappropriate description

	early	mid-season	late	n/a
full spray	✓	✓	✓	
half spray	✓	✓	✓	
no spray				✓

(b) Appropriate description

Fig. 1.5. Two descriptions of the treatments in Example 1.16

Notation In this book, general treatments are denoted by lower-case Latin letters, such as i, j. The whole set of treatments is denoted by $\mathcal{T}$, and the number of treatments by t.

The experimental protocol needs to contain a precise description of each treatment. This must include complete details, such as ingredients, proprietary name (if any), quantity, time of application, variety, species, etc. Then give each treatment a simple code like A, B, C or 1, 2, ... for reference later.

Definition *Treatment structure* means meaningful ways of dividing up $\mathcal{T}$.

Examples of treatment structure include:

unstructured This means that there is no structure to the treatments at all.

several new treatments plus control This is the structure in Example 1.14. It is examined further in Chapter 3.

all combinations of two factors See Example 1.10 and Chapter 5.

all combinations of two factors, plus control See Example 1.16 and Section 5.9.

all combinations of three factors See Examples 1.12 and 1.15 and Chapters 5, 12 and 13.

increasing doses of a quantitative factor This is not covered explicitly in this book, apart from the discussion of Example 5.10, because the relevant ideas can be found in most books on regression. Factors such as the amount of fertilizer in Example 1.7 are indeed quantitative, but not really on a continuous scale, because the farmer will use a whole number of bags of fertilizer.

Definition *Plot structure* means meaningful ways of dividing up the set Ω of plots, ignoring the treatments.

Examples of plot structure include:

unstructured There is no structure to the observational units at all.

experimental units containing observational units This is the structure in Example 1.2. It is discussed in Chapter 8.

1.4. Defining terms

	early	late	both	n/a
none				✓
A	✓	✓		
B	✓	✓		
C	✓	✓		
D	✓	✓		
X			✓	

Fig. 1.6. Treatment structure in Example 1.17

blocks This means local control: dividing the set of experimental units into homogeneous blocks. In Example 1.11 the housewives should be treated as blocks. See Chapter 4.

blocks containing subblocks containing plots This is the structure in Example 1.7. It is also discussed in Chapter 8.

blocks containing experimental units containing observational units This is mentioned in Chapter 8.

two different sorts of blocks, neither containing the other This plot structure occurs in Example 1.9, where the two sorts of block are patients and months. See Chapter 6.

All these structures (except for quantitative treatments) are described in a unified way in Chapter 10.

In principle, *any* type of treatment structure can occur with *any* type of plot structure. That is why it is neither possible nor sensible to give a short list of useful designs.

Definition The *design* is the allocation of treatments to plots.

Although we speak of allocating treatments to plots, mathematically the design is a function T from Ω to $\mathcal{T}$. Thus plot ω is allocated treatment $T(\omega)$. The function has to be this way round, because each plot can receive only one treatment. It may seem strange to apply a function T to an actual object such as a rat or a plot of land, but this is indeed what the design is.

We usually choose T to satisfy certain combinatorial properties. The design has theoretical plots (perhaps numbered $1, \ldots, N$) and coded treatments.

Definition The *plan* or *layout* is the design translated into actual plots.

Some randomization is usually involved in this translation process.

The actual plots must be labelled or drawn in such a way that the person applying the treatments can identify the plots uniquely. For example, in a field trial the North arrow will usually be shown on the plan.

Example 1.17 revisited (Oilseed rape) The North arrow was omitted from the plan for this experiment. The person applying chemicals at the later date held the plan upside down relative to the person applying them early. The effect of this is shown in Figure 1.7, where there are some quite unintended treatments.

Plot	Early	Late (intended)	Late (actual)
1	D	–	–
2	–	A	C
3	B	–	B
4	–	–	–
5	A	–	X
6	–	B	–
7	–	D	–
8	C	–	–
9	–	C	A
10	X	X	D
11	–	D	X
12	–	A	C
13	D	–	–
14	B	–	D
15	C	–	B
16	X	X	–
17	A	–	–
18	–	B	–
19	–	C	A
20	–	–	–

Fig. 1.7. Result of holding the plan upside down at the later date in Example 1.17

The treatments in the plan usually remain in coded form. Partly this is for brevity. Partly it is to prevent unconscious biases from the people applying the treatments or recording the data: see Chapter 7.

1.5 Linear model

The response on plot ω is a random variable Y_ω whose observed value after the experiment is y_ω. Thus we have a data vector $\mathbf{y}$ which is a realization of the random vector $\mathbf{Y}$.

We assume that we have measured on a suitable scale so that

$$Y_\omega = Z_\omega + \tau_{T(\omega)}, \qquad (1.1)$$

where $\tau_{T(\omega)}$ is a *constant*, depending on the treatment $T(\omega)$ applied to plot ω, and Z_ω is a *random variable*, depending on ω. Thus Z_ω can be thought of as the contribution of the plot ω to the response, while τ_i is the contribution of treatment i.

If we cannot assume the additive model (1.1) then we cannot extrapolate the conclusions of the experiment to any other plots. In such a case, is it worth doing the experiment? Even if we can assume model (1.1), it gives us more unknowns than measurements, so we have to assume something about the values of the Z_ω.

The probability space is taken to be the set of occasions and uncontrolled conditions under which the experiment might be carried out. The non-repeatability of the experiment gives Z_ω its randomness. If α and β are two different plots then Z_α and Z_β are different random variables on the same probability space, so they have a joint distribution, including a correlation, which may be zero.

What is the joint distribution of the $(Z_\omega)_{\omega \in \Omega}$? Here are some common assumptions:

simple textbook model the Z_ω are independent identically distributed normal random variables with mean 0 and variance σ^2;

fixed-effects model the Z_ω are independent normal random variables each with variance σ^2, and the mean μ_ω of Z_ω depends on the position of ω within Ω;

random-effects model the Z_ω have identical distributions, and the covariance $\text{cov}(Z_\alpha, Z_\beta)$ depends on how α and β are related in the plot structure;

randomization model the Z_ω have identical distributions, and $\text{cov}(Z_\alpha, Z_\beta)$ depends on the method of randomization.

Using our assumptions about the Z_ω, the analysis of the data should give

- minimum variance unbiased estimators of the treatment parameters τ_i and linear combinations thereof, such as $\tau_i - \tau_j$;
- estimates of the variances of those estimators;
- inferences about presence or absence of effects.

1.6 Summary

The top half of Figure 1.8 summarizes this chapter. The plot structure and the treatment structure must both be taken into account in choosing the design. Randomizing the design gives the plan. The method of randomization is usually dictated by the plot structure, but occasionally we use the method of randomization to define the plot structure.

The bottom half of the figure gives a preview of the rest of the book. The plot structure determines the null analysis, as we shall see in Chapter 10: this has no treatments and no data. Putting this together with the design gives the skeleton analysis, which has treatments but still no data. The skeleton analysis enables us to make educated guesses about variance and power using preliminary guesses about the size of σ^2. If the variance is too big or the power too low then we should go back and change the design (possibly changing the plot structure by adding plots or the treatment structure by removing treatments) and begin the cycle again.

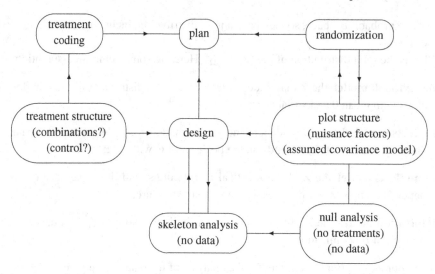

Fig. 1.8. Overview and preview

Questions for discussion

1.1 A professional apple-grower has written to you, making an appointment to discuss an experiment which he is proposing to conduct in the coming growing season. Part of his letter reads:

> There is a new type of chemical spray available, which is supposed to make the apple flowers more attractive to bees. Since bees are essential for setting the fruit, I want to investigate these chemicals. Two manufacturers are selling the new sprays, under the trade names Buzz!! and Attractabee.
>
> I propose dividing my orchard into three parts. I shall spray Attractabee onto one part, and Buzz!! onto the second part. The third part will be managed in the normal way, with none of these new sprays. I shall then see which part of the orchard does best.

Make notes on what you should discuss with him—and why!—at your meeting.

1.2 Several studies have suggested that drinking red wine gives some protection against heart disease, but it is not known whether the effect is caused by the alcohol or by some other ingredient of red wine. To investigate this, medical scientists enrolled 40 volunteers into a trial lasting 28 days. For the first 14 days, half the volunteers drank two glasses of red wine per day, while the other half had two standard drinks of gin. For the remaining 14 days the drinks were reversed: those who had been drinking red wine changed to gin, while those who had been drinking gin changed to red wine. On days 14 and 28, the scientists took a blood sample from each volunteer and measured the amount of inflammatory substance in the blood.

Identify the experimental units and observational units. How many are there of each? What is the plot structure?

What are the treatments? What is the treatment structure?

Questions for discussion

Table 1.1. *Data for Question 1.3*

Person	Day	Drug	Hours after drug administration					
			2	3	5	6	9	12
1	1	A	0.667	0.467	0.333	0.300	0.233	0.200
1	2	B	0.480	0.440	0.280	0.240	0.160	0.160
2	1	D	0.700	0.500	0.333	0.333	0.267	0.267
2	2	C	0.133	0.156	0.200	0.200	0.178	0.178
3	1	C	0.156	0.200	0.178	0.200	0.156	0.156
3	2	A	0.733	0.533	0.333	0.300	0.200	0.200
4	1	B	0.680	0.520	0.360	0.360	0.280	0.280
4	2	C	0.156	0.222	0.222	0.200	0.156	0.178
5	1	D	0.733	0.600	0.400	0.433	0.367	0.333
5	2	A	0.667	0.467	0.300	0.233	0.200	0.167
6	1	D	0.600	0.467	0.300	0.387	0.267	0.233
6	2	B	0.680	0.520	0.360	0.320	0.240	0.200
7	1	B	0.800	0.600	0.360	0.440	0.320	0.320
7	2	A	0.733	0.467	0.333	0.300	0.200	0.200
8	1	B	0.800	0.680	0.360	0.440	0.360	0.320
8	2	D	0.700	0.400	0.300	0.267	0.200	0.200
9	1	C	0.111	0.156	0.156	0.133	0.133	0.111
9	2	D	0.567	0.467	0.300	0.233	0.133	0.133
10	1	A	0.700	0.533	0.433	0.400	0.367	0.333
10	2	D	0.567	0.433	0.300	0.267	0.200	0.200
11	1	A	0.667	0.433	0.367	0.367	0.300	0.233
11	2	C	0.200	0.267	0.289	0.267	0.200	0.178
12	1	C	0.133	0.156	0.200	0.200	0.222	0.244
12	2	B	0.720	0.520	0.320	0.280	0.200	0.160

1.3 Twelve people took part in a study to compare four formulations of lithium carbonate. Formulations A, B and C were pills, while formulation D had the lithium carbonate dissolved in solution. Pills A and B were similar, but with different *excipients* (this is the name for the inactive substance that holds the active chemical). Pill C contained something to delay the release of the lithium carbonate. The formulations contained different quantities of lithium carbonate, as follows, in milligrams per dose.

$$\begin{array}{cccc} A & B & C & D \\ 300 & 250 & 450 & 300 \end{array}$$

On one day, each person was given a dose of one formulation. Blood samples were taken two, three, five, six, nine and twelve hours later. The quantity of lithium in the sample, after scaling to a common dose, was recorded. One week later, the procedure was repeated, with each person receiving a different formulation from the previous one. Table 1.1 shows the data.

What are the experimental units? What are the observational units? What are the treatments? How many are there of each?

Comment on the data recording.

1.4 Two types of concrete mix are to be compared.

Ingredient	Type (a)	Type (b)
cement	220 kg/m^3	390 kg/m^3
water	172 kg/m^3	210 kg/m^3
aggregate	1868 kg/m^3	1839 kg/m^3

Each of five operators mixes two batches, one of each type, then casts three cylindrical samples from each batch. The samples are left to harden for 21 days. After 7 days, each operator randomly chooses one cylinder of each type and ascertains the breaking load of each by using a standard concrete-testing machine. After 14 days, one of the remaining two cylinders of each type is chosen at random, and measured in the same way. The final cylinder of each type is measured after 21 days.

What are the treatments? How many are there? What is the treatment structure?

Identify the experimental units and observational units. What is the plot structure?

1.5 Read the paper 'The importance of experimental design in proteomic mass spectrometry experiments: some cautionary tales' written by Jianhua Hu, Kevin R. Coombes, Jeffrey S. Morris and Keith A. Baggerly, which was published in the journal *Briefings in Functional Genomics and Proteomics* in 2005 (Volume 3, pages 322–331). How far do you think their recommendations apply to experiments in other areas of science?

Chapter 2

Unstructured experiments

2.1 Completely randomized designs

If there is no reason to group the plots into blocks then we say that Ω is *unstructured*.

Suppose that treatment i is applied to r_i plots, in other words that i is replicated r_i times. Then
$$\sum_{i=1}^{t} r_i = |\Omega| = N.$$
Treatments should be allocated to the plots at random. Then the design is said to be *completely randomized*.

To construct and randomize the design, proceed as follows.

(i) Number the plots $1, 2, \ldots, N$.

(ii) Apply treatment 1 to plots $1, \ldots, r_1$; apply treatment 2 to plots $r_1 + 1, \ldots, r_1 + r_2$, and so on, to obtain a systematic design.

(iii) Choose a random permutation of $\{1, 2, \ldots, N\}$ and apply it to the design.

Example 2.1 (Fictitious) Suppose that there are three treatments coded A, B and C with $r_A = 5$ and $r_B = r_C = 4$. Then there are 13 plots. The systematic design is as follows.

plot	1	2	3	4	5	6	7	8	9	10	11	12	13
treatment	A	A	A	A	A	B	B	B	B	C	C	C	C

Suppose that the random permutation is
$$\begin{pmatrix} 1 & 2 & 3 & 4 & 5 & 6 & 7 & 8 & 9 & 10 & 11 & 12 & 13 \\ 6 & 2 & 8 & 11 & 13 & 1 & 12 & 5 & 7 & 4 & 9 & 3 & 10 \end{pmatrix},$$
where we are using the usual two-line way of displaying a permutation, which here indicates that 1 and 6 are interchanged, 2 does not move, 3 is moved to 8, and so on. In applying this permutation to the design we move the treatment A on plot 1 to plot 6, leave the treatment A on plot 2 still on plot 2, move the treatment A on plot 3 to plot 8, and so on. This gives the following plan.

plot	1	2	3	4	5	6	7	8	9	10	11	12	13
treatment	B	A	C	C	B	A	B	A	C	C	A	B	A

2.2 Why and how to randomize

Why do we randomize? It is to avoid

systematic bias for example, doing all the tests on treatment A in January then all the tests on treatment B in March;

selection bias for example, choosing the most healthy patients for the treatment that you are trying to prove is best;

accidental bias for example, using the first rats that the animal handler takes out of the cage for one treatment and the last rats for the other;

cheating by the experimenter.

Cheating is not always badly intentioned. For example, an experimenter may decide to give the extra milk rations to those schoolchildren who are most undernourished or she may choose to put a patient in a trial if she thinks that the patient will particularly benefit from the new treatment. This sort of cheating is for the benefit of the (small) number of people in the experiment but, by biasing the results, may be to the disadvantage of the (large) number of people who could benefit in future from a treatment which has been demonstrated, without bias, to be superior. As another example, she may be secretly trying to remove bias by trying to balance numbers over some nuisance factor without troubling the statistician, but this too can produce false results unless this 'balancing' is taken into account in the analysis. Yet again, she may be trying to make life a little easier for the technician by telling him to do all of one treatment first.

Thus doing an objective randomization and presenting the experimenter with the plan has the added benefit that it may force her to tell you something which she had previously thought unnecessary, such as 'We cannot do it that way because ...' or 'that will put all replicates of treatment A in the shady part of the field'.

How do we choose a random permutation? The process must be objective, so that you have no chance to cheat either. Simply writing down the numbers $1, \ldots, N$ in an apparently haphazard order is not good enough.

One excellent way to randomize is to shuffle a pack of cards. A normal pack of 52 playing cards should be thoroughly randomized after seven riffle shuffles. In Example 2.1 one can deal out the shuffled pack, noting the number (with Jack $= 11$, etc.) but not the suit, and ignoring repeated numbers.

Another good method is to ask a computer to produce a (pseudo-)random order of the numbers $1, \ldots, N$. Even a palm-top can do this. The permutation in Example 2.1 corresponds to the random order

$$6 \quad 2 \quad 8 \quad 11 \quad 13 \quad 1 \quad 12 \quad 5 \quad 7 \quad 4 \quad 9 \quad 3 \quad 10.$$

Two other methods of randomizing use sequences of random numbers, which can be generated by computers or calculators or found in books of tables. Both methods will be illustrated here for a random permutation of the numbers $1, \ldots, 13$.

My calculator produces random numbers uniformly distributed between 0 and 1. We need 13 different numbers, so record the random numbers to two decimal places. These are shown

2.3. The treatment subspace

0.72	0.34	0.65	0.99	0.01	0.23	0.30	0.57	0.17	0.63	0.14	1.00	0.94	0.29	0.05	0.38	0.86
12	14	5	19	1	3	10	17	17	3	14	20	14	9	5	18	6
12	X	5	X	1	3	10	X	X	R	X	X	X	9	R	X	6

	0.59	0.51	0.22	0.30	0.05	0.24	0.78	0.63	0.35	0.13	0.12	0.54	0.96	0.67
	19	11	2	10	5	4	18	3	15	13	12	14	16	7
	X	11	2	R	R	4	X	R	X	13	R	X	X	7

Fig. 2.1. Using a sequence of random numbers to generate a random permutation of 13 numbers

0.71	0.66	0.31	0.71	0.78	0.11	0.11	0.41	0.91	0.24	0.99	0.20	0.53	0.68	0.24	0.03
10	8	5	X	11	2	X	6	12	4	13	3	7	9	X	1

Fig. 2.2. Second method of using a sequence of random numbers to generate a random permutation of 13 numbers

in the top row in Figure 2.1. To turn these into the numbers 1–13 in a simple way, keeping the uniform distribution, multiply each random number by 100 and subtract multiples of 20 to leave a number in the range 1–20. Cross out numbers bigger than 13: these are marked X in the figure. Remove any number that has already occurred: these are marked R in the figure. Continue to produce random numbers until 12 different numbers have been listed. Then the final number must be the missing one, so it can be written down without further ado.

Thus the sequence of random numbers given in Figure 2.1 gives us the random order

$$12 \quad 5 \quad 1 \quad 3 \quad 10 \quad 9 \quad 6 \quad 11 \quad 2 \quad 4 \quad 13 \quad 7 \quad 8.$$

Although this process is simple, it usually requires the generation of far more random numbers than the number of plots being permuted. In the second method we generate only a few more random numbers than the number of plots, because only the exact repeats are crossed out. Then place a 1 under the smallest number, a 2 under the second smallest, and so on, replacing each random number by its rank. This process is shown in Figure 2.2. It gives us the random order

$$10 \quad 8 \quad 5 \quad 11 \quad 2 \quad 6 \quad 12 \quad 4 \quad 13 \quad 3 \quad 7 \quad 9 \quad 1.$$

2.3 The treatment subspace

Definition The function $T: \Omega \to \mathcal{T}$ is called the *treatment factor*.

There is an N-dimensional vector space associated with Ω. It consists of N-tuples of real numbers, with each place in the N-tuple associated with one plot. Formally this vector space is $\mathbb{R}^\Omega$, but it will be called V for the remainder of this book. Because data sheets usually have one row for each observational unit, with each variable as a column, vectors in this book are usually written as column vectors.

Definition The *treatment subspace* of V consists of those vectors in V which are constant on each treatment.

Table 2.1. *Some vectors in Example 2.1*

Ω	T	Typical vector $\mathbf{v}$	Data vector $\mathbf{y}$	Some vectors in V_T			Orthogonal basis for V_T			Unknown treatment parameters τ	Fitted values $\hat{\tau}$
				$\mathbf{v}_1$	$\mathbf{v}_2$	$\mathbf{v}_3$	$\mathbf{u}_A$	$\mathbf{u}_B$	$\mathbf{u}_C$		
1	B	v_1	y_1	1	0	$-\frac{1}{4}$	0	1	0	τ_B	$\hat{\tau}_B$
2	A	v_2	y_2	0	$\frac{1}{5}$	$\frac{1}{5}$	1	0	0	τ_A	$\hat{\tau}_A$
3	C	v_3	y_3	3	0	0	0	0	1	τ_C	$\hat{\tau}_C$
4	C	v_4	y_4	3	0	0	0	0	1	τ_C	$\hat{\tau}_C$
5	B	v_5	y_5	1	0	$-\frac{1}{4}$	0	1	0	τ_B	$\hat{\tau}_B$
6	A	v_6	y_6	0	$\frac{1}{5}$	$\frac{1}{5}$	1	0	0	τ_A	$\hat{\tau}_A$
7	B	v_7	y_7	1	0	$-\frac{1}{4}$	0	1	0	τ_B	$\hat{\tau}_B$
8	A	v_8	y_8	0	$\frac{1}{5}$	$\frac{1}{5}$	1	0	0	τ_A	$\hat{\tau}_A$
9	C	v_9	y_9	3	0	0	0	0	1	τ_C	$\hat{\tau}_C$
10	C	v_{10}	y_{10}	3	0	0	0	0	1	τ_C	$\hat{\tau}_C$
11	A	v_{11}	y_{11}	0	$\frac{1}{5}$	$\frac{1}{5}$	1	0	0	τ_A	$\hat{\tau}_A$
12	B	v_{12}	y_{12}	1	0	$-\frac{1}{4}$	0	1	0	τ_B	$\hat{\tau}_B$
13	A	v_{13}	y_{13}	0	$\frac{1}{5}$	$\frac{1}{5}$	1	0	0	τ_A	$\hat{\tau}_A$

Notation Since the treatment factor is T, the treatment subspace will be denoted V_T.

Definition A vector $\mathbf{v}$ in V is a *treatment vector* if $\mathbf{v} \in V_T$; it is a *treatment contrast* if $\mathbf{v} \in V_T$ and $\sum_{\omega \in \Omega} v_\omega = 0$.

Example 2.1 revisited (Fictitious) The left side of Table 2.1 shows the set Ω, the treatment factor T, a typical vector $\mathbf{v}$ in V and the data vector $\mathbf{y}$. Beyond these are some vectors in the treatment subspace V_T. The vectors are shown as columns. The vector $\mathbf{v}_3$ is a treatment contrast.

Notation For each treatment i let $\mathbf{u}_i$ be the vector whose value on plot ω is equal to

$$\begin{cases} 1 & \text{if } T(\omega) = i \\ 0 & \text{otherwise.} \end{cases}$$

Table 2.1 shows the special vectors $\mathbf{u}_A$, $\mathbf{u}_B$ and $\mathbf{u}_C$.

It is clear that every vector in V_T is a unique linear combination of the vectors $\mathbf{u}_1, \ldots, \mathbf{u}_t$. In Table 2.1, $\mathbf{v}_1 = \mathbf{u}_A + 3\mathbf{u}_C$, $\mathbf{v}_2 = (1/5)\mathbf{u}_A$ and $\mathbf{v}_3 = (1/5)\mathbf{u}_A - (1/4)\mathbf{u}_B$.

Recall the *scalar product*: for vectors $\mathbf{v}$ and $\mathbf{w}$ in V, the scalar product $\mathbf{v} \cdot \mathbf{w}$ of $\mathbf{v}$ and $\mathbf{w}$ is defined by

$$\mathbf{v} \cdot \mathbf{w} = \sum_{\omega \in \Omega} v_\omega w_\omega = v_1 w_1 + v_2 w_2 + \cdots + v_N w_N = \mathbf{v}^\top \mathbf{w},$$

where $^\top$ denotes transpose. In particular, $\mathbf{v} \cdot \mathbf{v} = \sum_{\omega \in \Omega} v_\omega^2$, which is sometimes called the *sum of squares* of $\mathbf{v}$ and sometimes the *squared length* of $\mathbf{v}$; it is also written as $\|\mathbf{v}\|^2$.

We say that the vector **v** is *orthogonal* to the vector **w** (written $\mathbf{v} \perp \mathbf{w}$) if $\mathbf{v} \cdot \mathbf{w} = 0$.

In Table 2.1, $\mathbf{u}_A \cdot \mathbf{u}_A = 5$ and $\mathbf{u}_B \cdot \mathbf{u}_B = \mathbf{u}_C \cdot \mathbf{u}_C = 4$. Also, $\mathbf{u}_A \cdot \mathbf{u}_B = 0$, so $\mathbf{u}_A$ is orthogonal to $\mathbf{u}_B$.

Proposition 2.1 *For each treatment i, $\mathbf{u}_i \cdot \mathbf{u}_i = r_i$.*

Proposition 2.2 *If $i \ne j$ then $\mathbf{u}_i \perp \mathbf{u}_j$ and so the set of vectors $\{\mathbf{u}_i : i \in \mathcal{T}\}$ is an orthogonal basis for V_T.*

Corollary 2.3 *If there are t treatments then $\dim(V_T) = t$.*

Also in the treatment subspace is the vector $\boldsymbol{\tau}$ of unknown treatment parameters. Its value on plot ω is equal to $\tau_{T(\omega)}$. Under the linear model assumed in Section 2.5, $\mathbb{E}(Y_\omega) = \tau_{T(\omega)}$, so the *fitted value* on plot ω is equal to the estimated value $\hat{\tau}_{T(\omega)}$ of $\tau_{T(\omega)}$. Thus we have a vector $\hat{\boldsymbol{\tau}}$ of fitted values, which is also in the treatment subspace. These two vectors are also shown in Table 2.1.

2.4 Orthogonal projection

Orthogonality is important in statistics, partly because orthogonal vectors often correspond to random variables with zero correlation. Many procedures in estimation and analysis of variance are nothing more than the decomposition of the data vector into orthogonal pieces.

Definition If W is a subspace of V then the *orthogonal complement* of W is the set

$$\{\mathbf{v} \in V : \mathbf{v} \text{ is orthogonal to } \mathbf{w} \text{ for all } \mathbf{w} \text{ in } W\}.$$

Notation The orthogonal complement of W is written $W^\perp$, which is pronounced 'W perp'.

Theorem 2.4 *Let W be a subspace of V. Then the following hold.*

(i) *$W^\perp$ is also a subspace of V.*

(ii) *$(W^\perp)^\perp = W$.*

(iii) *$\dim(W^\perp) = \dim V - \dim W$.*

(iv) *V is the* internal direct sum *$W \oplus W^\perp$; this means that given any vector $\mathbf{v}$ in V there is a unique vector $\mathbf{x}$ in W and a unique vector $\mathbf{z}$ in $W^\perp$ such that $\mathbf{v} = \mathbf{x} + \mathbf{z}$. We call $\mathbf{x}$ the* orthogonal projection *of $\mathbf{v}$ onto W, and write $\mathbf{x} = \mathbf{P}_W(\mathbf{v})$. See Figure 2.3. Orthogonal projection is a linear transformation, so $\mathbf{P}_W$ is effectively an $N \times N$ matrix, so we often write $\mathbf{P}_W(\mathbf{v})$ as $\mathbf{P}_W \mathbf{v}$.*

(v) *$\mathbf{P}_{W^\perp}(\mathbf{v}) = \mathbf{z} = \mathbf{v} - \mathbf{x} = \mathbf{v} - \mathbf{P}_W \mathbf{v}$.*

(vi) *For a fixed vector $\mathbf{v}$ in V and vector $\mathbf{w}$ in W, $\sum_{\omega \in \Omega}(v_\omega - w_\omega)^2 = \|\mathbf{v} - \mathbf{w}\|^2$. As $\mathbf{w}$ varies over W, this sum of squares of differences is minimized when $\mathbf{w} = \mathbf{P}_W \mathbf{v}$.*

(vii) *If $\{\mathbf{u}_1, \ldots, \mathbf{u}_n\}$ is an orthogonal basis for W then*

$$\mathbf{P}_W \mathbf{v} = \left(\frac{\mathbf{v} \cdot \mathbf{u}_1}{\mathbf{u}_1 \cdot \mathbf{u}_1}\right)\mathbf{u}_1 + \left(\frac{\mathbf{v} \cdot \mathbf{u}_2}{\mathbf{u}_2 \cdot \mathbf{u}_2}\right)\mathbf{u}_2 + \cdots + \left(\frac{\mathbf{v} \cdot \mathbf{u}_n}{\mathbf{u}_n \cdot \mathbf{u}_n}\right)\mathbf{u}_n.$$

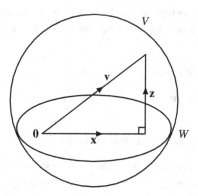

Fig. 2.3. The vector **x** is the orthogonal projection of the vector **v** onto the subspace W

2.5 Linear model

For unstructured plots we assume that

$$\mathbf{Y} = \boldsymbol{\tau} + \mathbf{Z},$$

where $\boldsymbol{\tau} \in V_T$, $\mathbb{E}(\mathbf{Z}) = \mathbf{0}$, $\mathrm{Var}(Z_\omega) = \sigma^2$ for all ω in Ω, and $\mathrm{cov}(Z_\alpha, Z_\beta) = 0$ for different plots α and β. In other words, $\mathbb{E}(\mathbf{Y}) = \boldsymbol{\tau}$, which is an unknown vector in V_T, and $\mathrm{Cov}(\mathbf{Y}) = \sigma^2 \mathbf{I}$, where $\mathbf{I}$ is the $N \times N$ identity matrix.

Under these assumptions, standard linear model theory gives the following results.

Theorem 2.5 *Assume that $\mathbb{E}(\mathbf{Y}) = \boldsymbol{\tau}$ and that $\mathrm{Cov}(\mathbf{Y}) = \sigma^2 \mathbf{I}$. Let W be a d-dimensional subspace of V. Then*

(i) $\mathbb{E}(\mathbf{P}_W \mathbf{Y}) = \mathbf{P}_W(\mathbb{E}(\mathbf{Y})) = \mathbf{P}_W \boldsymbol{\tau}$;

(ii) $\mathbb{E}(\|\mathbf{P}_W \mathbf{Y}\|^2) = \|\mathbf{P}_W \boldsymbol{\tau}\|^2 + d\sigma^2$.

Theorem 2.6 *Assume that $\mathbb{E}(\mathbf{Y}) = \boldsymbol{\tau} \in V_T$ and that $\mathrm{Cov}(\mathbf{Y}) = \sigma^2 \mathbf{I}$. Let $\mathbf{x}$ and $\mathbf{z}$ be any vectors in V_T. Then*

(i) *the best (that is, minimum variance) linear unbiased estimator of the scalar $\mathbf{x} \cdot \boldsymbol{\tau}$ is $\mathbf{x} \cdot \mathbf{Y}$;*

(ii) *the variance of the estimator $\mathbf{x} \cdot \mathbf{Y}$ is $\|\mathbf{x}\|^2 \sigma^2$;*

(iii) *the covariance of $\mathbf{x} \cdot \mathbf{Y}$ and $\mathbf{z} \cdot \mathbf{Y}$ is $(\mathbf{x} \cdot \mathbf{z})\sigma^2$.*

2.6 Estimation

Proposition 2.1 shows that, if the $\mathbf{u}_i$ are the vectors defined in Section 2.3, then $\mathbf{u}_i \cdot \mathbf{u}_i = r_i$. Moreover, if $\mathbf{v}$ is any other vector in V then $\mathbf{u}_i \cdot \mathbf{v}$ is equal to the sum of the values of $\mathbf{v}$ on those plots with treatment i. In particular $\mathbf{u}_i \cdot \boldsymbol{\tau} = r_i \tau_i$. Write $\mathrm{SUM}_{T=i}$ for the sum of the values

2.6. Estimation

of $\mathbf{Y}$ on the plots with treatment i, and $\text{sum}_{T=i}$ for the sum of the values of $\mathbf{y}$ on the plots with treatment i. Then
$$\mathbf{u}_i \cdot \mathbf{Y} = \text{SUM}_{T=i} \quad \text{and} \quad \mathbf{u}_i \cdot \mathbf{y} = \text{sum}_{T=i}.$$

Also, write the means $\text{SUM}_{T=i}/r_i$ and $\text{sum}_{T=i}/r_i$ as $\text{MEAN}_{T=i}$ and $\text{mean}_{T=i}$ respectively.

Similarly, let $\mathbf{u}_0$ be the all-1 vector; that is, $\mathbf{u}_0 = \sum_{i=1}^{t} \mathbf{u}_i$. For every vector $\mathbf{v}$ in V, write
$$\bar{v} = \sum_{\omega \in \Omega} v_\omega / N.$$

Then $\mathbf{u}_0 \cdot \mathbf{v} = \sum_{\omega \in \Omega} v_\omega = N\bar{v}$ for all $\mathbf{v}$ in V. In particular, $\mathbf{u}_0 \cdot \mathbf{u}_0 = N$,
$$\mathbf{u}_0 \cdot \mathbf{Y} = \text{SUM} = N\bar{Y} \quad \text{and} \quad \mathbf{u}_0 \cdot \mathbf{y} = \text{sum} = N\bar{y},$$

where SUM and sum are the grand totals $\sum_{\omega \in \Omega} Y_\omega$ and $\sum_{\omega \in \Omega} y_\omega$ respectively and $\bar{Y}$ and $\bar{y}$ are the grand means SUM/N and sum/N respectively.

To estimate the treatment parameter τ_i, put $\mathbf{x} = (1/r_i)\mathbf{u}_i$. Then $\mathbf{x} \cdot \boldsymbol{\tau} = \mathbf{x} \cdot \mathbf{u}_i/r_i = \tau_i$ and $\mathbf{x} \cdot \mathbf{Y} = \text{SUM}_{T=i}/r_i = \text{MEAN}_{T=i}$. Therefore Theorem 2.6(i) shows that the best linear unbiased estimator of τ_i is $\text{MEAN}_{T=i}$, with corresponding estimate $\hat{\tau}_i$ equal to $\text{mean}_{T=i}$.

Similarly, to estimate a linear combination such as $\sum_{i=1}^{t} \lambda_i \tau_i$, put $\mathbf{x} = \sum_{i=1}^{t} (\lambda_i/r_i)\mathbf{u}_i$. Now
$$\mathbf{x} \cdot \boldsymbol{\tau} = \sum_{i=1}^{t} \lambda_i \left(\frac{1}{r_i} \mathbf{u}_i \cdot \boldsymbol{\tau} \right) = \sum_{i=1}^{t} \lambda_i \tau_i$$

and
$$\mathbf{x} \cdot \mathbf{Y} = \sum_{i=1}^{t} \lambda_i \left(\frac{1}{r_i} \mathbf{u}_i \cdot \mathbf{Y} \right) = \sum_{i=1}^{t} \lambda_i \text{MEAN}_{T=i},$$

so $\sum \lambda_i \hat{\tau}_i$ is the best linear unbiased estimate of $\sum \lambda_i \tau_i$.

In particular, put $\bar{\tau} = \sum_{i=1}^{t} r_i \tau_i / N$, which is the linear combination of $\tau_1, \ldots, \tau_t$ which has $\lambda_i = r_i/N$. Then $\mathbf{x} = (1/N)\mathbf{u}_0$, so
$$\mathbf{x} \cdot \mathbf{Y} = \frac{\text{SUM}}{N} = \bar{Y},$$

and this is the best linear unbiased estimator of $\bar{\tau}$.

Now we look at Theorem 2.5 with $W = V_T$. Since $\boldsymbol{\tau} \in V_T$, we have
$$\mathbf{P}_{V_T} \boldsymbol{\tau} = \boldsymbol{\tau} = \sum_{i=1}^{t} \tau_i \mathbf{u}_i.$$

Proposition 2.2 and Theorem 2.4(vii) show that
$$\mathbf{P}_{V_T} \mathbf{Y} = \sum_{i=1}^{t} \left(\frac{\mathbf{Y} \cdot \mathbf{u}_i}{\mathbf{u}_i \cdot \mathbf{u}_i} \right) \mathbf{u}_i = \sum_{i=1}^{t} \frac{\text{SUM}_{T=i}}{r_i} \mathbf{u}_i = \sum_{i=1}^{t} \text{MEAN}_{T=i} \mathbf{u}_i.$$

Theorem 2.5(i) confirms that this is an unbiased estimator of $\boldsymbol{\tau}$.

Example 2.2 (Milk production) An experiment to compare the effects of three different diets on milk production used 32 Holstein dairy cows at similar points in their lactation cycles. They were fed the diets for three weeks. During the third week, the average daily milk production was recorded for each cow. These data, in pounds per day, are shown in Table 2.2.

Thus $r_A = 11$, $r_B = 13$ and $r_C = 8$. Furthermore, $\text{sum}_{T=A} = 660.0$, so $\hat{\tau}_A = 60.00$; $\text{sum}_{T=B} = 723.4$, so $\hat{\tau}_B = 55.65$; and $\text{sum}_{T=C} = 468.4$, so $\hat{\tau}_C = 58.55$. Therefore the best estimate of the difference in daily milk production between cows on diets A and B is 4.35.

Table 2.2. *Data for Example 2.2: milk yield per cow in pounds per day*

Diet A	60.7	59.7	61.9	61.8	62.6	62.5	57.9	59.0	59.6	57.1	57.2		
Diet B	55.6	52.9	52.7	53.1	60.5	58.0	60.5	53.3	59.2	56.2	50.5	56.0	54.9
Diet C	62.8	55.8	56.0	62.3	60.1	54.9	60.6	55.9					

2.7 Comparison with matrix notation

Some authors use matrices to express linear models. Let $\mathbf{X}$ be the $N \times t$ matrix defined by

$$\mathbf{X} = [\ \mathbf{u}_1 \quad \mathbf{u}_2 \quad \ldots \quad \mathbf{u}_t\ ],$$

where $\mathbf{u}_1, \ldots, \mathbf{u}_t$ are regarded as column vectors. Thus, in Example 2.1, $\mathbf{X}$ consists of the three columns of Table 2.1 headed $\mathbf{u}_A$, $\mathbf{u}_B$ and $\mathbf{u}_C$. Let $\boldsymbol{\beta}$ be the vector $[\tau_1, \tau_2, \ldots, \tau_t]^\top$. Note that $\boldsymbol{\beta}$ is not the same as $\boldsymbol{\tau}$, because $\boldsymbol{\beta}$ has t entries while $\boldsymbol{\tau}$ has N. Then $\boldsymbol{\tau} = \mathbf{X}\boldsymbol{\beta}$, and so the matrix formulation of the expectation model is $\mathbb{E}(\mathbf{Y}) = \mathbf{X}\boldsymbol{\beta}$, whose least-squares solution is

$$\hat{\boldsymbol{\beta}} = \left(\mathbf{X}^\top \mathbf{X}\right)^{-1} \mathbf{Y}. \tag{2.1}$$

Now,

$$\mathbf{X}^\top \mathbf{X} = \begin{bmatrix} \mathbf{u}_1 \cdot \mathbf{u}_1 & \mathbf{u}_1 \cdot \mathbf{u}_2 & \cdots & \mathbf{u}_1 \cdot \mathbf{u}_t \\ \mathbf{u}_2 \cdot \mathbf{u}_1 & \mathbf{u}_2 \cdot \mathbf{u}_2 & \cdots & \mathbf{u}_2 \cdot \mathbf{u}_t \\ \vdots & \vdots & \ddots & \vdots \\ \mathbf{u}_t \cdot \mathbf{u}_1 & \mathbf{u}_t \cdot \mathbf{u}_2 & \cdots & \mathbf{u}_t \cdot \mathbf{u}_t \end{bmatrix} = \begin{bmatrix} r_1 & 0 & \cdots & 0 \\ 0 & r_2 & \cdots & 0 \\ \vdots & \vdots & \ddots & 0 \\ 0 & 0 & \cdots & r_t \end{bmatrix},$$

and so

$$\left(\mathbf{X}^\top \mathbf{X}\right)^{-1} = \begin{bmatrix} \frac{1}{r_1} & 0 & \cdots & 0 \\ 0 & \frac{1}{r_2} & \cdots & 0 \\ \vdots & \vdots & \ddots & 0 \\ 0 & 0 & \cdots & \frac{1}{r_t} \end{bmatrix}.$$

Moreover,

$$\mathbf{X}^\top \mathbf{Y} = \begin{bmatrix} \mathbf{u}_1^\top \\ \mathbf{u}_2^\top \\ \vdots \\ \mathbf{u}_t^\top \end{bmatrix} \mathbf{Y} = \begin{bmatrix} \mathbf{u}_1 \cdot \mathbf{Y} \\ \mathbf{u}_2 \cdot \mathbf{Y} \\ \vdots \\ \mathbf{u}_t \cdot \mathbf{Y} \end{bmatrix} = \begin{bmatrix} \mathrm{SUM}_{T=1} \\ \mathrm{SUM}_{T=2} \\ \vdots \\ \mathrm{SUM}_{T=t} \end{bmatrix}.$$

Hence Equation (2.1) gives $\hat{\tau}_i = \mathrm{mean}_{T=i}$ for $i = 1, \ldots, t$, which agrees with Section 2.6.

2.8 Sums of squares

Definition Let W be any subspace of V. The *sum of squares* for W means either $\|\mathbf{P}_W \mathbf{Y}\|^2$ or $\|\mathbf{P}_W \mathbf{y}\|^2$. The *degrees of freedom* for W is another name for $\dim W$. The *mean square* for W is

$$\frac{\text{sum of squares for } W}{\text{degrees of freedom for } W},$$

2.8. Sums of squares

for either sense of sum of squares. The *expected mean square* for W, written $\text{EMS}(W)$, is the expectation of the mean square in the random sense; that is

$$\text{EMS}(W) = \frac{\mathbb{E}(\|\mathbf{P}_W\mathbf{Y}\|^2)}{\dim W}.$$

First we apply these ideas with $W = V_T$. Since

$$\mathbf{P}_{V_T}\mathbf{Y} = \sum_{i=1}^{t} \frac{\text{SUM}_{T=i}}{r_i}\mathbf{u}_i,$$

the sum of squares for V_T is equal to

$$\left(\sum_{i=1}^{t} \frac{\text{SUM}_{T=i}}{r_i}\mathbf{u}_i\right) \cdot \left(\sum_{j=1}^{t} \frac{\text{SUM}_{T=j}}{r_j}\mathbf{u}_j\right).$$

Now, $\mathbf{u}_i \cdot \mathbf{u}_j = 0$ whenever $i \neq j$, so this sum of squares

$$= \sum_{i=1}^{t} \frac{(\text{SUM}_{T=i})^2}{r_i^2}\mathbf{u}_i \cdot \mathbf{u}_i = \sum_{i=1}^{t} \frac{(\text{SUM}_{T=i})^2}{r_i}.$$

The quantity $\sum_i (\text{sum}_{T=i}^2/r_i)$ is called the *crude sum of squares for treatments*, which may be abbreviated to CSS(treatments).

The number of degrees of freedom for V_T is simply the dimension of V_T, which is equal to t.

The mean square for V_T is equal to

$$\sum_{i=1}^{t} \frac{(\text{SUM}_{T=i})^2}{r_i} \bigg/ t.$$

Theorem 2.5(ii) shows that

$$\mathbb{E}(\|\mathbf{P}_{V_T}\mathbf{Y}\|^2) = \|\mathbf{P}_{V_T}\boldsymbol{\tau}\|^2 + t\sigma^2 = \|\boldsymbol{\tau}\|^2 + t\sigma^2 = \sum_{i=1}^{t} r_i \tau_i^2 + t\sigma^2,$$

because $\boldsymbol{\tau} = \sum_{i=1}^{t} \tau_i \mathbf{u}_i$. Hence the expected mean square for V_T is equal to $\sum r_i \tau_i^2 / t + \sigma^2$.

Secondly we apply the ideas with $W = V_T^\perp$. By Theorem 2.4(v),

$$\begin{aligned}\mathbf{P}_W \mathbf{y} &= \mathbf{y} - \mathbf{P}_{V_T}\mathbf{y} = \mathbf{y} - \sum_{i=1}^{t} \hat{\tau}_i \mathbf{u}_i \\ &= \text{data vector} - \text{vector of fitted values} \\ &= \text{residual vector},\end{aligned}$$

so $\|\mathbf{P}_W \mathbf{y}\|^2$ is equal to the sum of the squares of the residuals. For this reason, all the quantities associated with W are named 'residual'. (The word 'error' is sometimes used, but this can be confusing to non-statisticians, who tend to interpret it as 'mistake'.)

Now, **y** is the sum of the orthogonal vectors $\mathbf{P}_{V_T}\mathbf{y}$ and $\mathbf{P}_W\mathbf{y}$, so Pythagoras's Theorem shows that

$$\sum_{\omega \in \Omega} y_\omega^2 = \|\mathbf{y}\|^2 = \|\mathbf{P}_{V_T}\mathbf{y}\|^2 + \|\mathbf{P}_W\mathbf{y}\|^2.$$

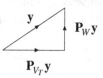

The quantity $\sum_{\omega \in \Omega} y_\omega^2$ is just the *total sum of squares*, so the sum of squares for residual is equal to the difference between the total sum of squares and the crude sum of squares for treatments; indeed, this is usually the easiest way to calculate it.

The number of degrees of freedom for residual, which may be written as df(residual), is equal to the dimension of W, which is $N-t$, by Theorem 2.4(iii). Hence the mean square for residual is equal to

$$\frac{\text{sum of squares for residual}}{N-t}.$$

This will be denoted MS(residual).

We know that $\mathbf{P}_W \boldsymbol{\tau} = \mathbf{0}$ because $\boldsymbol{\tau} \in V_T$. Thus Theorem 2.5(ii) shows that $\mathbb{E}(\|\mathbf{P}_W\mathbf{Y}\|^2) = (N-t)\sigma^2$ and hence

$$\text{EMS(residual)} = \sigma^2. \quad (2.2)$$

Example 2.2 revisited (Milk production) Here the crude sum of squares for treatments is

$$\frac{660.0^2}{11} + \frac{723.4^2}{13} + \frac{468.4^2}{8} = 107279.2477.$$

The total sum of squares is $60.7^2 + \cdots + 55.9^2 = 107515.62$. Therefore the residual sum of squares is $107515.6200 - 107279.2477 = 236.3723$. The number of degrees of freedom for residual is $32 - 3 = 29$, so the residual mean square is $236.3723/29 = 8.1508$.

2.9 Variance

Section 2.6 showed that the best linear unbiased estimator of the linear combination $\sum \lambda_i \tau_i$ is $\mathbf{x} \cdot \mathbf{Y}$, where $\mathbf{x} = \sum (\lambda_i/r_i)\mathbf{u}_i$. By Theorem 2.6(ii), the variance of this estimator is equal to $\|\mathbf{x}\|^2 \sigma^2$. Now

$$\|\mathbf{x}\|^2 \sigma^2 = \left(\sum_{i=1}^t \frac{\lambda_i}{r_i}\mathbf{u}_i\right) \cdot \left(\sum_{j=1}^t \frac{\lambda_j}{r_j}\mathbf{u}_j\right)\sigma^2 = \sum_{i=1}^t \frac{\lambda_i^2}{r_i^2}(\mathbf{u}_i \cdot \mathbf{u}_i)\sigma^2 = \left(\sum_{i=1}^t \frac{\lambda_i^2}{r_i}\right)\sigma^2.$$

Two cases are particularly important. To estimate the treatment parameter τ_i, for a fixed treatment i, put $\lambda_i = 1$ and $\lambda_j = 0$ for $j \neq i$. Then the variance is σ^2/r_i. To estimate the simple difference $\tau_i - \tau_j$ for fixed different i and j, put $\lambda_i = 1$, $\lambda_j = -1$ and $\lambda_k = 0$ if $k \neq i$ and $k \neq j$. Then the variance is equal to

$$\sigma^2\left(\frac{1}{r_i} + \frac{1}{r_j}\right). \quad (2.3)$$

Equation (2.2) shows that MS(residual) is an unbiased estimator of σ^2, so

$$\sum_{i=1}^t \frac{\lambda_i^2}{r_i} \times \text{MS(residual)} \quad (2.4)$$

is an unbiased estimator of the variance of the estimator of $\sum \lambda_i \tau_i$.

2.9. Variance

Definition The *standard error* for the estimate $\sum \lambda_i \hat{\tau}_i$ is the square root of the estimate of the variance given by (2.4); that is, the standard error is equal to

$$\sqrt{\sum_{i=1}^{t} \frac{\lambda_i^2}{r_i} \times \text{MS(residual)}}.$$

Thus the standard error for $\hat{\tau}_i$ is $\sqrt{\text{MS(residual)}/r_i}$. This is called the *standard error of a mean*, which may be abbreviated to s.e.m. Similarly, the standard error for $\hat{\tau}_i - \hat{\tau}_j$ is

$$\sqrt{\text{MS(residual)} \left(\frac{1}{r_i} + \frac{1}{r_j} \right)},$$

which is called the *standard error of a difference* and abbreviated to s.e.d.

Example 2.1 revisited (Fictitious) Here the estimate $\hat{\tau}_A$ of τ_A is equal to $\text{sum}_{T=A}/5$, with variance $\sigma^2/5$. The simple difference $\tau_A - \tau_B$ is estimated by $\text{sum}_{T=A}/5 - \text{sum}_{T=B}/4$, with variance $9\sigma^2/20$.

The estimators $\hat{\tau}_A$, $\hat{\tau}_B$ and $\hat{\tau}_C$ are mutually uncorrelated, but

$$\text{Cov}(\hat{\tau}_A - \hat{\tau}_B, \hat{\tau}_A - \hat{\tau}_C) = \sigma^2/5.$$

Furthermore, the estimators $\hat{\tau}_B - \hat{\tau}_C$ and $\tau_A - (\hat{\tau}_B + \hat{\tau}_C)/2$ are uncorrelated.

Example 2.2 revisited (Milk production) The standard error of the mean for diet A is equal to $\sqrt{8.1508/11}$, which is 0.86. The standard errors of the means for diets B and C are 0.79 and 1.01 respectively.

The standard error of the difference $\hat{\tau}_A - \hat{\tau}_B$ is $\sqrt{8.1508 \times 24/143}$, which is 1.17. Similarly, the standard errors of the differences $\hat{\tau}_A - \hat{\tau}_C$ and $\hat{\tau}_B - \hat{\tau}_C$ are 1.33 and 1.28 respectively.

The theory from Theorem 2.5 onwards has made no assumptions about the distribution of $\mathbf{Y}$ apart from its expectation and covariance. Assuming multivariate normality enables us to say more about the distributions of some statistics.

Theorem 2.7 *Suppose that the distribution of $\mathbf{Y}$ is multivariate normal, that $\mathbb{E}(\mathbf{Y}) = \boldsymbol{\tau} \in V_T$ and that $\text{Cov}(\mathbf{Y})$ is a scalar matrix. Then the following hold.*

(i) *If $\mathbf{x} = \sum_{i=1}^{t} (\lambda_i/r_i) \mathbf{u}_i$ then*

$$\frac{\mathbf{x} \cdot \mathbf{Y} - \sum \lambda_i \tau_i}{\sqrt{\left(\sum \frac{\lambda_i^2}{r_i} \right) \times \text{MS(residual)}}}$$

has a t-distribution on $N - t$ degrees of freedom.

(ii) *If $\mathbf{x}$ and $\mathbf{z}$ are in V and $\mathbf{x} \cdot \mathbf{z} = 0$ then $\mathbf{x} \cdot \mathbf{Y}$ and $\mathbf{z} \cdot \mathbf{Y}$ are independent estimators.*

2.10 Replication: equal or unequal?

If the treatments are unstructured then we assume that all estimates of simple treatment differences are equally important. Thus the variances of all these estimators should be as small as possible. Equation (2.3) shows that the average variance of these estimators is equal to

$$\frac{1}{t(t-1)}\sum_{i=1}^{t}\sum_{j\neq i}\left(\frac{1}{r_i}+\frac{1}{r_j}\right)\sigma^2 = \frac{1}{t}\sum_{i=1}^{t}\frac{1}{r_i}\sigma^2.$$

Proposition 2.8 *If positive numbers $r_1, \ldots, r_t$ have a fixed sum R then $\sum(1/r_i)$ is minimized when $r_1 = r_2 = \cdots = r_t = R/t$.*

This proposition, with $R = N$, shows that the average variance of the estimators of simple differences is minimized when the replications $r_1, \ldots, r_t$ are as equal as possible. That is why most designs are equireplicate.

Suppose that there is one treatment which is not sufficiently available for it to have replication N/t. If there are only two treatments this is often not a problem, because replacing replications of $N/2$ and $N/2$ by $N/3$ and $2N/3$ increases the variance from $4\sigma^2/N$ to $9\sigma^2/2N$, an increase of only 12.5%. You may even be able to increase the number of plots slightly, to maintain the variance, if there is an unlimited supply of the second treatment.

Example 2.3 (Limited availability) Suppose that there are two treatments and that 16 plots are available. Ideally, each treatment is applied to eight plots, in which case the variance of the difference is $\sigma^2/4$. Suppose that there is only enough of the first treatment for six plots but that there are unlimited supplies of the second treatment. Keeping 16 plots we can have replications 6 and 10, giving a variance of $4\sigma^2/15$. If we can use two more plots then we can increase the replication of the second treatment to 12, in which case the variance returns to $\sigma^2/4$.

Life is not so simple when there are more than two treatments. Here the best that you can do is use the maximum amount of any treatment(s) whose availability is less than N/t, and make the remaining replications as equal as possible. The number of plots is rarely specified exactly in advance, so it will usually be possible to make all the remaining replications equal by including a few extra plots.

2.11 Allowing for the overall mean

In Section 2.8 we saw that

$$\text{EMS}(V_T) = \frac{\sum_{i=1}^{t} r_i \tau_i^2}{t} + \sigma^2 = \frac{\sum_{i=1}^{t} r_i \tau_i^2}{t} + \text{EMS}(\text{residual}).$$

Thus the difference between $\text{EMS}(V_T)$ and $\text{EMS}(\text{residual})$ is non-negative, and is equal to zero if and only if $\tau_1 = \tau_2 = \cdots = \tau_t = 0$.

Usually we do not measure on a scale that makes it plausible that $\tau_1 = \tau_2 = \cdots = \tau_t = 0$. However, it is often plausible that $\tau_1 = \tau_2 = \cdots = \tau_t$. This is called the *null model*, in which

$$\mathbb{E}(Y_\omega) = \kappa \quad \text{for all } \omega \text{ in } \Omega,$$

2.11. Allowing for the overall mean

where κ is an unknown constant. In other words, $\mathbb{E}(\mathbf{Y})$ is a scalar multiple of $\mathbf{u}_0$.

Let V_0 be the subspace of V which consists of scalar multiples of $\mathbf{u}_0$. Then $\{\mathbf{u}_0\}$ is a basis for V_0 and $\dim V_0 = 1$. Theorem 2.4(vii) gives the following.

Proposition 2.9 *If* $\mathbf{v} \in V$ *then*

$$\mathbf{P}_{V_0}\mathbf{v} = \left(\frac{\mathbf{v}\cdot\mathbf{u}_0}{\mathbf{u}_0\cdot\mathbf{u}_0}\right)\mathbf{u}_0 = \left(\frac{\text{grand total of }\mathbf{v}}{N}\right)\mathbf{u}_0 = \bar{v}\mathbf{u}_0$$

and

$$\|\mathbf{P}_{V_0}\mathbf{v}\|^2 = \left(\frac{\text{grand total of }\mathbf{v}}{N}\right)^2 \mathbf{u}_0\cdot\mathbf{u}_0 = \frac{(\text{grand total of }\mathbf{v})^2}{N} = N\bar{v}^2.$$

Proposition 2.9 shows that

$$\mathbf{P}_{V_0}\mathbf{Y} = \bar{Y}\mathbf{u}_0 = \frac{\text{SUM}}{N}\mathbf{u}_0$$

so the sum of squares for V_0 is

$$\|\mathbf{P}_{V_0}\mathbf{Y}\|^2 = \frac{\text{SUM}^2}{N^2}\mathbf{u}_0\cdot\mathbf{u}_0 = \frac{\text{SUM}^2}{N}.$$

The sum of squares for V_0 is called the *crude sum of squares for the mean*, or just the *sum of squares for the mean*, which will be written as SS(mean). Substituting data gives

$$\text{SS(mean)} = \frac{\text{sum}^2}{N}$$

and

$$\text{MS(mean)} = \frac{\text{SS(mean)}}{1} = \frac{\text{sum}^2}{N}.$$

This version, with data rather than random variables, is sometimes called the *correction for the mean*, which suggests, rather misleadingly, that there is something incorrect about the full data.

Proposition 2.9 also shows that $\mathbf{P}_{V_0}\boldsymbol{\tau} = \bar{\tau}\mathbf{u}_0$. Then Theorem 2.5 gives

$$\mathbb{E}(\mathbf{P}_{V_0}\mathbf{Y}) = \mathbf{P}_{V_0}(\mathbb{E}(\mathbf{Y})) = \mathbf{P}_{V_0}\boldsymbol{\tau} = \bar{\tau}\mathbf{u}_0$$

and

$$\mathbb{E}\left(\|\mathbf{P}_{V_0}(\mathbf{Y})\|^2\right) = (\bar{\tau}\mathbf{u}_0)\cdot(\bar{\tau}\mathbf{u}_0) + \sigma^2 = N\bar{\tau}^2 + \sigma^2.$$

Thus

$$\text{EMS(mean)} = \frac{N\bar{\tau}^2 + \sigma^2}{1} = N\bar{\tau}^2 + \sigma^2.$$

Now, $\mathbf{u}_0 = \mathbf{u}_1 + \cdots + \mathbf{u}_t$, which is in V_T, and so V_0 is a subspace of V_T. We can apply the ideas of Section 2.4 with V_0 in place of W and V_T in place of V. Thus we define

$$W_T = \{\mathbf{v}\in V_T : \mathbf{v}\text{ is orthogonal to }V_0\}$$
$$= V_T \cap V_0^\perp,$$

and find that

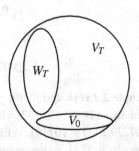

(i) $\dim W_T = \dim V_T - \dim V_0 = t - 1$;

(ii) $\mathbf{P}_{W_T}\mathbf{v} = \mathbf{P}_{V_T}\mathbf{v} - \mathbf{P}_{V_0}\mathbf{v}$ for all $\mathbf{v}$ in V;

(iii) $\left\|\mathbf{P}_{W_T}\mathbf{v}\right\|^2 + \left\|\mathbf{P}_{V_0}\mathbf{v}\right\|^2 = \left\|\mathbf{P}_{V_T}\mathbf{v}\right\|^2$ for all $\mathbf{v}$ in V.

Applying (ii) and (iii) with $\mathbf{v} = \boldsymbol{\tau}$ gives

$$\mathbf{P}_{W_T}\boldsymbol{\tau} = \mathbf{P}_{V_T}\boldsymbol{\tau} - \mathbf{P}_{V_0}\boldsymbol{\tau} = \boldsymbol{\tau} - \bar{\tau}\mathbf{u}_0 = \sum_{i=1}^{t}(\tau_i - \bar{\tau})\mathbf{u}_i \qquad (2.5)$$

and

$$\sum_{i=1}^{t} r_i(\tau_i - \bar{\tau})^2 = \left\|\mathbf{P}_{W_T}\boldsymbol{\tau}\right\|^2 = \left\|\boldsymbol{\tau}\right\|^2 - \left\|\bar{\tau}\mathbf{u}_0\right\|^2 = \sum_{i=1}^{t} r_i\tau_i^2 - N\bar{\tau}^2,$$

which is zero if and only if all the τ_i are equal.

Applying (ii) with $\mathbf{v} = \mathbf{y}$ gives

$$\begin{aligned}
\mathbf{P}_{W_T}\mathbf{y} &= \sum_{i=1}^{t}\left(\frac{\text{sum}_{T=i}}{r_i}\right)\mathbf{u}_i - \frac{\text{sum}}{N}\mathbf{u}_0 \\
&= \text{fitted values for treatments} - \text{fit for null model} \\
&= \sum_{i=1}^{t}(\hat{\tau}_i - \bar{y})\mathbf{u}_i.
\end{aligned}$$

The coefficients $\hat{\tau}_i - \bar{y}$ are called *treatment effects*. Taking sums of squares gives

$$\left\|\mathbf{P}_{W_T}\mathbf{y}\right\|^2 = \sum_{i=1}^{t}\frac{(\text{sum}_{T=i})^2}{r_i} - \frac{\text{sum}^2}{N}.$$

The sum of squares for W_T is called the *sum of squares for treatments*. We may abbreviate this to SS(treatments), so we have

$$\text{sum of squares for treatments} = \text{crude sum of squares for treatments}$$
$$- \text{sum of squares for the mean}.$$

Correspondingly, the *mean square for treatments*, MS(treatments), is given by

$$\text{MS(treatments)} = \frac{\text{SS(treatments)}}{t-1}.$$

Now, $\mathbb{E}(\mathbf{P}_{W_T}\mathbf{Y}) = \mathbf{P}_{W_T}\boldsymbol{\tau}$ and $\left\|\mathbf{P}_{W_T}\boldsymbol{\tau}\right\|^2 = \sum_i r_i\tau_i^2 - N\bar{\tau}^2$, so Theorem 2.5 shows that

$$\mathbb{E}\left(\left\|\mathbf{P}_{W_T}\mathbf{Y}\right\|^2\right) = \sum_i r_i\tau_i^2 - N\bar{\tau}^2 + (t-1)\sigma^2.$$

Hence

$$\text{EMS(treatments)} = \frac{\sum_i r_i\tau_i^2 - N\bar{\tau}^2}{t-1} + \sigma^2.$$

Example 2.2 revisited (Milk production) The grand total is 1851.8, so the sum of squares for the mean is $1851.8^2/32$, which is 107161.3513. Hence the sum of squares for treatments is $107279.2477 - 107161.3513$, which is 117.8964. The number of degrees of freedom for treatments is $3 - 1$, so the mean square for treatments is $117.8964/2$, which is 58.9482.

2.12 Hypothesis testing

The previous section shows how to decompose the vector space V into the sum of three orthogonal pieces:
$$V = V_T \oplus V_T^\perp = V_0 \oplus W_T \oplus V_T^\perp.$$

Correspondingly, the overall dimension N, the data vector $\mathbf{y}$ and its sum of squares can all be shown as the sum of three pieces. The sums of squares have their corresponding mean squares and expected mean squares, although there is no longer any sense in adding the three pieces.

	V	$=$	V_0	$\oplus$	W_T	$\oplus$	$V_T^\perp$
dimension	N	$=$	1	$+$	$(t-1)$	$+$	$(N-t)$
data	$\mathbf{y}$	$=$	$\bar{y}\mathbf{u}_0$	$+$	$\left(\sum_i \text{mean}_{T=i}\mathbf{u}_i - \bar{y}\mathbf{u}_0\right)$	$+$	$\left(\mathbf{y} - \sum_i \text{mean}_{T=i}\mathbf{u}_i\right)$
					treatment effects		residual
sum of squares	$\sum_{\omega\in\Omega} Y_\omega^2$	$=$	$\dfrac{\text{SUM}^2}{N}$	$+$	SS(treatments)	$+$	SS(residual)
mean square			$\dfrac{\text{SUM}^2}{N}$		$\dfrac{\text{SS(treatments)}}{t-1}$		$\dfrac{\text{SS(residual)}}{N-t}$
expected mean square			$N\bar{\tau}^2 + \sigma^2$		$\dfrac{\sum_i r_i\tau_i^2 - N\bar{\tau}^2}{t-1} + \sigma^2$		σ^2

To test the null hypothesis
$$H_0 : \bar{\tau} = 0$$
against the alternative hypothesis
$$H_1 : \bar{\tau} \neq 0,$$
look at MS(mean). If MS(mean) $\approx$ MS(residual) then we can conclude that $\bar{\tau}$ may well be zero. However, we are not usually interested in this.

To test the null hypothesis
$$H_0 : \tau_1 = \tau_2 = \cdots = \tau_t$$
against the alternative hypothesis
$$H_1 : \boldsymbol{\tau} \text{ is not a constant vector,}$$
look at MS(treatments). If MS(treatments) $\approx$ MS(residual) then we can conclude that $\boldsymbol{\tau}$ may well be constant, in other words that there are no treatment differences.

Note that both of the above tests are one-sided, because the differences EMS(mean) $- \sigma^2$ and EMS(treatments) $- \sigma^2$ are both non-negative.

The calculations are shown in an analysis-of-variance table (usually abbreviated to 'anova table'). There is one row for each 'source'; that is, subspace. The quantity in the column

Table 2.3. *Analysis-of-variance table for unstructured plots and unstructured treatments under the simple model*

Source	Sum of squares	Degrees of freedom	Mean square	Variance ratio
mean	$\dfrac{\text{sum}^2}{N}$	1	SS(mean)	$\dfrac{\text{MS(mean)}}{\text{MS(residual)}}$
treatments	$\sum_i \dfrac{(\text{sum}_{T=i})^2}{r_i} - \dfrac{\text{sum}^2}{N}$	$t-1$	$\dfrac{\text{SS(treatments)}}{t-1}$	$\dfrac{\text{MS(treatments)}}{\text{MS(residual)}}$
residual	$\leftarrow \cdots\cdots$ by subtraction $\cdots\cdots \rightarrow$		$\dfrac{\text{SS(residual)}}{\text{df(residual)}}$	—
Total	$\sum_\omega y_\omega^2$	N		

Table 2.4. *Analysis of variance in Example 2.2*

Source	SS	df	MS	VR
mean	107161.3513	1	107161.3513	13147.39
diets	117.8964	2	58.9482	7.23
residual	236.3723	29	8.1508	—
Total	107515.62	32		

headed 'variance ratio' is the ratio of two mean squares whose expectations are equal under some null hypothesis to be tested: the numerator is the mean square for the current row, while the denominator is another mean square. Table 2.3 is the analysis-of-variance table for unstructured plots and unstructured treatments.

Example 2.2 revisited (Milk production) We have already calculated everything except the variance ratio. Table 2.4 gives the analysis of variance.

Comparing the size of a mean square with the mean square for residual gives an indication that some parameter of interest is nonzero. However, a proper significance test cannot be done without knowing the distribution of the variance ratio under the null hypothesis.

Theorem 2.10 *Suppose that the distribution of* $\mathbf{Y}$ *is multivariate normal, that* $\mathbb{E}(\mathbf{Y}) = \boldsymbol{\tau} \in V_T$ *and that* $\text{Cov}(\mathbf{Y}) = \sigma^2 \mathbf{I}$. *Let* W_1 *and* W_2 *be subspaces of* V *with dimensions* d_1 *and* d_2. *Then the following hold.*

(i) *If* $\mathbf{P}_{W_1}\boldsymbol{\tau} = \mathbf{0}$ *then* $\text{SS}(W_1)/\sigma^2$ *has a* χ^2-*distribution with* d_1 *degrees of freedom.*

(ii) *If* W_1 *is orthogonal to* W_2 *and* $\mathbf{P}_{W_1}\boldsymbol{\tau} = \mathbf{P}_{W_2}\boldsymbol{\tau} = \mathbf{0}$ *then the ratio* $\text{MS}(W_1)/\text{MS}(W_2)$ *has an F-distribution with* d_1 *and* d_2 *degrees of freedom.*

Of course, there are many experiments where the response variable is manifestly not normally distributed: for example, the observations may be counts (Example 1.1), percentages

$$\begin{array}{lll}
\mathbb{E}(Y_\omega) = \tau_{T(\omega)} & \mathbb{E}(\mathbf{Y}) \in V_T & V_T \\
& & \quad | \\
\mathbb{E}(Y_\omega) = \kappa & \mathbb{E}(\mathbf{Y}) \in V_0 & V_0 \\
& & \quad | \\
\mathbb{E}(Y_\omega) = 0 & \mathbb{E}(\mathbf{Y}) \in \{\mathbf{0}\} & \{\mathbf{0}\}
\end{array} \left.\begin{array}{l} \\ \\ \end{array}\right\} W_T \quad \left.\begin{array}{l} \\ \\ \end{array}\right\} V_0$$

Fig. 2.4. Three models for the expectation of **Y**

(Example 1.3) or ordinal scores (Example 1.11). Nonetheless, an F-test will usually give the qualitatively correct conclusion.

Most textbooks and computer output do not show the line for the mean in the analysis-of-variance table, as I have done in Table 2.3. There are three reasons why I do so. In the first place, I think that the calculations are more transparent when the 'sum of squares' and 'degrees of freedom' columns add up to the genuine totals rather than to adjusted totals.

In the second place, fitting V_0 as a submodel of V_T is a first taste of what we shall do many times with structured treatments: fit submodels and see what is left over. Figure 2.4 shows the chain of vector subspaces $\{\mathbf{0}\} \subseteq V_0 \subseteq V_T$ according to the usual convention that a smaller space is shown below a larger space, with a vertical line to indicate that the smaller space is contained in the larger. This is the opposite convention to the one used in the analysis-of-variance table, where the subspaces at the bottom of the figure are shown at the top of the table. Each hypothesis test compares the size of the extra fit in one subspace (compared to the fit in the subspace below) against the residual mean square. I see no benefit in treating V_0 differently from any other submodel.

The third reason will become clear in Section 2.14, where we shall see that retaining a line for the mean points to the need to split up the rows of the analysis-of-variance table by strata even in the simplest case.

2.13 Sufficient replication for power

Suppose that there are two treatments and that $\tau_1 - \tau_2 = \delta$. Let Δ be the estimator for δ given in Section 2.6. Then $\Delta - \delta$ has expectation 0 and variance $\sigma^2 v$, where

$$v = \frac{1}{r_1} + \frac{1}{r_2},$$

from Equation (2.3). Let $\Gamma = \text{MS}(\text{residual})$, which is the estimator for σ^2 given in Section 2.9. Put

$$X = \frac{(\Delta - \delta)}{\sqrt{v\Gamma}}.$$

Under normality, Theorem 2.7(i) shows that X has a t-distribution on d degrees of freedom, where d is the number of residual degrees of freedom.

Let a be the 0.975 point of the t-distribution on d degrees of freedom, as shown in Figure 2.5. Some values of a are shown in Table 2.5. If a two-sided t-test with significance level 0.05 is performed on the data then we will conclude that $\delta \neq 0$ if $|\Delta|/\sqrt{v\Gamma} > a$. Thus

Table 2.5. *Values of a such that $\Pr[X \le a] = 0.975$, where X has a t-distribution on d degrees of freedom*

d	2	4	6	10	15	20	30	60
a	4.30	2.78	2.45	2.23	2.13	2.09	2.04	2.00
$\Pr[X > 2a]$	< 0.01	< 0.003	0.001	0.0005	0.0003	0.0002	0.0001	0.0001

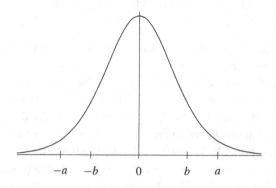

Fig. 2.5. Random variable X with a t-distribution: $\Pr[X > a] = \Pr[X < -a] = 0.025$ and $\Pr[X > b] = \Pr[X < -b] = 0.1$.

the probability p of not finding enough evidence to conclude that $\delta \ne 0$ is given by

$$p = \Pr\left[-a\sqrt{v\Gamma} < \Delta < a\sqrt{v\Gamma}\right] = \Pr\left[-a - \frac{\delta}{\sqrt{v\Gamma}} < X < a - \frac{\delta}{\sqrt{v\Gamma}}\right].$$

Consider the real function h defined by

$$h(x) = \Pr[-a - x < X < a - x],$$

so that $p = h(\delta/\sqrt{v\Gamma})$. The probability density function of each t-distribution takes larger values between 0 and a than it does between $-2a$ and $-a$, so $h(x)$ decreases from $x = 0$ to $x = a$ and also decreases from $x = 0$ to $x = -a$. Therefore if $-a < \delta/\sqrt{v\Gamma} < a$ then $p = h(\delta/\sqrt{v\Gamma}) > h(a) = h(-a) = \Pr[0 < X < 2a] = 0.5 - \Pr[X > 2a]$. Table 2.5 shows that $\Pr[X > 2a]$ is negligibly small, so p is unacceptably high.

If $\delta/\sqrt{v\Gamma}$ is bigger than a then $-a - \delta/\sqrt{v\Gamma} < -2a$. Therefore

$$0 \le \Pr[X < -a - \delta/\sqrt{v\Gamma}] < \Pr[X < -2a] = \Pr[X > 2a] \approx 0,$$

and so $p = \Pr[X < a - \delta/\sqrt{v\Gamma}] - \Pr[X < -a - \delta/\sqrt{v\Gamma}] \approx \Pr[X < a - \delta/\sqrt{v\Gamma}]$. If we want p to be at most 0.1 then we need to have $a - \delta/\sqrt{v\Gamma} < -b$, where $-b$ is the 0.1 point of the t-distribution with d degrees of freedom (and so b is the 0.9 point): see Figure 2.5. Thus $a + b < \delta/\sqrt{v\Gamma}$. If $\delta/\sqrt{v\Gamma}$ is less than $-a$ then a similar argument shows that we must have $\delta/\sqrt{v\Gamma} < -(a+b)$. Thus, in either case, we need

$$(a+b)^2 v\Gamma < \delta^2.$$

2.13. Sufficient replication for power

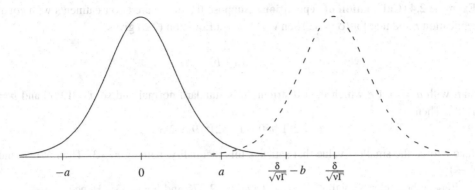

Fig. 2.6. Solid curve defines the interval $[-a,a]$ used for the hypothesis test; dashed curve gives the probability density function of $\Delta/\sqrt{v\Gamma}$

Figure 2.6 shows this pictorially when $\delta/\sqrt{v\Gamma} > a$. The solid curve shows the probability density function of a t random variable on d degrees of freedom. This defines the interval $[-a,a]$ used in the hypothesis test. The dashed curve shows the probability density function of $\Delta/\sqrt{v\Gamma}$. With probability 0.9, $\Delta/\sqrt{v\Gamma} > \delta/\sqrt{v\Gamma} - b$, and we want this lower limit to be outside the interval $[-a,a]$. Hence $a+b < \delta/\sqrt{v\Gamma}$.

Replacing Γ by its expectation σ^2 gives

$$(a+b)^2 v < (\delta/\sigma)^2. \qquad (2.6)$$

Consider the ingredients in Equation (2.6). We assume that $|\delta|$ is a known quantity, the size of the smallest difference that we want to detect. The variance σ^2 is assumed unknown, but previous experiments on similar material may give a rough estimate for its value. In more complicated experiments (see Sections 2.14 and 4.6 and Chapters 8 and 10), we shall need to replace σ^2 by the appropriate stratum variance. In many experiments the variance $v\sigma^2$ of the estimator of δ is given by $v = 1/r_1 + 1/r_2$, but Chapter 11 shows that a more complicated formula is needed in non-orthogonal designs. The values a and b depend partly on the number of degrees of freedom, which depends on the design. They also depend on some other choices: a on the significance level of the t-test; b on the upper limit of acceptability for p.

If we have even a rough idea of the size of $|\delta|/\sigma$, and have set the value of a by choosing a significance level for the t-test, then Equation (2.6) gives an inequality to be satisfied by v and b. There are two ways in which this can be used.

In some areas, such as agricultural research, it is typical to propose the number of treatments and their replications first, according to resources available. This gives a value for v, from which Equation (2.6) gives an upper bound for b, from which t-tables give a value for p. If this value is acceptably low the experiment proceeds. If not, a modified experiment is proposed with a smaller value of v, usually by increasing resources or omitting less interesting treatments.

In other areas, such as clinical trials, it is more common to set both the significance level and the power (which is equal to $1 - p$) in advance. Then, assuming equal replication r, Equation (2.6) is used to update values of r and d alternately until convergence is achieved.

Example 2.4 (Calculation of replication) Suppose that there are two treatments with equal replication r and that $|\delta|/\sigma = 3$. Then $v = 2/r$ and Equation (2.6) gives

$$r > \frac{2}{9}(a+b)^2.$$

Start with $d = \infty$, for which the t-distribution is standard normal and so $a = 1.960$ and $b = 1.282$. Then

$$r > 2(1.960+1.282)^2/9 \approx 2.3.$$

Take r to be the smallest value that satisfies this inequality, namely $r = 3$. Then $N = 6$ and $d = 4$.

Repeat the cycle. Now that $d = 4$ we have $a = 2.776$ and $b = 1.533$. Hence

$$r > 2(2.776+1.533)^2/9 \approx 4.1$$

so put $r = 5$. Then $d = 8$.

This new value of d gives $a = 2.306$ and $b = 1.397$, so

$$r > 2(2.306+1.397)^2/9 \approx 3.04.$$

Thus we put $r = 4$. Then $d = 6$.

Now $a = 2.447$ and $b = 1.440$ so

$$r > 2(2.447+1.440)^2/9 \approx 3.4.$$

This is satisfied by the current value of r, and we have already seen that the value immediately below does not satisfy Equation (2.6), so we stop. We conclude that eight experimental units should suffice.

Note that power can be increased by including extra treatments (because this increases d) but that this does not alter the variance of the estimator of a difference between two treatments.

2.14 A more general model

The chapter concludes with a slightly more general model than the one in Section 2.5.

As before, we assume that

$$\mathbb{E}(\mathbf{Y}) = \boldsymbol{\tau} \in V_T.$$

However, we change the assumption about covariance to

$$\operatorname{cov}(Z_\alpha, Z_\beta) = \begin{cases} \sigma^2 & \text{if } \alpha = \beta \\ \rho\sigma^2 & \text{if } \alpha \neq \beta. \end{cases}$$

In other words, the correlation between responses on pairs of different plots is ρ, which may not be zero. Complete randomization justifies this assumption, as we shall show in Chapter 14. Write $\mathbf{C} = \operatorname{Cov}(\mathbf{Y})$. Under our assumptions,

$$\mathbf{C} = \sigma^2 \mathbf{I} + \rho\sigma^2(\mathbf{J}-\mathbf{I}) = \sigma^2\left[(1-\rho)\mathbf{I} + \rho\mathbf{J}\right],$$

2.14. A more general model

where $\mathbf{J}$ is the $N \times N$ all-1 matrix.

Note that
$$\mathbf{J} = \begin{bmatrix} \mathbf{u}_0^\top \\ \mathbf{u}_0^\top \\ \vdots \\ \mathbf{u}_0^\top \end{bmatrix},$$

so if $\mathbf{x}$ is any vector in V then every entry of $\mathbf{Jx}$ is equal to $\mathbf{u}_0 \cdot \mathbf{x}$. In particular, $\mathbf{Ju}_0 = N\mathbf{u}_0$, and $\mathbf{Jx} = \mathbf{0}$ if $\mathbf{x} \perp \mathbf{u}_0$.

Of course, $\mathbf{Iu}_0 = \mathbf{u}_0$. Therefore

$$\mathbf{Cu}_0 = \sigma^2(1 - \rho + N\rho)\mathbf{u}_0,$$

so $\mathbf{u}_0$ is an eigenvector of $\mathbf{C}$ with eigenvalue $\sigma^2(1 - \rho + N\rho)$.

If $\mathbf{x} \in V$ and $\mathbf{x} \perp \mathbf{u}_0$ then $\mathbf{Ix} = \mathbf{x}$ and $\mathbf{Jx} = \mathbf{0}$, so

$$\mathbf{Cx} = \sigma^2(1 - \rho)\mathbf{x},$$

and therefore $\mathbf{x}$ is an eigenvector of $\mathrm{Cov}(\mathbf{Y})$ with eigenvalue $\sigma^2(1 - \rho)$.

The results from Theorem 2.5 onwards have assumed that $\mathbf{C}$ is a scalar matrix. Changing this assumption makes no difference to expectations of linear functions of $\mathbf{Y}$, but it does change the expectation of quadratic functions of $\mathbf{Y}$, that is, sums of squares. If $\mathbf{C} = \sigma^2 \mathbf{I}$ then all formulas for variance or expected mean square involve σ^2. If $\mathbf{x}$ is an eigenvector of $\mathbf{C}$ with eigenvalue ξ then $\mathbf{C}$ acts on $\mathbf{x}$ just like $\xi \mathbf{I}$. Thus careful replacement of σ^2 by the relevant eigenvalue gives the correct results. There is one possible difficulty in generalizing Theorem 2.6(iii) to the case when $\mathbf{x}$ and $\mathbf{z}$ are eigenvectors of $\mathbf{C}$ with different eigenvalues. However, $\mathbf{C}$ is a symmetric matrix, so eigenvectors with different eigenvalues are orthogonal to each other, so that $\mathbf{x} \cdot \mathbf{z} = 0$ in this case. The other places where different eigenvalues might occur are the generalizations of Theorem 2.7(i) and Theorem 2.10(ii): we deal with both of these by restricting the results to eigenvectors with the same eigenvalue.

Theorem 2.11 *Suppose that $\mathbb{E}(\mathbf{Y}) = \boldsymbol{\tau} \in V_T$ and $\mathrm{Cov}(\mathbf{Y}) = \mathbf{C}$. Then the following hold.*

(i) *If W is any subspace of V then $\mathbb{E}(\mathbf{P}_W(\mathbf{Y})) = \mathbf{P}_W \boldsymbol{\tau}$.*

(ii) *If W consists entirely of eigenvectors of $\mathbf{C}$ with eigenvalue ξ, and if $\dim W = d$, then*

$$\mathbb{E}(\|\mathbf{P}_W(\mathbf{Y})\|^2) = \|\mathbf{P}_W(\boldsymbol{\tau})\|^2 + d\xi.$$

(iii) *If $\mathbf{x} \in V_T$ and $\mathbf{x}$ is an eigenvector of $\mathbf{C}$ then the best linear unbiased estimator of $\mathbf{x} \cdot \boldsymbol{\tau}$ is $\mathbf{x} \cdot \mathbf{Y}$.*

(iv) *If $\mathbf{x}$ is an eigenvector of $\mathbf{C}$ with eigenvalue ξ then the variance of $\mathbf{x} \cdot \mathbf{Y}$ is $\|\mathbf{x}\|^2 \xi$.*

(v) *Suppose that $\mathbf{x}$ and $\mathbf{z}$ are eigenvectors of $\mathbf{C}$ with eigenvalues ξ and η respectively. If $\xi = \eta$ then $\mathrm{cov}(\mathbf{x} \cdot \mathbf{Y}, \mathbf{z} \cdot \mathbf{Y}) = (\mathbf{x} \cdot \mathbf{z})\xi$; if $\xi \neq \eta$ then $\mathrm{cov}(\mathbf{x} \cdot \mathbf{Y}, \mathbf{z} \cdot \mathbf{Y}) = 0$.*

Table 2.6. *Analysis-of-variance table for unstructured plots and unstructured treatments under the more general model*

Stratum	Source	df	EMS
V_0 'mean'	mean	1	$N\bar{\tau}^2 + \xi_0$
$V_0^\perp$ 'plots'	treatments	$t-1$	$\dfrac{\sum_i r_i(\tau_i - \bar{\tau})^2}{t-1} + \xi_1$
	residual	$N-t$	ξ_1
Total		N	

(vi) Suppose that $\mathbf{x}$ *is an eigenvector of* $\mathbf{C}$ *with eigenvalue* ξ, *that* $\mathbf{x} = \sum_i (\lambda_i / r_i) \mathbf{u}_i$, *that* W *is a d-dimensional subspace consisting of eigenvectors of* $\mathbf{C}$ *with eigenvalue* ξ, *and that* W *is orthogonal to* V_T. *If* $\mathbf{Y}$ *has a multivariate normal distribution then*

$$\frac{\mathbf{x} \cdot \mathbf{Y} - \sum \lambda_i \tau_i}{\sqrt{\left(\sum \frac{\lambda_i^2}{r_i}\right) \times \mathrm{MS}(W)}}$$

has a t-distribution on d degrees of freedom and $\mathrm{SS}(W)/\xi$ *has a* χ^2-*distribution on d degrees of freedom.*

(vii) If W_1 *and* W_2 *are subspaces with dimensions* d_1 *and* d_2, *both consisting of eigenvectors of* $\mathbf{C}$ *with eigenvalue* ξ, *orthogonal to each other, with* $\mathbf{P}_{W_1}\tau = \mathbf{P}_{W_2}\tau = \mathbf{0}$, *and if* $\mathbf{Y}$ *has a multivariate normal distribution then* $\mathrm{MS}(W_1)/\mathrm{MS}(W_2)$ *has an F-distribution on* d_1 *and* d_2 *degrees of freedom.*

Definition A *stratum* is an eigenspace of $\mathrm{Cov}(\mathbf{Y})$ (note that this is not the same as a stratum in sampling).

The analysis of variance proceeds just as before, except that we first decompose V into the different strata. Under the assumptions of this section, V_0 is one stratum, with dimension 1 and eigenvalue $\sigma^2(1 - \rho + N\rho)$, while $V_0^\perp$ is the other stratum, with dimension $N - 1$ and eigenvalue $\sigma^2(1 - \rho)$. Call these eigenvalues ξ_0 and ξ_1 respectively. We then obtain the analysis-of-variance table shown in Table 2.6.

Now we calculate the variance ratio only for terms with the same eigenvalue.

There is no way of estimating ξ_0, and hence no way of assessing whether $\bar{\tau}$ is (statistically significantly) different from zero, and no way of estimating the variance of the estimator of any treatment parameter τ_i. However, all treatment contrasts are in $V_0^\perp$, so their linear combinations and their variances may be estimated just as before. Experiments in which we are interested only in treatment contrasts are called *comparative experiments*.

Questions for discussion

2.1 A psychology course has the 21 students shown in Table 2.7. The professor wants to use the students to test two new types of pill for keeping people awake, called Wakey-Wakey and Zizzaway. He has only six pills of Wakey-Wakey and five pills of Zizzaway. He plans to use eleven students. Each student will be shut alone in the observation room, swallow their allocated pill, and then follow a set programme of activities until they fall asleep. A hidden watcher will record when they swallow the pill and when they fall asleep.

Design the experiment for the professor, to the extent of giving him a plan allocating pills to students.

2.2 A marine engineer is investigating ways of treating the standard metal components used in the construction of underwater structures at sea, such as piers and oil-drilling platforms. He wants to protect them against corrosion.

A colleague has developed a new sort of paint for the components. The engineer would like to see whether two coats of this paint give better protection than a single coat. So he will paint some metal components once, some twice, then immerse them all in his experimental tank of sea water. After three months, he will remove all the metal components from the tank, and measure the amount of corrosion on each.

He has a virtually unlimited supply of the metal components. The tank has room for up to 30 components. However, the paint is new, and there is enough for only 24 coats of paint.

Advise the engineer how best to use his resources in his experiment.

2.3 A technician has to measure the acidity of four soils. You give him three samples of each soil and ask him to make the twelve measurements in random order. He says that a random order will confuse him and that it will be better if he measures the acidity of all three samples of soil A, then all three samples of soil B, and so on. Make notes on arguments you will use to persuade him that a random order is better.

2.4 A physician wants to test a new drug to compare it with the current standard drug. He would like to have 90% power for detecting a difference of 15 units if he does a hypothesis test at the 5% level of significance. He believes that the value of σ^2, in the population, is about 100, for measurements in those units. If he uses r people for each drug, what is the smallest value of r that he should use?

Table 2.7. *Available students in Question 2.1*

Name	Sex	Age	Name	Sex	Age	Name	Sex	Age
Adrian	M	19	Helen	F	20	Olivia	F	20
Belinda	F	20	Ingrid	F	20	Peter	M	20
Caroline	F	19	James	M	20	Quentin	M	27
David	M	20	Katherine	F	19	Ruth	F	20
Esther	F	28	Linda	F	28	Sarah	F	19
Fiona	F	20	Michael	M	20	Trixie	F	20
Gregory	M	19	Naomi	F	19	Ursula	F	20

2	1	6	4	6	7	5	3
9	12	18	10	24	17	30	16
1	5	4	3	5	1	1	6
10	7	4	10	21	24	29	12
2	7	3	1	3	7	2	4
9	7	18	30	18	16	16	4
5	1	7	6	1	4	1	2
9	18	17	19	32	5	26	4

Fig. 2.7. Field plan in Question 2.5

2.5 A completely randomized experiment was conducted to compare seven treatments for their effectiveness in reducing scab disease in potatoes. The field plan is shown in Figure 2.7. The upper figure in each plot denotes the treatment, coded 1–7. The lower figure denotes an index of scabbiness of potatoes in that plot: 100 potatoes were randomly sampled from the plot, for each one the percentage of the surface area infected with scabs was assessed by eye and recorded, and the average of these 100 percentages was calculated to give the scabbiness index.

(a) Give the analysis-of-variance table for these data.

(b) Is there any evidence that the mean scabbiness is different according to different treatments? Justify your answer.

(c) Estimate the mean scabbiness produced by each treatment.

(d) What is the standard error of the above estimates?

(e) What is the standard error of the differences between means?

Chapter 3
Simple treatment structure

3.1 Replication of control treatments

Suppose that treatment 1 is a control treatment and that treatments $2, \ldots, t$ are new treatments which we want to compare with the control. Then we want to estimate $\tau_i - \tau_1$ for $i = 2, \ldots, t$. From Equation (2.3), the average variance of these estimators is equal to

$$\frac{1}{t-1} \sum_{i=2}^{t} \left(\frac{1}{r_1} + \frac{1}{r_i} \right) \sigma^2 = \left(\frac{1}{r_1} + \frac{1}{t-1} \sum_{i=2}^{t} \frac{1}{r_i} \right) \sigma^2.$$

For given values of r_1 and N, Proposition 2.8, with $R = N - r_1$, shows that this average variance is minimized when $r_2 = r_3 = \cdots = r_t = (N - r_1)/(t-1)$.

Put $r_1 = r$ and

$$g(r) = \frac{1}{r} + \frac{1}{t-1} \sum_{i=2}^{t} \frac{t-1}{N-r} = \frac{1}{r} + \frac{t-1}{N-r},$$

so that we need to choose r to minimize $g(r)$. Now, g is differentiable on $(0, N)$ and increases without limit as $r \to 0$ and as $r \to N$. Moreover,

$$g'(r) = -\frac{1}{r^2} + \frac{(t-1)}{(N-r)^2},$$

which is zero when, and only when, $r = (N-r)/\sqrt{t-1}$. Thus $g(r)$ is minimized when, and only when, $r = (N-r)/\sqrt{t-1}$; that is

$$r_1 = r = (t-1)r_2/\sqrt{t-1} = \sqrt{t-1}\, r_2. \tag{3.1}$$

In practice we have to use approximate solutions to Equation (3.1) because all the replications must be integers.

Sometimes there is more than one control treatment, and we want to compare every new treatment with every control treatment. Proposition 2.8 shows that all the control treatments should have the same replication as each other, say r_1, while all the new treatments should have the same replication as each other, say r_t. If there are n control treatments and m new treatments we then have $nr_1 + mr_t = N$, and need to minimize

$$\frac{1}{r_1} + \frac{1}{r_t}$$

subject to this constraint. Put
$$g(r) = \frac{1}{r} + \frac{m}{N-nr}.$$
Then
$$g'(r) = -\frac{1}{r^2} + \frac{nm}{(N-nr)^2},$$
which is zero when $r = (N-nr)/\sqrt{nm}$. Thus the average variance of estimators of differences between control treatments and new treatments is minimized when
$$\sqrt{n}\,r_1 = \sqrt{n}\,r = (N-nr)/\sqrt{m} = \sqrt{m}\,r_t. \tag{3.2}$$

Example 3.1 (Example 1.17 continued: Oilseed rape) In this experiment there were two control treatments and eight new treatments. Equation (3.2) gives $\sqrt{2}\,r_1 = \sqrt{8}\,r_t$, so the replication of the controls should have been twice that of the new treatments. In fact, all treatments were applied to two plots each. Perhaps the comparisons between pairs of new treatments were deemed as interesting as those between new treatments and controls?

There are some experiments with a single control treatment where it is known in advance that the control treatment is ineffective. In such circumstances, comparisons between new treatments are more informative then any comparison between a new treatment and the control. This suggests that the new treatments should have higher replication than the control. However, the person who wants to include the control treatment probably wants to compare all new treatments with it, rather than among themselves, so will want higher replication for the control. You may have to compromise on equal replication.

3.2 Comparing new treatments in the presence of a control

Suppose that treatment 1 is a control treatment and that treatments $2, \ldots, t$ are new treatments. Rather than asking

is $\tau_i = \tau_1$?

for $i = 2, \ldots, t$, we could ask the two questions

(i) is $\tau_2 = \tau_3 = \cdots = \tau_t$?

(ii) is τ_1 equal to the average of $\tau_2, \ldots, \tau_t$?

To test the null hypothesis
$$H_0 : \tau_2 = \tau_3 = \cdots = \tau_t$$
against the alternative hypothesis
$$H_1 : \tau_2, \ldots, \tau_t \text{ are not all equal,}$$
we calculate the mean square for the non-control treatments. This is equal to $\mathrm{SS}(\mathrm{new})/(t-2)$, where
$$\mathrm{SS}(\mathrm{new}) = \sum_{i=2}^{t} \frac{(\mathrm{sum}_{T=i})^2}{r_i} - \frac{(\sum_{i=2}^{t} \mathrm{sum}_{T=i})^2}{N - r_1}. \tag{3.3}$$

3.2. Comparing new treatments in the presence of a control

$$
\begin{array}{llll}
& \mathbb{E}(Y_\omega) = \tau_{T(\omega)} & \mathbb{E}(\mathbf{Y}) \in V_T & V_T \\
\text{H}_0\ ?\ \downarrow & & & \quad\quad \Big\} W'_T \\
& \mathbb{E}(Y_\omega) = \begin{cases} \tau_1 & \text{if } T(\omega) = 1 \\ \phi & \text{otherwise} \end{cases} & \mathbb{E}(\mathbf{Y}) \in V_C & V_C \\
\text{H}'_0\ ?\ \downarrow & & & \quad\quad \Big\} W_C \\
& \mathbb{E}(Y_\omega) = \kappa & \mathbb{E}(\mathbf{Y}) \in V_0 & V_0 \\
& & & \quad\quad \Big\} V_0 \\
& \mathbb{E}(Y_\omega) = 0 & \mathbb{E}(\mathbf{Y}) \in \{\mathbf{0}\} & \{\mathbf{0}\}
\end{array}
$$

Fig. 3.1. Four models for the expectation of **Y** when one treatment is a control

This is then compared with the residual mean square. If we decide that there is a constant ϕ such that $\tau_i = \phi$ for $i = 2, \ldots, t$ then it is reasonable to go on and test the null hypothesis

$$\text{H}'_0 : \tau_1 = \phi$$

against the alternative hypothesis

$$\text{H}'_1 : \tau_1 \neq \phi.$$

We obtain the sum of squares for this by pretending that all the new treatments are a single treatment: it is equal to

$$\frac{(\text{sum}_{T=1})^2}{r_1} + \frac{\sum_{i=2}^{t}(\text{sum}_{T=i})^2}{N - r_1} - \frac{\text{sum}^2}{N}. \tag{3.4}$$

These tests can be related to a chain of vector subspaces similar to the one in Figure 2.4. Let C be the 'control' factor on Ω: it is defined by

$$C(\omega) = \begin{cases} 1 & \text{if } T(\omega) = 1 \\ 2 & \text{otherwise.} \end{cases}$$

Define the subspace V_C of V to consist of all those vectors $\mathbf{v}$ for which $v_\alpha = v_\beta$ whenever $C(\alpha) = C(\beta)$. Then $\dim V_C = 2$ and $V_0 \subset V_C \subset V_T$. Further, define

$$W_C = V_C \cap V_0^\perp,$$

which has dimension 1, and

$$W'_T = V_T \cap V_C^\perp,$$

which has dimension $t - 2$. Then the null hypothesis H_0 corresponds to the model

$$\mathbb{E}(\mathbf{Y}) \in V_C.$$

If we have accepted H_0 then H'_0 corresponds to

$$\mathbb{E}(\mathbf{Y}) \in V_0,$$

which is

$$\mathbf{P}_{W_C}(\mathbb{E}(\mathbf{Y})) = \mathbf{0}.$$

Thus we obtain the chain of models shown in Figure 3.1.

Table 3.1. *First three columns of the analysis-of-variance table when there is one control treatment and no other structure*

Source	Sum of squares	Degrees of freedom
mean	CSS(mean)	1
control	CSS(control) − CSS(mean)	1
new treatments	CSS(treatments) − CSS(control)	$t-2$
residual	←················by subtraction················→	
Total	$\sum_\omega y_\omega^2$	N

To test H_0, we look at the size of $\mathbf{P}_{V_T}\mathbf{y} - \mathbf{P}_{V_C}\mathbf{y}$, which is equal to $\mathbf{P}_{W'_T}\mathbf{y}$. Since

$$\mathbf{P}_{V_C}\mathbf{y} = \mathrm{mean}_{C=1}\mathbf{u}_1 + \mathrm{mean}_{C=2}(\mathbf{u}_0 - \mathbf{u}_1),$$

we have

$$\begin{aligned}
\mathrm{SS}(W'_T) = \left\|\mathbf{P}_{W'_T}\mathbf{y}\right\|^2 &= \left\|\mathbf{P}_{V_T}\mathbf{y}\right\|^2 - \left\|\mathbf{P}_{V_C}\mathbf{y}\right\|^2 \\
&= \sum_{i=1}^{t} \frac{(\mathrm{sum}_{T=i})^2}{r_i} - \frac{(\mathrm{sum}_{C=1})^2}{r_1} - \frac{(\mathrm{sum}_{C=2})^2}{N-r_1} \\
&= \sum_{i=2}^{t} \frac{(\mathrm{sum}_{T=i})^2}{r_i} - \frac{(\mathrm{sum}_{C=2})^2}{N-r_1},
\end{aligned}$$

as in Equation (3.3). In words, the

sum of squares for new treatments = crude sum of squares for treatments
− crude sum of squares for control.

Similarly, to test H'_0, we look at the size of $\mathbf{P}_{W_C}\mathbf{y}$, which is equal to $\mathbf{P}_{V_C}\mathbf{y} - \mathbf{P}_{V_0}\mathbf{y}$. We have

$$\begin{aligned}
\mathrm{SS}(W_C) = \left\|\mathbf{P}_{W_C}\mathbf{y}\right\|^2 &= \left\|\mathbf{P}_{V_C}\mathbf{y}\right\|^2 - \left\|\mathbf{P}_{V_0}\mathbf{y}\right\|^2 \\
&= \frac{(\mathrm{sum}_{C=1})^2}{r_1} + \frac{(\mathrm{sum}_{C=2})^2}{N-r_1} - \frac{\mathrm{sum}^2}{N},
\end{aligned}$$

as in Equation (3.4). In words, the

sum of squares for control = crude sum of squares for control
− sum of squares for the mean.

The first three columns of the analysis-of-variance table are shown in Table 3.1.

Note that SS(control) + SS(new treatments) = SS(treatments), so we have decomposed the sum of squares for treatments into two meaningful parts, each of which is used to investigate a relevant hypothesis.

3.3. Other treatment groupings

Table 3.2. *Data in Example 3.2*

					Total
Fungicide A	29.3	33.6	33.7	29.6	126.2
Fungicide B	35.2	33.3	36.1	31.5	136.1
Fungicide C	35.7	39.3	36.8	28.0	139.8
Fungicide D	34.6	30.6	31.2	37.1	133.5
No fungicide	20.9	22.7	27.8	18.5	89.9
					625.5

Table 3.3. *Analysis-of-variance table for Example 3.2*

Source	SS	df	MS	VR
mean	19562.5125	1	19562.5125	1667.43
control	387.2000	1	387.2000	33.00
fungicides	24.7750	3	8.2583	0.70
residual	175.9825	15	11.7322	—
Total	20150.4700	20		

Example 3.2 (Fungicide on potatoes) Four fungicides (*A*–*D*) were compared with each other and with a no-fungicide control for their effect on potatoes. Each treatment was applied to four plots. Table 3.2 shows the yields of potatoes in tonnes/hectare. The data have been rearranged into treatment order for ease of showing the treatment totals.

The total sum of squares is 20150.47.

$$\text{SS(control)} = \frac{535.6^2}{16} + \frac{89.9^2}{4} - \frac{625.5^2}{20} = 387.2000;$$

$$\text{SS(fungicides)} = \frac{126.2^2}{4} + \frac{136.1^2}{4} + \frac{139.8^2}{4} + \frac{133.5^2}{4} - \frac{535.6^2}{16} = 24.7750;$$

the residual sum of squares is $20150.47 - 625.5^2/20 - 387.2000 - 24.7750 = 175.9825$. Table 3.3 gives the analysis of variance.

It is evident from the table that the fungicides differ very much from the control in their effect on the yield of potatoes, but that there is little difference between the four fungicides. The relevant means are

Fungicide	No fungicide
33.475	22.475

and the standard error of their difference is $\sqrt{11.7322 \times 5/16}$, which is 1.915.

3.3 Other treatment groupings

The treatments may be grouped into two or more types for reasons other than that one type consists of controls. The consequences for replication and analysis depend on the reasons for including the different types in the experiment. The following examples illustrate some of the possibilities.

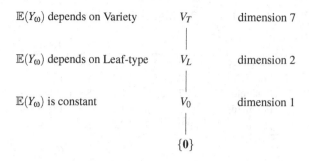

Fig. 3.2. Four models for expectation in Example 3.3

Example 3.3 (Rubber trees) Seven varieties of rubber tree are planted in an experiment to compare yields of the varieties. It happens that the varieties have two visually different leaf-types: three of the varieties have round green leaves (leaf-type 1) while the other four have long serrated grey leaves (leaf-type 2).

Variety	1	2	3	4	5	6	7
Leaf-type	2	1	2	1	2	2	1

The main purpose of the experiment is to compare the seven varieties, so they should be equally replicated. However, it is useful to ask if any differences among the varieties can be explained as differences between the two leaf-types. Thus the analysis should fit the chain of models shown in Figure 3.2, where V_L is the subspace of V_T consisting of vectors which are constant on each leaf-type.

Calculations similar to those in Section 3.2 give the partial analysis-of-variance table shown in Table 3.4, where r denotes the common replication. There are five degrees of freedom to investigate the question 'Do varieties differ within leaf-type?' and one degree of freedom for the question 'Do the leaf-types differ?'

Example 3.4 (Drugs at different stages of development) A company which develops and manufactures pharmaceuticals wants to compare six treatments for a certain disease. The initial trial will use healthy volunteers, simply to measure the amount of certain chemicals released into the blood two hours after the treatments are administered. Three of the treatments are three different doses of a formulation (coded A) that has been under development for some time. The other three are three different doses of a new formulation (coded B) that has not been so extensively studied. The main aim of the trial is to compare the doses of formulation A; the secondary aim is to compare the new formulation with the old one; and the lowest priority is given to comparing doses of the new formulation.

The main effort in the experiment should go into comparing the doses of the old formulation A. Sufficient replication must be used for these three treatments, guided by the principles in Sections 2.10 and 2.13. The company decides to use 12 volunteers for each dose, thus using 36 volunteers. However, it has sufficient resources to use 48 volunteers, so the three doses of formulation B are assigned to four volunteers each.

Compared to the more limited design with three doses of A and only 36 volunteers, the design with the extra 12 volunteers increases the precision of the estimate of σ^2. It also gives

3.3. Other treatment groupings

Table 3.4. *First three columns of the analysis-of-variance table in Example 3.3*

Source	Sum of squares	Degrees of freedom
mean	CSS(mean)	1
leaf-types	CSS(leaf-types) − CSS(mean)	1
varieties	CSS(varieties) − CSS(leaf-types)	5
residual	←·········· by subtraction ··········→	
Total	$\sum_\omega y_\omega^2$	$7r$

more residual degrees of freedom, and hence more power for detecting differences between the doses of A. In addition, it gives some information about doses of B.

Variances of estimators of some contrasts are as follows.

between two doses of A $\qquad \dfrac{2}{12}\sigma^2 = \dfrac{1}{6}\sigma^2$

between a dose of A and a dose of B $\qquad \left(\dfrac{1}{12}+\dfrac{1}{4}\right)\sigma^2 = \dfrac{1}{3}\sigma^2$

between the average effect of A and the average effect of B $\qquad \dfrac{1}{9}\left(\dfrac{3}{12}+\dfrac{3}{4}\right)\sigma^2 = \dfrac{1}{9}\sigma^2$

between two doses of B $\qquad \dfrac{2}{4}\sigma^2 = \dfrac{1}{2}\sigma^2$

If the company is correct in its judgement that replication 12 is sufficient for the doses of A, then comparisons among these are sufficiently precise, as is the comparison between the average effect of A and the average effect of B. Comparisons between doses of B are less precise, but may yield useful information if there are large differences, which should help the company decide whether to proceed with the development of formulation B.

Now it is sensible to split the sum of squares for treatments into three parts:

(i) the sum of squares for differences between formulations, which is calculated like the sum of squares for leaf-types in Example 3.3;

(ii) the sum of squares for differences between doses of A, which is calculated as in Equation (3.3) under the pretence that all doses of B are a single control;

(iii) the sum of squares for differences between doses of B, which is calculated as in Equation (3.3) under the pretence that all doses of A are a single control.

Define treatment factors F, A and B as follows.

Treatment	Old formulation			New formulation		
	1	2	3	4	5	6
F	1	1	1	2	2	2
A	1	2	3	0	0	0
B	0	0	0	1	2	3

Table 3.5. *Models for* $\mathbb{E}(Y_\omega)$ *in Example 3.4*

Coordinate (and parameters)	Vector (and subspace)
$\mathbb{E}(Y_\omega) = \tau_{T(\omega)}$	$\mathbb{E}(\mathbf{Y}) \in V_T$
$\mathbb{E}(Y_\omega) = \begin{cases} \lambda & \text{if } T(\omega) \text{ is old} \\ \tau_{T(\omega)} & \text{if } T(\omega) \text{ is new} \end{cases}$	$\mathbb{E}(\mathbf{Y}) \in V_B$
$\mathbb{E}(Y_\omega) = \begin{cases} \tau_{T(\omega)} & \text{if } T(\omega) \text{ is old} \\ \mu & \text{if } T(\omega) \text{ is new} \end{cases}$	$\mathbb{E}(\mathbf{Y}) \in V_A$
$\mathbb{E}(Y_\omega) = \begin{cases} \lambda & \text{if } T(\omega) \text{ is old} \\ \mu & \text{if } T(\omega) \text{ is new} \end{cases}$	$\mathbb{E}(\mathbf{Y}) \in V_F$
$\mathbb{E}(Y_\omega) = \kappa$	$\mathbb{E}(\mathbf{Y}) \in V_0$
$\mathbb{E}(Y_\omega) = 0$	$\mathbb{E}(\mathbf{Y}) \in \{\mathbf{0}\}$

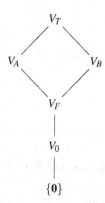

Fig. 3.3. Relationships among the subspaces in Example 3.4

These define vector spaces V_F, V_A and V_B analogous to V_T and V_C. The relationships between these spaces are shown in Figure 3.3. The expectation model corresponding to each of these is given in Table 3.5.

Example 3.5 (Reducing feed for chickens) An experiment is to be conducted to see if chickens can be fed a slightly inferior diet in the 16 weeks before slaughter without affecting their final weight. The chickens are housed in 40 cages of 20 birds each, and feeds are applied to whole cages. The ten treatments are shown in Table 3.6. Thus the non-control treatments can be grouped into three different methods, each of which has several variants.

The main interest in the experiment is probably in comparing each method with the control treatment, which is the normal feed. Thus Equation (3.1) suggests that we should have r_1 cages with the control and r cages with each method, where $r_1 \approx \sqrt{3}r$. With 40 cages, we could have $r = 8$ and $r_1 = 16$ or $r = 9$ and $r_1 = 13$. All the treatments pertaining to

3.3. Other treatment groupings

Table 3.6. *Treatments in the chicken-feeding experiment in Example 3.5*

		Treatment
Control		1
Reduce protein content by	5%	2
	10%	3
	15%	4
Change diet to a given cheaper one after	4 weeks	5
	8 weeks	6
	12 weeks	7
Replace 5% of the protein by an equal volume of roughage of type	A	8
	B	9
	C	10

each method should be equally replicated. For the ten treatments in Table 3.6 this suggests replication

    3   for all non-control treatments
   13   for the control treatment.

However, if we replace one 'quantity of protein' treatment by another 'roughage' treatment, then we might choose replication

    3   for the 3 time treatments
    4   for the 2 quantity treatments
    2   for the 4 roughage treatments
   15   for the control treatment.

In practice, such a range of different replications will make the design impossible if there is any blocking (see Chapter 4), so it may be better to opt for equal replication, at least for the non-control treatments.

In this example the sum of squares for treatments should be split into five parts, which can all be calculated following the principles given earlier in this chapter. For generality, suppose that there are n_i treatments pertaining to method i, for $i = 1, 2, 3$:

(i) the sum of squares for the difference between the control treatment and the rest (1 degree of freedom);

(ii) the sum of squares for the differences between methods (2 degrees of freedom);

(iii) the sum of squares for the differences between treatments of the quantity type ($n_1 - 1$ degrees of freedom);

(iv) the sum of squares for the differences between treatments of the time type ($n_2 - 1$ degrees of freedom);

(v) the sum of squares for the differences between treatments of the roughage type ($n_3 - 1$ degrees of freedom).

We shall return to these examples in Chapters 5 and 10.

Questions for discussion

3.1 Suppose that there are n control treatments, each replicated r_1 times, and m new treatments, each replicated r_t times. Find the optimal ratio r_1/r_t if all treatment comparisons are of interest except those between control treatments.

3.2 Consider the scabbiness experiment in Question 2.5. The seven coded treatments consisted of one 'do nothing' control and six spray treatments, as shown below. The amounts of sulphur are given in pounds per acre.

Treatment	1	2	3	4	5	6	7
Amount of sulphur	0	300	600	1200	300	600	1200
Timing	N/A	autumn	autumn	autumn	spring	spring	spring

Give two plausible reasons for the particular choice of unequal replication made in this experiment.

3.3 Re-analyse the data from Question 2.5, taking account of the fact that treatment 1 is a control.

3.4 Write down formulas for the various sums of squares in Example 3.4. For each sum of squares, also write down the null hypothesis for which it is appropriate.

3.5 A clinical trial on asthma compared three doses of a new formulation of a bronchodilator, three unrelated doses of a standard formulation of the bronchodilator, and placebo.

(a) Define three relevant factors on the treatments.

(b) Hence specify relevant models for the expectation of the response, in the format given in Table 3.5.

(c) Show the relationships between the corresponding model subspaces, in a diagram like the one in Figure 3.3.

(d) Hence explain how the treatment sum of squares should be decomposed, and give the number of degrees of freedom for each part.

(e) For each part of the treatment sum of squares, write down the corresponding question about the treatments.

Chapter 4

Blocking

4.1 Types of block

If the plots are not all reasonably similar, we should group them together into *blocks* in such a way that plots within each block are alike.

There are three main types of block.

4.1.1 Natural discrete divisions

These divisions between the experimental units are already present. If the experimental units are new-born animals then litters make natural blocks. In an experiment on people or animals, the two sexes make obvious blocks. In testing tags on cows' ears, the ears are the experimental units and cows are the blocks. In an industrial process, a block could be a batch of chemical or ore used for part of the process.

Example 4.1 (Insect repellent) Midges in Scotland are a severe irritant in July and August. A researcher wants to try out some insect repellents, which are applied to people's skin. Twelve people volunteer for the experiment. It is known that people vary widely in their attractiveness to midges, so the researcher uses people as blocks, applying different repellents to each arm. After a fixed period of exposure to midges, the number of midge bites on each arm of each volunteer is recorded.

Sometimes there is more than one type of natural discrete block. If the experimental units are half-leaves of tobacco plants then whole leaves make one sort of block while the plants make another. In a consumer experiment, such as Example 1.11, testers and weeks are both natural blocks. In an experiment in the laboratory, technicians, benches and days may all be blocks.

If an experiment is carried out on plots that had previously been used for another experiment then you should consider whether to deem the previous treatments to be blocks. This is because the previous treatments may have left some residue that may affect the responses in the new experiment. This type of block is particularly important in experiments on trees, which may have to be used for different experiments year after year.

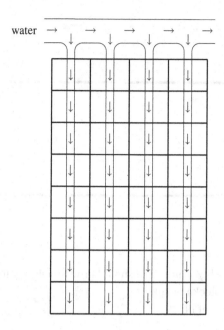

Fig. 4.1. Irrigation channels in the rice experiment

Example 4.2 (Irrigated rice) Rice is usually grown on irrigated land. Figure 4.1 shows 32 plots in a rice paddy to be used for an experiment. Irrigation channels branch off the main irrigation channel, each one watering a long strip of plots. These strips, or irrigation groupings, should be considered as blocks.

Example 4.3 (Road signs) A road transport research laboratory in Wales wanted to investigate whether the proposed introduction of bilingual road signs would prove distracting to drivers. They made two sets of road signs, one with Welsh on top and English underneath, the other with English on top and Welsh underneath. On each of several afternoons they erected one set of signs around an off-road test driving track, and invited volunteers to do a test drive. During each drive, a researcher sat in the car with the driver and asked a series of questions about their driving, designed to evaluate whether the driver's concentration was lowered when near one of the bilingual road signs. Since it took some time to erect a set of signs, it was not practicable to ask volunteers to drive round the track twice, once with each set of signs.

When the statistician came to analyse the data, he noticed that all the volunteers were either retired people or university students. This in itself was not surprising, because these are the classes of people who can easily be free in the normal working day. Unfortunately, the person designing the experiment had not noticed this, and had unwittingly allocated all of the retired volunteers to the signs with Welsh on top and all of the student volunteers to the signs with English on top. There was, therefore, no way of telling whether the recorded differences in concentration were due to an inherent difference between retired people and university students or were caused by the relative positions of the two languages on the signs.

4.1. Types of block

If the researchers had thought about this problem in advance, they should have used the two classes of volunteers as blocks.

4.1.2 Continuous gradients

If an experiment is spread out in time or space then there will probably be continuous underlying trends but no natural boundaries. In such cases the plots can be grouped into blocks of plots which are contiguous in time or space. To some extent the positioning of the block boundaries is arbitrary.

In an experiment on people or animals, age, weight and state of health are continuous variables which are often suitable for determining blocks. To be in the same block two people do not have to have exactly the same weight, but weight ranges can be chosen so that blocks have a suitable size. Similarly, severity of disease can be used to block patients in a clinical trial.

Example 4.4 (Laboratory measurement of samples) Consider the technician measuring soil samples in Question 2.3. His experimental units follow one another in time. As time goes on, he may get more accurate, or he may get tired. Outside factors, such as temperature or humidity, may change. Dividing up the sequence of experimental units into three or four blocks of consecutive plots should remove these unnecessary sources of variation from the conclusions of the experiment.

Example 4.5 (Field trial) The plots in an agricultural field trial may cover quite a large area, encompassing changes in fertility. Sometimes it is possible to form natural blocks by marking out a stony area, a shady area and so on. More often it is simply assumed that plots close to each other are more likely to respond similarly than plots far apart, so small compact areas are chosen as blocks.

In Example 4.2, the distance from the main irrigation channel gives a continuous source of variability that should also be used for blocking, but now there is some freedom to choose how large a distance each block should cover.

4.1.3 Choice of blocking for trial management

Some aspects of trial management force differences between the plots. As far as possible, these differences should match (some of) the block boundaries.

In a clinical trial, patients may have to be divided into groups to be attended to by different doctors or nurses. These groups should be blocks.

In a laboratory experiment, technicians may be thought of as natural blocks if their times and places of work are already fixed. However, if technicians can be allocated to tasks as part of the management of the experiment, then it may be possible to adjust their work so that, for example, the number of samples analysed by one person in one session is equal to the number of treatments.

There are many experiments where one or more treatment factors can be applied only to large areas: see Example 1.7. These large areas form a sort of block.

Example 4.5 revisited (Field trial) In the developed world, most agricultural operations are by tractor. Typically a tractor is driven as far as possible in a straight line before being turned round. This suggests that blocks in field trials should be long thin areas corresponding to a few passes of the tractor.

Example 4.6 (Citrus orchards) Similarly, citrus orchards are planted with the trees in a rectangular grid. The space between rows is bigger than the space between columns, so that lorries can drive along the rows for operations such as applying pesticides or harvesting. Therefore, both contiguity and management considerations suggest that rows should be blocks.

4.1.4 How and when to block

If possible,

(i) blocks should all have the same size;

(ii) blocks should be big enough to allow each treatment to occur at least once in each block.

Natural discrete blocks should always be used once they have been recognized. If possible, choose plots and blocks to satisfy (i).

Example 4.7 (Piglets) If the experimental units are piglets then litters are natural blocks. Litters are not all of the same size, typically being in the range 8–12, depending on the breed. It would be sensible to use only some fixed number, say nine, of piglets from each litter. Then you need an objective rule for which piglets to choose from the larger litters, such the heaviest piglets. Alternatively, if larger blocks are needed, start with more sows than necessary and use only those litters large enough to give, say, ten piglets.

Natural blocks have an upper limit on their size, so it may be impossible to satisfy (ii). In the cows' ears example, blocks have size two no matter how many treatments there are.

Blocks should always be used for management. Then all trial operations—sowing, harvesting, interim applications of treatments, measuring—are done block by block, in case there are interruptions, improvements in technique, replacement of staff, etc. This ensures that any extra variation caused by changing conditions is already accounted for by the blocking. Management blocks can usually be chosen to satisfy both (i) and (ii).

Example 4.8 (Weed control) Field trials to compare methods of controlling weeds are sometimes planted by hand. If a trial is too large for one person to do all the planting, then it is recommended that either each person should plant one or more whole blocks, or each person should plant the same number of rows in every plot.

To eliminate the effects of a continuous trend, blocks can also be chosen to satisfy both (i) and (ii). Usually such blocking is helpful, but it may be better not to use this sort of block if doing so would make the number of residual degrees of freedom very small: see Example 4.15.

As noted in Example 4.5, the requirements of blocking for trial management may conflict with those of blocking to remove a continuous trend. You may have to decide which is more important for the experiment at hand.

4.2. Orthogonal block designs

We have also noted examples where more than one sort of block is needed. This point will be developed further in Chapters 6 and 8.

Example 4.9 (Mushrooms) Mushrooms are grown indoors in tunnels or sheds. In an experiment to compare different strains of mushroom it therefore seems obvious that we should use the tunnels as natural discrete blocks. However, different strains of mushroom are normally grown at different temperatures and it may not be practicable to vary the temperature within a tunnel. As a result, strains will grow at different rates, so that harvesting by strains, rather than by tunnels, becomes inevitable.

In such a situation, it may be better to treat the tunnels as experimental units, so that each strain can be grown in its appropriate temperature. This may lead to issues concerning replication and correct data analysis: see Chapter 8.

4.2 Orthogonal block designs

For the rest of this chapter we suppose that Ω consists of b blocks of equal size k. We thus have a *block factor B* which is defined by

$$B(\omega) = \text{the block containing } \omega.$$

The *block subspace* V_B consists of those vectors in V which take a constant value on each block. For $j = 1, \ldots, b$, let $\mathbf{v}_j$ be the vector whose entry on plot ω is equal to

$$\begin{cases} 1 & \text{if } \omega \text{ is in block } j; \\ 0 & \text{otherwise.} \end{cases}$$

Then $\mathbf{v}_j \cdot \mathbf{v}_j = k$, while $\mathbf{v}_j \cdot \mathbf{v}_l = 0$ if $j \neq l$. Therefore $\{\mathbf{v}_j : j = 1, \ldots, b\}$ is an orthogonal basis for V_B, and $\dim V_B = b$.

Now, $\mathbf{u}_0 = \sum_{j=1}^{b} \mathbf{v}_j \in V_B$, so $V_0 \subset V_B$. Just as we defined W_T, we put

$$W_B = \{\mathbf{v} \in V_B : \mathbf{v} \text{ is orthogonal to } V_0\} = V_B \cap V_0^\perp.$$

Example 4.10 (More fiction) Table 4.1 shows a block design for three treatments in two blocks of size four. The treatment vectors $\mathbf{u}_A$, $\mathbf{u}_B$, $\mathbf{u}_C$ and the block vectors $\mathbf{v}_1$ and $\mathbf{v}_2$ are shown. Treatment A occurs twice in block 1, so $\mathbf{u}_A \cdot \mathbf{v}_1 = 2$. Two further vectors are shown: $\mathbf{x} = \mathbf{u}_A - 2\mathbf{u}_B \in V_T$ and $\mathbf{w} = \mathbf{v}_1 - \mathbf{v}_2 \in V_B$. Moreover, $\mathbf{x} \cdot \mathbf{u}_0 = \mathbf{w} \cdot \mathbf{u}_0 = 0$, so $\mathbf{x} \in W_T$ and $\mathbf{w} \in W_B$. Note that $\mathbf{x} \cdot \mathbf{w} = 0$, so $\mathbf{x} \perp \mathbf{w}$.

Definition A block design is *orthogonal* if the spaces W_T and W_B are orthogonal to each other.

Theorem 4.1 *Given a block design for t treatments in b blocks of size k, let s_{ij} be the number of times that treatment i occurs in block j, for $i = 1, \ldots, t$ and $j = 1, \ldots, b$. Then the block design is orthogonal if and only if $s_{ij} = r_i/b$ for $i = 1, \ldots, t$ and $j = 1, \ldots, b$.*

Table 4.1. *Some vectors in Example 4.10*

Plot	Block	T	$\mathbf{u}_0$	$\mathbf{u}_A$	$\mathbf{u}_B$	$\mathbf{u}_C$	$\mathbf{v}_1$	$\mathbf{v}_2$	$\mathbf{x}$ in W_T	$\mathbf{w}$ in W_B
1	1	A	1	1	0	0	1	0	1	1
2	1	A	1	1	0	0	1	0	1	1
3	1	B	1	0	1	0	1	0	-2	1
4	1	C	1	0	0	1	1	0	0	1
5	2	C	1	0	0	1	0	1	0	-1
6	2	A	1	1	0	0	0	1	1	-1
7	2	B	1	0	1	0	0	1	-2	-1
8	2	A	1	1	0	0	0	1	1	-1

Proof First note that $s_{ij} = \mathbf{u}_i \cdot \mathbf{v}_j$.

Since W_T is orthogonal to V_0, $W_T \perp W_B$ if and only if $W_T \perp V_B$, which happens if and only if

$$\left(\sum_{i=1}^{t} a_i \mathbf{u}_i \right) \cdot \mathbf{v}_j = 0 \quad \text{for } j = 1, \ldots, b$$

whenever $\sum_i a_i r_i = 0$; that is, $\sum_i a_i s_{ij} = 0$ for all j whenever $\sum_i a_i r_i = 0$.

If $s_{ij} = r_i/b$ for each i then $\sum_i a_i s_{ij} = \sum_i (a_i r_i)/b$, which is zero whenever $\sum_i a_i r_i = 0$. This is true for all j, so $W_T \perp W_B$.

Conversely, suppose that $W_T \perp W_B$. Fix i different from 1, and put $a_1 = 1/r_1$, $a_i = -1/r_i$ and $a_l = 0$ if $l \notin \{1, i\}$. Then $\sum_{l=1}^{t} a_l r_l = 0$ while

$$\sum_{l=1}^{t} a_l s_{lj} = \frac{s_{1j}}{r_1} - \frac{s_{ij}}{r_i}$$

for all j. Since $W_T \perp W_B$, it follows that

$$\frac{s_{1j}}{r_1} = \frac{s_{ij}}{r_i}$$

for all j. This is true for all i, including $i = 1$, so counting the plots in block j gives

$$k = \sum_{i=1}^{t} s_{ij} = \frac{s_{1j}}{r_1} \sum_{i=1}^{t} r_i = \frac{s_{1j}}{r_1} N = \frac{s_{1j}}{r_1} bk.$$

Therefore $s_{1j} = r_1/b$ and hence $s_{ij} = r_i/b$ for all i. ∎

Definition A *complete-block design* has blocks of size t, with each treatment occurring once in each block.

Corollary 4.2 *Complete-block designs are orthogonal.*

We consider only orthogonal block designs for the remainder of this chapter.

4.3 Construction and randomization

Construct and randomize an orthogonal block design as follows.

(i) Apply treatment i to r_i/b plots in block 1, for $i = 1, \ldots, t$, and randomize, just as for a completely randomized design.

(ii) Repeat for each block, using a fresh randomization each time, independent of the preceding randomizations.

Example 4.11 (Wine tasting) Four wines are tasted and evaluated by each of eight judges. A plot is one tasting by one judge; judges are blocks. So there are eight blocks and 32 plots. Plots within each judge are identified by order of tasting.

The systematic design is the same for each judge.

Judge j				
Tasting	1	2	3	4
Wine	1	2	3	4

To randomize this design we need eight independent random permutations of four objects. Here we use the method described at the end of Section 2.2, using a stream of random digits and taking as many as are needed for each successive block. The random digits are shown in the top row of Figure 4.2 and the randomized plan in Figure 4.3.

Example 4.12 (Example 1.11 continued: Detergents) Suppose that there are three detergents to be tested, but that the research organization has only limited quantities of the two new ones (A and B), while it has effectively unlimited quantities of its standard detergent (C). Each housewife participating in the trial will do four washloads, so the researchers decide that every housewife will wash two loads with detergent C and one with each of A and B.

The systematic design is

Housewife j				
Washload	1	2	3	4
Detergent	A	B	C	C

for each of the ten housewives. After randomization, the plan could appear like the one in Figure 4.4.

4.4 Models for block designs

Recall that $Y_\omega = \tau_{T(\omega)} + Z_\omega$, where Z_ω is the effect of plot ω. There are two common models for how the blocks affect Z_ω.

In the first model, the blocks affect the expectation but not the covariance. Thus

$$\mathbb{E}(Z_\omega) = \zeta_{B(\omega)},$$

where $\zeta_{B(\omega)}$ is an unknown constant depending on the block $B(\omega)$ containing ω. However, the covariance still has its simplest form; that is

$$\operatorname{cov}(Z_\alpha, Z_\beta) = \begin{cases} \sigma^2 & \text{if } \alpha = \beta \\ 0 & \text{otherwise.} \end{cases}$$

```
5 4 6 7 | 8 6 2 0 | 8 0 2 2 4 |
2 1 3 4 | 4 3 2 1 | 4 1 2 X 3 |

2 7 4 1 | 5 9 8 5 7 | 4 7 4 7 2 6 |
2 4 3 1 | 1 4 3 X 2 | 2 4 X X 1 3 |

9 3 3 1 5 | 6 4 1 6 3
4 2 X 1 3 | 4 3 1 X 2
```

Fig. 4.2. Stream of random digits, used to randomize the design in Example 4.11

Judge 1				
Tasting	1	2	3	4
Wine	2	1	3	4

Judge 2				
Tasting	1	2	3	4
Wine	4	3	2	1

Judge 3				
Tasting	1	2	3	4
Wine	4	1	2	3

Judge 4				
Tasting	1	2	3	4
Wine	2	4	3	1

Judge 5				
Tasting	1	2	3	4
Wine	1	4	3	2

Judge 6				
Tasting	1	2	3	4
Wine	2	4	1	3

Judge 7				
Tasting	1	2	3	4
Wine	4	2	1	3

Judge 8				
Tasting	1	2	3	4
Wine	4	3	1	2

Fig. 4.3. Randomized plan in Example 4.11

In this model, we say that the blocks have *fixed effects*.

In the second model the blocks make no contribution to the expectation, so that $\mathbb{E}(Z_\omega) = 0$. However, the covariance between the responses on plots α and β depends on whether $\alpha = \beta$, α and β are different but in the same block, or α and β are in different blocks. Thus

$$\text{cov}(Z_\alpha, Z_\beta) = \begin{cases} \sigma^2 & \text{if } \alpha = \beta \\ \rho_1 \sigma^2 & \text{if } \alpha \neq \beta \text{ but } B(\alpha) = B(\beta) \\ \rho_2 \sigma^2 & \text{if } B(\alpha) \neq B(\beta). \end{cases}$$

Of course, $1 \geq \rho_1$ and $1 \geq \rho_2$. Usually we expect that $\rho_1 > \rho_2$, because plots in the same block should respond in a more alike manner than plots in different blocks. Now we say that the blocks have *random effects*.

Let $\mathbf{J}_B$ be the $N \times N$ matrix whose (α, β)-entry is equal to

$$\begin{cases} 1 & \text{if } B(\alpha) = B(\beta) \\ 0 & \text{otherwise}. \end{cases}$$

4.5. Analysis when blocks have fixed effects

Housewife 1				
Washload	1	2	3	4
Detergent	C	C	B	A

Housewife 2				
Washload	1	2	3	4
Detergent	C	A	B	C

Housewife 3				
Washload	1	2	3	4
Detergent	A	C	B	C

Housewife 4				
Washload	1	2	3	4
Detergent	B	C	C	A

Housewife 5				
Washload	1	2	3	4
Detergent	C	A	C	B

Housewife 6				
Washload	1	2	3	4
Detergent	A	C	C	B

Housewife 7				
Washload	1	2	3	4
Detergent	A	C	C	B

Housewife 8				
Washload	1	2	3	4
Detergent	C	B	A	C

Housewife 9				
Washload	1	2	3	4
Detergent	B	C	A	C

Housewife 10				
Washload	1	2	3	4
Detergent	C	C	B	A

Fig. 4.4. Randomized plan in Example 4.12

For example, if $b = 2$ and $k = 4$ and the first four plots are in block 1 then

$$\mathbf{J}_B = \begin{bmatrix} 1 & 1 & 1 & 1 & 0 & 0 & 0 & 0 \\ 1 & 1 & 1 & 1 & 0 & 0 & 0 & 0 \\ 1 & 1 & 1 & 1 & 0 & 0 & 0 & 0 \\ 1 & 1 & 1 & 1 & 0 & 0 & 0 & 0 \\ 0 & 0 & 0 & 0 & 1 & 1 & 1 & 1 \\ 0 & 0 & 0 & 0 & 1 & 1 & 1 & 1 \\ 0 & 0 & 0 & 0 & 1 & 1 & 1 & 1 \\ 0 & 0 & 0 & 0 & 1 & 1 & 1 & 1 \end{bmatrix} = \begin{bmatrix} \mathbf{v}_1^\perp \\ \mathbf{v}_1^\perp \\ \mathbf{v}_1^\perp \\ \mathbf{v}_1^\perp \\ \mathbf{v}_2^\perp \\ \mathbf{v}_2^\perp \\ \mathbf{v}_2^\perp \\ \mathbf{v}_2^\perp \end{bmatrix}.$$

Then, in the random-effects model,

$$\begin{aligned} \mathrm{Cov}(\mathbf{Y}) &= \sigma^2 \mathbf{I} + \rho_1 \sigma^2 (\mathbf{J}_B - \mathbf{I}) + \rho_2 \sigma^2 (\mathbf{J} - \mathbf{J}_B) \\ &= \sigma^2 [(1-\rho_1)\mathbf{I} + (\rho_1 - \rho_2)\mathbf{J}_B + \rho_2 \mathbf{J}]. \end{aligned}$$

Some natural discrete classifications with a small number of possibilities (such as sex) are best considered as fixed. For example, 20-year-old human males might always be heavier than 20-year-old human females and we might want to find out how much heavier. Most other classifications are just a nuisance and are best thought of as random. For example, plots at the top end of the field may do better than plots at the bottom end in wet years and worse in dry years, but, on the whole, plots at the top end will tend to perform more similarly to each other than to plots at the bottom end.

4.5 Analysis when blocks have fixed effects

When blocks have fixed effects, the expectation part of the model is that

$$\mathbb{E}(Y_\omega) = \tau_{T(\omega)} + \zeta_{B(\omega)}. \tag{4.1}$$

In vector terms, this is
$$\mathbb{E}(\mathbf{Y}) = \boldsymbol{\tau} + \boldsymbol{\zeta},$$
where $\boldsymbol{\tau} \in V_T$ and $\boldsymbol{\zeta} \in V_B$. Equation (2.5) shows that $\boldsymbol{\tau} = \boldsymbol{\tau}_0 + \boldsymbol{\tau}_T$, where $\boldsymbol{\tau}_0 = \bar{\tau} \mathbf{u}_0 \in V_0$ and $\boldsymbol{\tau}_T = \boldsymbol{\tau} - \bar{\tau} \mathbf{u}_0 \in W_T$. Similarly, $\boldsymbol{\zeta} = \boldsymbol{\zeta}_0 + \boldsymbol{\zeta}_B$, where $\boldsymbol{\zeta}_0 = \bar{\zeta} \mathbf{u}_0 \in V_0$ and $\boldsymbol{\zeta}_B = \boldsymbol{\zeta} - \bar{\zeta} \mathbf{u}_0 \in W_B$. Thus
$$\mathbb{E}(\mathbf{Y}) = (\boldsymbol{\tau}_0 + \boldsymbol{\zeta}_0) + \boldsymbol{\tau}_T + \boldsymbol{\zeta}_B$$
with $\boldsymbol{\tau}_0 + \boldsymbol{\zeta}_0$ in V_0, $\boldsymbol{\tau}_T$ in W_T and $\boldsymbol{\zeta}_B$ in W_B.

Now, $\boldsymbol{\tau}_0$ and $\boldsymbol{\zeta}_0$ are both multiples of the all-1 vector $\mathbf{u}_0$ and so they cannot be distinguished, either in the model (4.1) or from the data. This can be seen in another way. We could replace τ_i by $(\tau_i + c)$ for some constant c, for all i, and replace ζ_j by $(\zeta_j - c)$, for all j, without changing Equation (4.1). This implies that neither $\boldsymbol{\tau}_0$ nor $\boldsymbol{\zeta}_0$ can be estimated. However, we can estimate treatment contrasts, and we can estimate block contrasts.

The definition of the sum of two vector subspaces gives
$$V_T + V_B = \{\mathbf{v} + \mathbf{w} : \mathbf{v} \in V_T, \mathbf{w} \in V_B\}.$$

Thus Equation (4.1) can be rewritten as
$$\mathbb{E}(\mathbf{Y}) \in V_T + V_B.$$

Suppose that $\mathbf{x} \in W_T$. Then $\mathbf{x} \in V_T + V_B$. Applying Theorem 2.6 with $V_T + V_B$ in place of V_T shows that $\mathbf{x} \cdot \mathbf{Y}$ is the best linear unbiased estimator of $\mathbf{x} \cdot (\boldsymbol{\tau}_0 + \boldsymbol{\zeta}_0 + \boldsymbol{\tau}_T + \boldsymbol{\zeta}_B)$. Now, $\mathbf{x} \cdot \boldsymbol{\tau}_0 = \mathbf{x} \cdot \boldsymbol{\zeta}_0 = 0$ because $\mathbf{x} \in V_0^\perp$, and $\mathbf{x} \cdot \boldsymbol{\zeta}_B = 0$ because $\mathbf{x} \in W_B^\perp$: that is why we restrict attention to orthogonal designs throughout this chapter. Therefore
$$\mathbf{x} \cdot (\boldsymbol{\tau}_0 + \boldsymbol{\zeta}_0 + \boldsymbol{\tau}_T + \boldsymbol{\zeta}_B) = \mathbf{x} \cdot \boldsymbol{\tau}_T,$$
whose best linear unbiased estimator is $\mathbf{x} \cdot \mathbf{Y}$ with variance $\|\mathbf{x}\|^2 \sigma^2$. Similarly, if $\mathbf{z} \in W_B$ then $\mathbf{z} \cdot \mathbf{Y}$ is the best linear unbiased estimator of $\mathbf{z} \cdot \boldsymbol{\zeta}_B$, with variance $\|\mathbf{z}\|^2 \sigma^2$.

Likewise, we have $\mathbf{P}_{W_T}(\boldsymbol{\tau}_0 + \boldsymbol{\zeta}_0) = \mathbf{0}$ because $\boldsymbol{\tau}_0 + \boldsymbol{\zeta}_0$ is orthogonal to W_T. Similarly, $\mathbf{P}_{W_T}(\boldsymbol{\zeta}_B) = \mathbf{0}$ because $\boldsymbol{\zeta}_B$ is in W_B, which is orthogonal to W_T. Now Theorem 2.5 shows that
$$\mathbb{E}\left(\mathbf{P}_{W_T}(\mathbf{Y})\right) = \mathbf{P}_{W_T}(\mathbb{E}(\mathbf{Y})) = \mathbf{P}_{W_T}(\boldsymbol{\tau}_0 + \boldsymbol{\zeta}_0 + \boldsymbol{\tau}_T + \boldsymbol{\zeta}_B) = \boldsymbol{\tau}_T$$
and that $\mathbb{E}\left(\mathbf{P}_{W_B}(\mathbf{Y})\right) = \boldsymbol{\zeta}_B$.

Put $W_E = (V_T + V_B)^\perp$. This is going to be the residual subspace: the reason for the notation E will be explained in Chapter 10. Then V is the following direct sum of orthogonal subspaces:
$$V = V_0 \oplus W_T \oplus W_B \oplus W_E.$$

We have constructed both W_T and W_B to be orthogonal to V_0. The subspaces W_T and W_B are orthogonal to each other because we have assumed that the design is orthogonal. Finally, we have constructed W_E to be orthogonal to the previous three subspaces because $V_T + V_B = V_0 \oplus W_T \oplus W_B$.

Figure 4.5 may be helpful. Note that this is not a Venn diagram. For example, V_0 and W_T are complementary *subspaces* of V_T, not complementary subsets.

4.5. Analysis when blocks have fixed effects

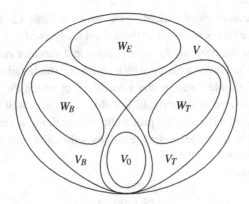

Fig. 4.5. Orthogonal subspaces for an orthogonal block design

Just as in Section 2.12, the orthogonal decomposition of V leads to (orthogonal) decompositions of the dimension, expectation, data and sum of squares, as follows:

$$V \quad = \quad V_0 \quad \oplus \quad W_T \quad \oplus \quad W_B \quad \oplus \quad W_E$$

dimension $\quad N = bk \quad = \quad 1 \quad + \quad (t-1) \quad + \quad (b-1) \quad + \quad (N-b-t+1)$

expectation $\quad \mathbb{E}(\mathbf{Y}) \quad = \quad (\tau_0 + \zeta_0) \quad + \quad \tau_T \quad + \quad \zeta_B \quad + \quad \mathbf{0}$

data $\quad \mathbf{y} \quad = \quad \bar{y}\mathbf{u}_0 \quad + \quad \mathbf{y}_T \quad + \quad \mathbf{y}_B \quad + \quad \text{residual}$

sum of squares $\quad \sum_{\omega \in \Omega} y_\omega^2 \quad = \quad \dfrac{\text{sum}^2}{N} \quad + \quad \text{SS(treatments)} \quad + \quad \text{SS(blocks)} \quad + \quad \text{SS(residual)}$

where

$$\mathbf{y}_T = \mathbf{P}_{W_T}\mathbf{y} = \sum_{i=1}^{t} (\text{mean}_{T=i})\,\mathbf{u}_i - \bar{y}\mathbf{u}_0,$$

$$\mathbf{y}_B = \mathbf{P}_{W_B}\mathbf{y} = \sum_{j=1}^{b} (\text{mean}_{B=j})\,\mathbf{v}_j - \bar{y}\mathbf{u}_0,$$

$$\text{SS(treatments)} = \sum_{i=1}^{t} \frac{(\text{sum}_{T=i})^2}{r_i} - \frac{\text{sum}^2}{N},$$

$$\text{SS(blocks)} = \sum_{j=1}^{b} \frac{(\text{sum}_{B=j})^2}{k} - \frac{\text{sum}^2}{N},$$

$$\begin{aligned}\text{SS(residual)} &= \text{sum of squares of the residuals} \\ &= \sum_{\omega \in \Omega} y_\omega^2 - \text{SS(mean)} - \text{SS(treatments)} - \text{SS(blocks)},\end{aligned}$$

and $\text{sum}_{B=j}$ and $\text{mean}_{B=j}$ are the total and mean respectively of the values of y_ω for ω in block j.

Hence we obtain the analysis-of-variance table shown in Table 4.2. Of course, this is really two analysis-of-variance tables in one. The theoretical analysis-of-variance table, which tells us what to do, can omit the columns for mean square and variance ratio, but must show the column for EMS (expected mean square), which shows us which variance ratios to calculate. The analysis-of-variance table given by the actual data, in which the formulae are replaced by their values, does not need to show the EMS column, but may well include a final column headed 'F-probability', which gives the probability of obtaining a variance ratio at least as big as the one obtained in the table, under the null hypothesis of zero effect for that line, and assuming normality.

Use the variance ratio
$$\frac{\text{MS(treatments)}}{\text{MS(residual)}}$$
to test for treatment differences, and the variance ratio
$$\frac{\text{MS(blocks)}}{\text{MS(residual)}}$$
to test for block differences (if you are interested in them). Both tests are one-sided. It is reasonable to take the view that we expect blocks to differ, in which case there is no point in performing a test for block differences.

Example 4.13 (Metal cords) An experiment was conducted to compare two protective dyes for metal, both with each other and with 'no dye'. Ten braided metal cords were broken into three pieces. The three pieces of each cord were randomly allocated to the three treatments. Thus the cords were blocks. After the dyes had been applied, the cords were left to weather for a fixed time, then their strengths were measured. Table 4.3 shows the strengths as a percentage of the nominal strength specification. The data in this table are not shown in the randomized order; they have been put into a systematic order for ease of calculating totals. Here A denotes 'no dye', while B and C are the two dyes.

From these data,
$$\text{SS(mean)} = 2975.8^2/30 = 295179.52,$$
$$\text{CSS(treatments)} = \left(966.7^2 + 992.9^2 + 1016.2^2\right)/10 = 295302.17,$$
$$\text{CSS(cords)} = \left(317.7^2 + 282.7^2 + \cdots + 283.2^2\right)/3 = 295827.42,$$

and $\sum y^2 = 296231.92$. Therefore SS(treatments) = 295302.17 − 295179.52 = 122.65 and SS(cords) = 295827.42 − 295179.52 = 647.90. It follows that SS(residual) = 296231.92 − 295179.52 − 122.65 − 647.90 = 281.85. If the cords have fixed effects, this gives the analysis of variance in Table 4.4. Since the 95% point of F on 2 and 18 degrees of freedom is 3.55, we can conclude that the treatments differ. We report the treatment means as $\hat{\tau}_A = 96.67$, $\hat{\tau}_B = 99.29$ and $\hat{\tau}_C = 101.62$, on the understanding that it is the *differences* between these values that are meaningful. The standard error of a difference is $\sqrt{15.66 \times 2/10} = 1.77$.

Figure 4.6 shows boxplots of the raw data for each treatment followed by boxplots for the same data with the block means subtracted. These show that subtracting the block means reduces the variability of each treatment without affecting their relative locations. Thus an analysis of the data that does not allow for the differences between blocks is unlikely to draw such strong conclusions about treatment differences.

Table 4.2. *Analysis-of-variance table for blocks and unstructured treatments when blocks have fixed effects*

Source		Sum of squares	Degrees of freedom	Mean square	EMS	Variance ratio
V_0	mean	$\dfrac{\text{sum}^2}{N}$	1	SS(mean)	$\\|\boldsymbol{\tau}_0 + \boldsymbol{\zeta}_0\\|^2 + \sigma^2$	$\dfrac{\text{MS(mean)}}{\text{MS(residual)}}$
W_B	blocks	$\sum_j \dfrac{(\text{sum}_{B=j})^2}{k} - \dfrac{\text{sum}^2}{N}$	$b-1$	$\dfrac{\text{SS(blocks)}}{b-1}$	$\dfrac{\\|\boldsymbol{\zeta}_B\\|^2}{b-1} + \sigma^2$	$\dfrac{\text{MS(blocks)}}{\text{MS(residual)}}$
W_T	treatments	$\sum_i \dfrac{(\text{sum}_{T=i})^2}{r_i} - \dfrac{\text{sum}^2}{N}$	$t-1$	$\dfrac{\text{SS(treatments)}}{t-1}$	$\dfrac{\\|\boldsymbol{\tau}_T\\|^2}{t-1} + \sigma^2$	$\dfrac{\text{MS(treatments)}}{\text{MS(residual)}}$
	residual	←········ by subtraction ········→		$\dfrac{\text{SS(residual)}}{\text{df(residual)}}$	σ^2	—
Total		$\sum_\omega y_\omega^2$	N			

Table 4.3. *Data in Example 4.13*

Cord	Treatment			Cord total
	A	B	C	
1	102.4	108.5	106.8	317.7
2	93.7	92.3	96.7	282.7
3	97.4	93.1	100.6	291.1
4	96.1	106.9	101.9	304.9
5	102.5	92.0	103.3	297.8
6	87.8	95.5	94.9	278.2
7	102.6	108.4	106.5	317.5
8	95.2	94.6	101.2	291.0
9	96.9	103.4	111.4	311.7
10	92.1	98.2	92.9	283.2
Treatment total	966.7	992.9	1016.2	2975.8

Table 4.4. *Analysis-of-variance table for Example 4.13 when cords have fixed effects*

Source	SS	df	MS	VR
mean	295179.52	1	295179.52	18849.27
cords	647.90	9	72.00	4.60
treatments	122.65	2	61.32	3.92
residual	281.85	18	15.66	–
Total	2962.32	30		

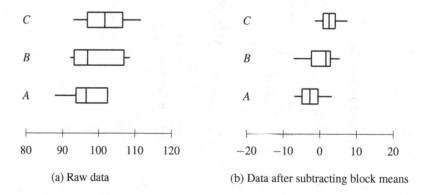

(a) Raw data   (b) Data after subtracting block means

Fig. 4.6. Boxplots of the data in Example 4.13

4.6. Analysis when blocks have random effects

Table 4.5. *Analysis-of-variance table for blocks and unstructured treatments when blocks have random effects*

Stratum	Source	df	EMS	VR
V_0 mean	mean	1	$\|\tau_0\|^2 + \xi_0$	–
W_B blocks	blocks	$b-1$	ξ_1	–
$V_B^\perp$ plots	treatments	$t-1$	$\dfrac{\|\tau_T\|^2}{t-1} + \xi_2$	$\dfrac{\text{MS(treatments)}}{\text{MS(residual)}}$
	residual	$b(k-1)-(t-1)$	ξ_2	–
Total		N		

4.6 Analysis when blocks have random effects

Put $\mathbf{C} = \text{Cov}(\mathbf{Y})$. Then we have

$$\mathbf{C} = \sigma^2[(1-\rho_1)\mathbf{I} + (\rho_1-\rho_2)\mathbf{J}_B + \rho_2\mathbf{J}]. \tag{4.2}$$

If plot ω is in block j then the ω-row of $\mathbf{J}_B$ is just $\mathbf{v}_j$. Hence if $\mathbf{x}$ is any vector in V then the ω-entry in $\mathbf{J}_B\mathbf{x}$ is equal to $\mathbf{v}_j \cdot \mathbf{x}$. In particular, if $\mathbf{x} = \mathbf{u}_0$ then $\mathbf{v}_j \cdot \mathbf{x} = k$ for all j and so $\mathbf{J}_B\mathbf{u}_0 = k\mathbf{u}_0$. Since $\mathbf{I}\mathbf{u}_0 = \mathbf{u}_0$ and $\mathbf{J}\mathbf{u}_0 = N\mathbf{u}_0$, we see that

$$\mathbf{C}\mathbf{u}_0 = \sigma^2[(1-\rho_1) + k(\rho_1-\rho_2) + N\rho_2]\mathbf{u}_0,$$

so that $\mathbf{u}_0$ is an eigenvector of $\mathbf{C}$ with eigenvalue ξ_0, where

$$\xi_0 = \sigma^2[(1-\rho_1) + k(\rho_1-\rho_2) + N\rho_2].$$

If $\mathbf{x} \in V_B$ then $\mathbf{x} = \sum_j \lambda_j \mathbf{v}_j$ for some scalars $\lambda_1, \ldots, \lambda_b$; hence $\mathbf{v}_j \cdot \mathbf{x} = k\lambda_j$ and so $\mathbf{J}_B\mathbf{x} = k\mathbf{x}$. In Section 2.14 we saw that $\mathbf{J}\mathbf{x} = \mathbf{0}$ if $\mathbf{x} \in V_0^\perp$. Hence if $\mathbf{x} \in W_B = V_B \cap V_0^\perp$ then

$$\mathbf{C}\mathbf{x} = \sigma^2[(1-\rho_1) + k(\rho_1-\rho_2)]\mathbf{x},$$

and so $\mathbf{x}$ is an eigenvector of $\mathbf{C}$ with eigenvalue ξ_1, where

$$\xi_1 = \sigma^2[(1-\rho_1) + k(\rho_1-\rho_2)].$$

Finally, if $\mathbf{x} \in V_B^\perp \subseteq V_0^\perp$ then $\mathbf{J}_B\mathbf{x} = \mathbf{0}$ and $\mathbf{J}\mathbf{x} = \mathbf{0}$ so $\mathbf{C}\mathbf{x} = \xi_2\mathbf{x}$, where $\xi_2 = \sigma^2(1-\rho_1)$.

Thus the eigenspaces of $\mathbf{C}$ (the strata) are V_0, W_B and $V_B^\perp$, with dimensions 1, $b-1$ and $N-b$ and eigenvalues ξ_0, ξ_1 and ξ_2 respectively. Usually we expect that $\xi_1 > \xi_2$, because $\xi_1 = \xi_2 + k\sigma^2(\rho_1 - \rho_2)$.

For an orthogonal block design, $W_T \subseteq V_B^\perp$. Hence Theorem 2.11 shows that the appropriate analysis-of-variance table is that shown in Table 4.5. The arithmetic calculations are identical to those for the fixed-effects model. Assess treatment differences just as before. For the effects of blocks, do a two-sided test using

$$\frac{\text{MS(blocks)}}{\text{MS(residual)}}.$$

Table 4.6. *Analysis-of-variance table for Example 4.13 when cords have random effects*

Stratum	Source	SS	df	MS	VR
mean	mean	295179.52	1	295179.52	–
cords	cords	647.90	9	72.00	–
plots	treatments	122.65	2	61.32	3.92
	residual	281.85	18	15.66	–
Total		2962.32	30		

If MS(blocks) >> MS(residual) then the choice of blocks was good: do it similarly next time. If MS(blocks) << MS(residual) then

either $\xi_1 < \xi_2$ because plots within a block compete (for example, if all plots in a chamber in a greenhouse share a single system of circulating liquid nutrients, so ρ_1 is negative)

or $\xi_1 < \xi_2$ and there is a better way of blocking

or trial management has not been by block (for example, if the plots are in a rectangle, columns have been used as blocks but harvesting has been done by rows)

or the scientist is fiddling the data, and is not expecting you to notice very low values of the variance ratio (such fiddling usually leads to too small estimates of variability, because the data are 'too good').

Example 4.13 revisited (Metal cords) Given that the metal cords used in the experiment are just ten out of many coming off the production line, it is more likely that they produce random effects than fixed effects. This gives the analysis of variance in Table 4.6. This contains exactly the same numbers as Table 4.4 (apart from the final column), but the layout and the interpretation are subtly different.

4.7 Why use blocks?

If we should use blocks and do not, what happens?

If the blocks contribute fixed effects then the vector ζ_B is almost certainly not zero. If the treatments are not allocated orthogonally to blocks then ζ_B will not be orthogonal to W_T, so $\mathbf{P}_{W_T}\zeta_B$ will be nonzero. The estimator of τ_T is $\mathbf{P}_{W_T}\mathbf{Y}$, whose expectation is $\tau_T + \mathbf{P}_{W_T}\zeta_B$. Thus the treatment estimators will be biased. It is most likely that ζ_B is also not orthogonal to $V_T^\perp$, so the estimator of σ^2 will also be biased. In fact, Theorem 2.5(ii) shows that the expectation of this estimator will be

$$\frac{\left\|\mathbf{P}_{V_T^\perp}\zeta_B\right\|^2}{(N-t)} + \sigma^2,$$

so that the variance will be overestimated.

Example 4.14 (Hay fever) Two new treatments for hay fever, from different manufacturers, are compared with a placebo, using 18 volunteers in the hay-fever season. For convenience, in one week the physicians give treatment *A* to six volunteers and the placebo to three; in the

4.8. Loss of power with blocking

second week they give treatment B to six more volunteers and the placebo to three more. They measure the respiratory function of each volunteer three hours after treatment.

Unfortunately, a change in the weather has the effect that different pollens are in the air in the second week, with the result that respiratory function increases by 4 units, on average. Then the expectation of the estimator of $\tau_B - \tau_A$ is equal to $\tau_B - \tau_A + 4$. If the result of this trial is that the physicians recommend treatment B rather than treatment A, the manufacturer of treatment A might well try to take the physicians to court. (Of course, a much larger trial would normally be run before such a recommendation is made, but the point about the danger of failing to block is still valid.)

Moreover, $\mathbf{P}_{V_T^\perp} \zeta_B$ now has coordinates zero on the two new treatments and coordinates ± 2 on the placebo, so the expectation of the residual mean square is $24/15 + \sigma^2$.

If the blocks contribute random effects then treatment estimators are unbiased but their variances are larger than they need be: on average, ξ_2 will be replaced by

$$\frac{(b-1)\xi_1 + (N-b)\xi_2}{N-1},$$

which is bigger than ξ_2 when $\xi_1 > \xi_2$.

If we do use blocks in the design but forget to include them in the analysis, what happens?

Now the treatment estimators are unbiased, but in both models our estimates of their variances are too high, so we may fail to detect genuine treatment differences. For fixed effects, the expectation of the estimator of σ^2 is equal to

$$\frac{\|\zeta_B\|^2}{(N-t)} + \sigma^2;$$

for random effects, the expectation of the estimator of ξ_2 is equal to

$$\frac{(b-1)\xi_1 + (N-b-t+1)\xi_2}{N-t} = \frac{(b-1)(\xi_1 - \xi_2) + (N-t)\xi_2}{N-t},$$

which is bigger than ξ_2 when $\xi_1 > \xi_2$.

Example 4.13 revisited (Metal cords) If the analysis ignores blocks then the residual mean square becomes $(647.90 + 281.85)/27 = 34.44$, so the variance ratio for treatments becomes only 1.78, which is well below the 95% point of F on 2 and 27 degrees of freedom. The analyst may well fail to detect treatment differences.

4.8 Loss of power with blocking

The following example, which is taken from a case where the manufacturer tried to sue the statisticians for using blocks, shows the only circumstances where blocking may be a disadvantage: there are no natural block boundaries, there are a small number of residual degrees of freedom, and the purpose of the experiment is (arguably) hypothesis testing rather than estimation.

Table 4.7. *Analysis of variance for the two designs in Example 4.15*

Stratum	Source	df	EMS
mean	mean	1	$6\bar{\tau}^2 + \xi_0$
plots	treatments	1	$\frac{6}{4}(\tau_1 - \tau_2)^2 + \xi$
	residual	4	ξ
Total		6	

Stratum	Source	df	EMS
mean	mean	1	$6\bar{\tau}^2 + \xi_0$
blocks	blocks	2	ξ_1
plots	treatments	1	$\frac{3}{2}(\tau_1 - \tau_2)^2 + \xi_2$
	residual	2	ξ_2
Total		6	

(a) Completely randomized design (b) Complete-block design

Example 4.15 (Pasture grass) A new additive is claimed to vastly improve the quality of pasture grass. Are farmers wasting their money in buying it? There are two treatments: the new additive, and nothing.

Plots must be large enough for several sheep to graze freely. Hence the replication cannot be large: replication 3 is chosen. Should the design be completely randomized or in three randomized complete blocks?

Put

$$\tau_1 = \text{response to nothing}$$
$$\tau_2 = \text{response to the new additive.}$$

The null hypothesis is $H_0 : \tau_1 = \tau_2 = \bar{\tau} = \left(\frac{\tau_1 + \tau_2}{2}\right)$. Now,

$$\tau_T = \begin{cases} \tau_1 - \bar{\tau} = \frac{\tau_1 - \tau_2}{2} & \text{on plots with nothing} \\ \tau_2 - \bar{\tau} = \frac{\tau_2 - \tau_1}{2} & \text{on plots with the new additive.} \end{cases}$$

Using the model for the completely randomized design from Section 2.14, but writing ξ in place of ξ_1 (to avoid confusion with the next model), we obtain the analysis of variance in Table 4.7(a).

Now we consider the complete-block design. There are no natural block boundaries, so the random-effects model is appropriate, and we obtain the analysis of variance in Table 4.7(b).

The completely randomized design mixes up the five degrees of freedom orthogonal to V_0, so $5\xi = 2\xi_1 + 3\xi_2$. Hence $\xi > \xi_2$ if $\xi_1 > \xi_2$.

The variance of the estimator of $\tau_1 - \tau_2$ is

$$\begin{cases} \frac{2}{3}\xi & \text{in the completely randomized design} \\ \frac{2}{3}\xi_2 & \text{in the complete-block design} \end{cases}$$

so the complete-block design gives smaller variance and so is better for estimation.

For hypothesis testing, we consider the one-sided alternative that τ_2 is bigger than τ_1. To test at the 5% significance level, we need the 0.95 point of the t-distribution, which is 2.920

on 2 degrees of freedom and 2.132 on 4 degrees of freedom. To have 90% power of detecting that $\tau_2 > \tau_1$, we also need the 0.90 points of these distributions, which are 1.886 and 1.533 respectively. The argument in Section 2.13 shows that to have probability at least 0.9 of detecting that $\tau_2 > \tau_1$ when doing a one-sided test at the 5% significance level we need

$$\begin{cases} \tau_2 - \tau_1 > (2.132 + 1.533) \times \sqrt{\frac{2}{3}\xi} & \text{in the completely randomized design} \\ \tau_2 - \tau_1 > (2.920 + 1.886) \times \sqrt{\frac{2}{3}\xi_2} & \text{in the complete-block design.} \end{cases}$$

Thus

$$\begin{aligned} \text{the block design is better} &\iff 4.806\sqrt{\xi_2} < 3.665\sqrt{\xi} \\ &\iff \xi > 1.720\xi_2 \\ &\iff \xi_1 > 2.8\xi_2. \end{aligned}$$

Typically we have $\xi_1 \approx 1.5\xi_2$ for such a trial, so smaller differences are more likely to be detected by the unblocked design. A scientist who is more interested in proving that the new additive is better (than in accurately estimating how much better) might complain if the experiment is conducted in blocks rather than in a completely randomized design.

Questions for discussion

4.1 A scientist at a horticultural research station is planning an experiment on cabbages. He wants to compare six different methods of keeping the cabbages free from slugs. The experimental area has 24 plots. He shows you the sketch in Figure 4.7. Advise him on how to design the experiment.

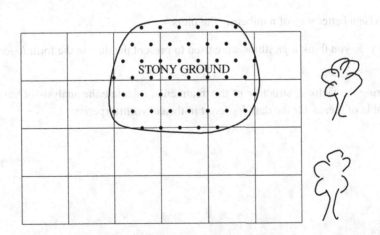

Fig. 4.7. Sketch of experimental area in Question 4.1

4.2 Redo Question 2.1 under the assumption that the professor has ten pills of Wakey-Wakey and ten of Zizzaway.

There is only one observation room, so only one pill can be tested per day. Your plan should show which student should take which pill on which day.

What information should you give the professor about the plan?

4.3 The plan in Figure 4.8 is the field layout of an experiment conducted in 1935 at Rothamsted Experimental Station (an agricultural research station founded in 1843). Each plot had a notice on it showing the block number and the plot number. These are the top two numbers given in each plot in the plan. The purpose of the experiment was to compare various types of fumigant, in single and double doses, for their ability to control eelworms in the soil where oats were being grown. A 'control' treatment (i.e. no fumigant) was included. In the plan, each plot shows, in order below the plot number, the level of a factor called Fumigant, then the dose, then the type of chemical. In early March, 400 gm of soil were sampled from each plot, and the number of eelworm cysts in each sample counted and recorded. The fumigants were ploughed into the soil in mid-March, and the oats were sown one week later. They were grown, and harvested in August. In October, the plots were sampled again in the same way, and the number of cysts recorded. The variable *logcount* was calculated as

$$logcount = \log(\text{number of eelworm cysts at harvest}) - \log(\text{number of eelworm cysts in spring before treatment}),$$

where the logarithms are to base e. This variable is shown at the bottom of each plot in the plan.

(a) How many treatments were there?

(b) How were the plots grouped into blocks?

(c) After sampling soil from plot 16 of block II, which plot should the scientist sample next?

(d) Devise a better way of numbering the plots.

(e) Why do you think logarithms were used to present the data in the form *logcount*?

4.4 Ignoring the factorial structure of the treatments, calculate the analysis-of-variance table and the table of means for the data *logcount* in the eelworm experiment.

Questions for discussion

I 1 1 0 Z 0.549	I 2 2 2 K −0.011	I 3 2 1 N 0.457	I 4 2 1 M 0.599	I 5 2 1 S 0.341	I 6 1 0 Z 0.784			
I 7 1 0 Z 0.759	I 8 2 2 M 0.365	I 9 2 2 S 0.277	I 10 2 1 K 0.107	I 11 1 0 Z 1.187	I 12 2 2 N 0.740	IV 37 2 2 M 0.739	IV 38 2 2 S 0.268	IV 39 2 2 K 0.574
II 13 2 1 K 0.771	II 14 1 0 Z 0.873	II 15 2 1 S 0.803	II 16 2 2 K 0.609	III 25 2 2 K 0.414	III 26 1 0 Z 0.521	III 27 2 1 K 0.191	III 28 2 1 M 1.088	IV 40 1 0 Z 1.482	IV 41 2 1 K 0.791	IV 42 2 1 N 1.316
II 17 1 0 Z 1.269	II 18 2 2 N 1.067	II 19 2 2 S 0.888	II 20 2 1 N 1.665	III 29 1 0 Z 2.170	III 30 2 2 N 2.325	III 31 2 2 S 0.499	III 32 1 0 Z 1.719	IV 43 2 1 M 1.457	IV 44 1 0 Z 0.616	IV 45 1 0 Z 1.398
II 21 2 2 M 0.812	II 22 1 0 Z 1.081	II 231 2 1 M 1.355	II 24 1 0 Z 1.618	III 33 2 1 S 1.247	III 34 2 1 N 1.792	III 35 1 0 Z 1.807	III 36 2 2 M 1.826	IV 46 1 0 Z 2.138	IV 47 2 2 N 1.992	IV 48 2 1 S 1.271

	type of fumigant		dose of fumigant
Z	no fumigant	0	no fumigant
N	chlorodinitrobenzene	1	single dose
M	'Cymag'	2	double dose
S	carbon disulphide jelly		
K	'Seekay'		

Fig. 4.8. Field layout for the experiment in Question 4.3

4.5 A group of ecologists is planning an experiment to compare 36 species of small insects for their effect in decomposing a certain mixture of leaves. Identical quantities of the leaf mixture will be put into 180 glass jars. Each species will be allocated to five of these jars: several insects from that species will be put into those jars. The jars will then be covered with transparent film and left on the bench in the laboratory for several weeks before the amount of leaf decomposition is measured.

Because of the shape of the available bench space, the jars must be arranged in a 5×36 rectangle. One row of jars is next to the window and receives the most sunlight; the fifth row receives the least. The ecologists suggest the following different methods for dealing with this difference in sunlight.

(a) Place the 180 jars in their positions in the rectangle in a haphazard manner. It is then likely that no species is especially favoured.

(b) Start as above. Each morning, rearrange the jars within each column by moving the jar nearest to the window to the position furthest from the window, and moving all the other jars in that column one place nearer to the window. That way, the differences in sunlight should be averaged out for each jar.

(c) Treat each row as a block. Put one jar of each species in each row. In each row independently, choose a random permutation of $1, \ldots, 36$ and use it to randomize the positions of the jars in that row. The differences in sunlight will be allowed for in the analysis.

(d) Put all five jars for each species into a single column, to ensure that each species is exposed to the full range of sunlight conditions. Choose a single random permutation of $1, \ldots, 36$ and use it to allocate species to columns.

Advise the ecologists of the advantages and disadvantages of these methods.

Chapter 5

Factorial treatment structure

5.1 Treatment factors and their subspaces

In this section we consider experiments where the treatment structure is that treatments consist of all combinations of levels of two treatment factors.

Example 5.1 (Example 1.7 continued: Rye-grass) Here the twelve treatments are all combinations of the levels of the following two treatment factors.

Factor	Levels
Cultivar (C)	Cropper, Melle, Melba
Fertilizer (F)	0, 80, 160, 240 kg/ha

The treatments may be labelled $1, \ldots, 12$ according to the following table.

	\multicolumn{4}{c}{F}			
C	0	80	160	240
Cropper	1	2	3	4
Melle	5	6	7	8
Melba	9	10	11	12

Thus treatment 6 is the ordered pair (Melle, 80). In particular,

$$T(\omega) = 6 = (\text{Melle}, 80) \iff \begin{cases} C(\omega) = \text{Melle} & \text{and} \\ F(\omega) = 80. \end{cases}$$

Notation If treatments are all combinations of levels of treatment factors F and G, write $T = F \wedge G$ to show that each level of T is a combination of levels of F and G.

In Example 5.1, $T = C \wedge F$.

Notation If F is a treatment factor, write n_F for the number of levels of F.

In Example 5.1, $n_C = 3$ and $n_F = 4$.

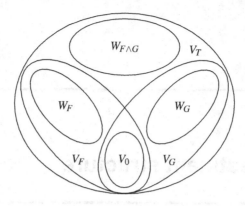

Fig. 5.1. Orthogonal subspaces of the treatment space when treatments are all combinations of the levels of two treatment factors

Now we define subspaces V_F, W_F, V_G and W_G of V analogous to those defined for treatments in Sections 2.3 and 2.11 and for blocks in Section 4.2. Thus we put

$$\begin{aligned} V_F &= \{\text{vectors in } V \text{ which are constant on each level of } F\}, \\ W_F &= V_F \cap V_0^\perp, \\ V_G &= \{\text{vectors in } V \text{ which are constant on each level of } G\}, \\ W_G &= V_G \cap V_0^\perp. \end{aligned}$$

Then $\dim V_F = n_F$, $\dim W_F = n_F - 1$, $\dim V_G = n_G$ and $\dim W_G = n_G - 1$.

Theorem 5.1 *If every combination of levels of factors F and G occurs on the same number of plots then $W_F \perp W_G$.*

Proof Similar to the proof of Theorem 4.1. ∎

Theorem 5.1 shows that if every combination of levels of the factors F and G occurs on the same number of plots then

$$V_F + V_G = V_0 \oplus W_F \oplus W_G \tag{5.1}$$

orthogonally, and hence that

$$\dim(V_F + V_G) = 1 + (n_F - 1) + (n_G - 1).$$

If the treatment factor T is defined by all combinations of levels of F and G, and all these combinations occur at least once, then $\dim V_T = t = n_F n_G$. Now, $V_F + V_G \subset V_T$, so put

$$W_{F \wedge G} = V_T \cap (V_F + V_G)^\perp. \tag{5.2}$$

Then

$$\dim W_{F \wedge G} = \dim V_T - \dim(V_F + V_G) = n_F n_G - 1 - (n_F - 1) - (n_G - 1) = (n_F - 1)(n_G - 1).$$

Figure 5.1 may be helpful. Compare this with Figure 4.5.

5.2. Interaction

Table 5.1. *Some models for $\mathbb{E}(Y_\omega)$ when treatments are all combinations of levels of F and G*

Coordinate (and parameters)	Vector (and subspace)	Model name
$\mathbb{E}(Y_\omega) = \tau_{T(\omega)}$	$\mathbb{E}(\mathbf{Y}) \in V_T$	full treatment model
$\mathbb{E}(Y_\omega) = \lambda_{F(\omega)} + \mu_{G(\omega)}$	$\mathbb{E}(\mathbf{Y}) \in V_F + V_G$	additive in F and G
$\mathbb{E}(Y_\omega) = \lambda_{F(\omega)}$	$\mathbb{E}(\mathbf{Y}) \in V_F$	F only
$\mathbb{E}(Y_\omega) = \mu_{G(\omega)}$	$\mathbb{E}(\mathbf{Y}) \in V_G$	G only
$\mathbb{E}(Y_\omega) = \kappa$	$\mathbb{E}(\mathbf{Y}) \in V_0$	null model
$\mathbb{E}(Y_\omega) = 0$	$\mathbb{E}(\mathbf{Y}) \in \{\mathbf{0}\}$	zero model

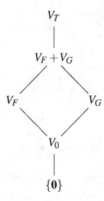

Fig. 5.2. Relationships among the subspaces listed as possible models in Table 5.1

5.2 Interaction

We continue to suppose that treatments consist of all combinations of levels of F and G. Table 5.1 shows some plausible models for the expectation of the response in this situation. The models are listed in decreasing order of their number of parameters. The relationships among the subspaces considered as models are shown in Figure 5.2.

If $\mathbb{E}(\mathbf{Y}) \in V_F + V_G$ then the difference between $\mathbf{Y}$ values for different levels of F does not depend on the level of G, and vice versa. The expected responses for four combinations are shown in Table 5.2. Thus plotting mean$_{F=i}$ against i for each level of G gives approximately parallel curves.

Example 5.1 revisited (Rye-grass) In this experiment the response on each plot was the percentage of water-soluble carbohydrate in the crop. If the twelve treatment means are as shown in Figure 5.3 then we have approximately parallel curves. Thus we can report, for example, that the percentage of water-soluble carbohydrate in the grain is 0.95 higher for Cropper than it is for Melle, irrespective of the amount of fertilizer.

If $\mathbb{E}(\mathbf{Y})$ is not in $V_F + V_G$ then we say that there is an *interaction* between F and G. In Example 5.1 there is no interaction.

Table 5.2. *Illustration of differences when factors are additive*

	Level 1 of G	Level 2 of G	Difference
Level 1 of F	$\lambda_1 + \mu_1$	$\lambda_1 + \mu_2$	$\mu_1 - \mu_2$
Level 2 of F	$\lambda_2 + \mu_1$	$\lambda_2 + \mu_2$	$\mu_1 - \mu_2$
Difference	$\lambda_1 - \lambda_2$	$\lambda_1 - \lambda_2$	

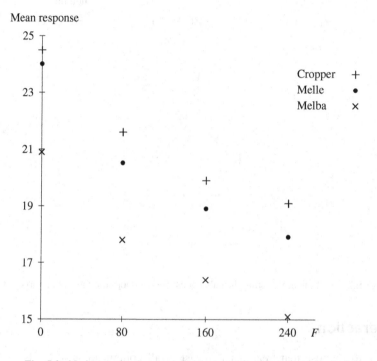

Fig. 5.3. No interaction between Cultivar and Fertilizer in Example 5.1

Example 5.2 (Herbicides) If the combination of two herbicides gives more weed control than would be expected by adding the effects of the two separate herbicides, this is called *synergism*. On the other hand, if the combination gives less control than the sum of the effects, then there is said to be *antagonism* between the herbicides. These are both forms of interaction.

The most extreme form of interaction occurs when the separate curves actually cross over each other.

Example 5.3 (Cow-peas) In an experiment in South Africa, the treatments consisted of five varieties of cow-pea in combination with three methods of cultivation. The mean yields, in tonnes/hectare, are shown in Figure 5.4. Here there is *crossover* interaction. It is important to report this interaction. Clearly cultivation method 1 is best for varieties C and D and worst for the other three, while cultivation method 2 is worst for varieties C and D and best for the

5.2. Interaction

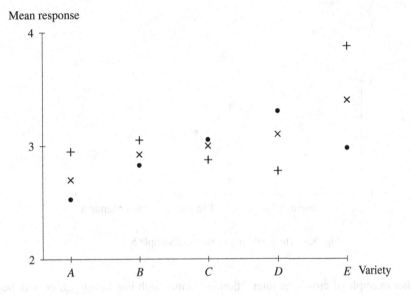

Fig. 5.4. Crossover interaction between Variety and Cultivation method in Example 5.3

other three. Cultivation method 3 is intermediate for every variety and so might be the safest one to use if a farmer is trying a new variety not among the five tested here.

Some people draw lines between successive points with the same symbol in diagrams like those in Figures 5.3 and 5.4. The lines certainly aid the human eye to see if the successions of points are parallel. However, they can also be misleading. In Figure 5.3 such a line would suggest a value for the percentage of water-soluble carbohydrate in the crop if nitrogen fertilizer were applied at 100 kg/ha, and the value might not be too far from the truth. In Figure 5.4 such a line would suggest a variety intermediate between varieties A and B, which may be nonsense.

If one treatment factor is quantitative and the other is qualitative, then people who draw lines would plot the levels of the quantitative factor along the x-axis, as in Figure 5.3, so that the intermediate values suggested by the lines might have some meaning. However, it is often easier to read the diagram if the factor with the higher number of levels is plotted along the x-axis, and the levels of the other factor are shown by symbols, whether or not either of them is quantitative.

Example 5.4 (Modern cereals) Modern cereals have been bred to produce high yields, but they often need large amounts of fertilizer and other additives in order to do so. For such a cereal, the yield rises rather steeply in response to extra fertilizer. By comparison, a traditional old variety will not be capable of producing the highest yields, but it may have a rather mild response to fertilizer, so that it does better than the new variety if not much fertilizer is applied.

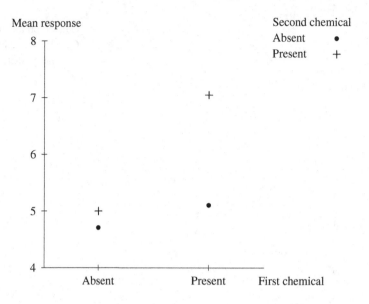

Fig. 5.5. Threshold interaction in Example 5.5

This is another example of crossover interaction. A farmer with low resources, or with poor soil, might do better to grow the traditional variety.

In psychology, a crossover interaction is called *disordinal*.

Example 5.5 (Enzyme in blood) Two chemical preparations were injected into mice to see if they affected the quantity of some enzyme in their blood. Thus there were two treatment factors, F and G, whose levels were absence and presence of each of the chemicals respectively. The mean responses are shown in Figure 5.5. It is clear that both chemicals are needed in order to increase the quantity of enzyme. Thus their effects are not additive, and there is interaction between them.

Example 5.5 shows *threshold* interaction: each treatment factor needs to be present at a certain level before the other can take effect. In more complicated cases the treatment factors may have more than two levels, and it may be that only one of them acts as a threshold for the other.

Example 5.6 (Saplings) At a forestry research station, poplar saplings were planted in an experiment to compare treatments for getting the young trees established. Four types of collar were put around the saplings to prevent them from predators and other damage: these were combined with five different heights of the saplings at planting time. After some months, the height was measured again. It turned out that one type of collar excluded so much light that all the plants died, so the response on those saplings was zero irrespective of the height at planting time. This is an extreme form of threshold interaction. Unless the qualitative factor Collar-type is at a level that permits the plants to grow, no other treatment factor can have any effect.

5.2. Interaction

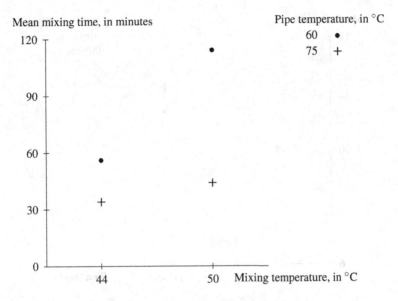

Fig. 5.6. Useful interaction in Example 5.7

An interaction with the above behaviour can be good if it implies that setting one factor to the correct level means that we can be slack about controlling the others.

Example 5.7 (Tablet manufacture) A pharmaceutical company can vary several factors during the manufacture of medicinal tablets. In one particular process, the staff can vary the temperature at which the ingredients are mixed; and, independently, they can vary the temperature in the pipe carrying the liquid ingredient into the mixture. They measure how long it takes for the mixing process to be complete—the shorter the better.

Figure 5.6 demonstrates that not only does the higher pipe temperature reduce the mixing time for each mixing temperature. It also reduces the difference between the mixing times for the two mixing temperatures. This suggests that, during production, it is worth being careful to maintain the higher pipe temperature: then it will not matter too much if the mixing temperature varies.

The opposite of a threshold interaction is a *trigger* interaction. Here at least one treatment factor needs to have the correct level to change the response but there is no further gain from having both at the correct level.

Example 5.8 (Catalysts in a chemical reaction) A chemical process in the production of an industrial chemical has several stages. At each stage a catalyst may be present to improve the reaction. Figure 5.7 shows the effect of having a catalyst present or absent at each of two stages. The response is a measure of the quality of the chemical produced. Relative to the residual mean square, the three responses with at least one catalyst present were judged to be the same.

Two further types of interaction are worth mentioning. If there is no crossover then the distances between the curves may increase as the general response increases. This can be

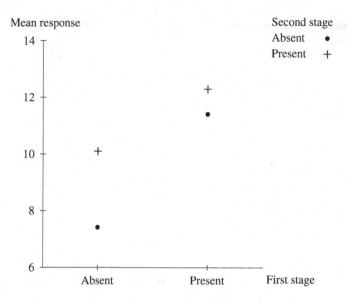

Fig. 5.7. Trigger interaction in Example 5.8

an indication that we are measuring on the wrong scale. The basic assumption (1.1) is that treatment i adds a constant τ_i to the response on each plot where it is applied. Similarly, we assume in Equation (4.1) that each block adds a fixed constant.

Example 5.9 (Counts of bacteria) If the purpose of the treatments is to reduce the number of bacteria in milk then it is much more likely that the effect of each treatment is to *multiply* the initial number of bacteria by a positive constant (which we hope is less than 1). Thus we would expect Equation (1.1) to apply to the logarithms of the counts rather than to the counts themselves.

Let Y_ω be the number of bacteria per millilitre in sample ω. Suppose that treatment factors F and G each act multiplicatively on the counts, so that $Y_\omega = Z_\omega \times \lambda_{F(\omega)}$ if only factor F is applied and $Y_\omega = Z_\omega \times \mu_{G(\omega)}$ if only factor G is applied. If these factors continue to act multiplicatively in the presence of each other, then $Y_\omega = Z_\omega \times \lambda_{F(\omega)} \times \mu_{G(\omega)}$, with the result that $\log(Y_\omega) = \log(Z_\omega) + \log(\lambda_{F(\omega)}) + \log(\mu_{G(\omega)})$: in other words, the model is additive in F and G on the log scale but not on the original count scale.

Similarly, we may measure volume but expect the linear model (1.1) to apply to the linear measurement, so that we need to take cube roots before analysing the data.

Suppose that there is a non-linear monotonic function f such that $f(Y)$ satisfies the linear model (1.1). If there is no interaction between treatment factors F and G when we consider the transformed data $f(y)$, then there *will* be interaction between them when we consider the untransformed data y. This is rather hard to spot graphically. If f is nearly linear over the range of the data, zero interaction on the transformed scale looks like zero interaction on the original scale. If f is not nearly linear over this range then calculating means of the data on the wrong scale gives a seriously misleading impression.

5.2. Interaction

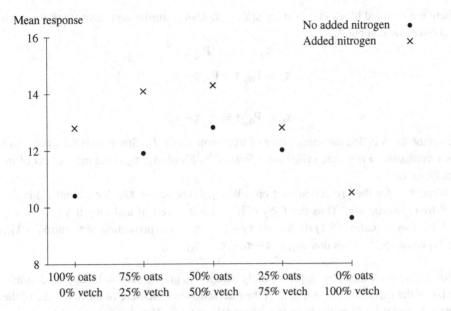

Fig. 5.8. The improvement due to added nitrogen declines as the proportion of vetch increases: see Example 5.10

Finally, the distances between the curves may increase as the levels of one of the treatment factors increases. There may be a simple explanation for this.

Example 5.10 (Vetch and oats) An experiment on forage crops compared five seed mixtures in the presence and absence of nitrogen fertilizer. Figure 5.8 shows the mean responses in tons per acre. The seed mixtures range from all oats, no vetch to no oats, all vetch. The yield is highest for the 50:50 mixture. Added nitrogen improves the yield of all of the mixtures, but the improvement declines as the proportion of vetch increases. This makes perfect sense. Cereal crops, such as oats, need to take nitrogen from the soil if they can, while legumes, such as vetch, actually fix nitrogen in the soil.

Examples 5.5–5.10 all exhibit interaction even though there is a simple explanation of the non-additivity in some cases.

Of course, there is no need to remember the names for all these types of interaction. For one thing, they are not really well defined. What is a crossover interaction from the point of view of one factor may or may not be a crossover interaction from the point of view of the other. Threshold and trigger interactions are, in fact, the same thing from different perspectives: in both cases there is at least one level of one factor at which the response does not vary with the levels of the other factor. What is important is to be aware that interaction can occur and that it sometimes, but not always, has a relatively simple explanation.

If there are no fixed block effects then $\mathbb{E}(\mathbf{Y}) = \boldsymbol{\tau}$. Using similar arguments to those given in Section 4.5, we can put

$$\boldsymbol{\tau}_0 = \bar{\tau}\mathbf{u}_0 = \mathbf{P}_{V_0}\boldsymbol{\tau},$$

$$\boldsymbol{\tau}_F = \mathbf{P}_{W_F}\boldsymbol{\tau} = \mathbf{P}_{V_F}\boldsymbol{\tau} - \boldsymbol{\tau}_0,$$

and

$$\boldsymbol{\tau}_G = \mathbf{P}_{W_G}\boldsymbol{\tau} = \mathbf{P}_{V_G}\boldsymbol{\tau} - \boldsymbol{\tau}_0.$$

The vector $\boldsymbol{\tau}_F$ is called the *main effect* of treatment factor F. Sometimes the entries in $\boldsymbol{\tau}_F$, which must sum to zero, are called the *effects* of F. Similarly, $\boldsymbol{\tau}_G$ is the main effect of treatment factor G.

Write $\boldsymbol{\tau}_{FG}$ for the projection of $\boldsymbol{\tau}$ onto $W_{F \wedge G}$. This vector $\boldsymbol{\tau}_{FG}$ (or its entries) is called the *F-by-G interaction*. Thus the F-by-G interaction is zero if and only if $\boldsymbol{\tau} \in V_F + V_G$. If $W_F \perp W_G$ then Equation (5.1) shows that $\boldsymbol{\tau}_0 + \boldsymbol{\tau}_F + \boldsymbol{\tau}_G$ is the projection of $\boldsymbol{\tau}$ onto $V_F + V_G$, so then Equation (5.2) shows that $\boldsymbol{\tau}_{FG} = \boldsymbol{\tau} - \boldsymbol{\tau}_0 - \boldsymbol{\tau}_F - \boldsymbol{\tau}_G$.

Of course, what we mean by 'approximately parallel' in graphs such as Figure 5.3 depends on the size of the variance. The estimate of the interaction is a measure of the departure of the fit in the full model V_T from the fit in the submodel $V_F + V_G$. The significance of this departure is assessed by comparing its size (divided by its degrees of freedom) with the residual mean square, as we show in Section 5.5.

5.3 Principles of expectation models

It is time to come clean over an issue that I have been fudging until now. In Chapter 1 I suggested that there was a clear dichotomy between estimation and testing. That may be true when the treatments are unstructured. However, we have now met several cases where we are interested in many different models for the expectation: see Section 2.11, Chapter 3 and Table 5.1. In these circumstances we usually do hypothesis tests to select the smallest model supported by the data, and then estimate the parameters of that model.

A collection of subspaces of V which are to serve as expectation models cannot be arbitrary, but should obey the following principles.

Principle 5.1 (Intersection Principle) If V_1 and V_2 are both expectation models then $V_1 \cap V_2$ should also be an expectation model.

The Intersection Principle is there to avoid ambiguity in model-fitting. If $\mathbf{y}$ is in (or is close to) $V_1 \cap V_2$, then our fitted model should be V_1 (or a subspace of it) and it should also be V_2 (or a subspace of it): if the Intersection Principle is not satisfied then we cannot fit the model $V_1 \cap V_2$ but there is no way of deciding between the models V_1 and V_2.

Principle 5.2 (Sum Principle) If V_1 and V_2 are both expectation models then $V_1 + V_2$ should also be an expectation model.

There are three reasons for the Sum Principle. First, it is a feature of all linear models that if $\mathbf{v}$ and $\mathbf{w}$ are allowable as vectors of fitted values then so should $\mathbf{v} + \mathbf{w}$ be: apply this with $\mathbf{v}$ in V_1 and $\mathbf{w}$ in V_2.

5.3. Principles of expectation models

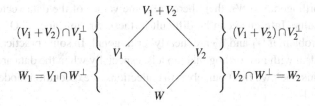

Fig. 5.9. Fitting submodels of $V_1 + V_2$

Secondly, it avoids ambiguity, just like the Intersection Principle. Suppose that $\mathbf{y} = \mathbf{v} + \mathbf{w}$, where $\mathbf{v}$ and $\mathbf{w}$ are nonzero vectors in V_1 and V_2 respectively. Unless $V_1 + V_2$ is an expectation model, we shall be forced to make an arbitrary choice between V_1 and V_2.

Thirdly, it implies that all the expectation models are contained in a single maximal model. For most of this book the maximal model is V_T. In Chapters 9, 12 and 13 we shall take a different maximal model when we can assume that some interactions are zero. The maximal model is the starting point for testing hypotheses about models. It also defines the residual: nothing in the maximal model is *ever* put into residual, even when we fit a smaller model.

The Intersection and Sum Principles are both automatically satisfied if the collection of expectation models forms a chain, as in Figures 2.4, 3.1 and 3.2. Moreover, neither is affected by the numbers of replications of the treatments.

The third principle can be affected by the numbers of replications. It is also more controversial.

Principle 5.3 (Orthogonality Principle) If V_1 and V_2 are both expectation models and if $W = V_1 \cap V_2$ then $V_1 \cap W^\perp$ should be orthogonal to $V_2 \cap W^\perp$.

Put $W_1 = V_1 \cap W^\perp$ and $W_2 = V_2 \cap W^\perp$. Then $V_1 = W + W_1$ and $V_2 = W + W_2$, and so $V_1 + V_2 = W + W_1 + W_2 = V_1 + W_2$. If W_2 is orthogonal to W_1 then W_2 is orthogonal to V_1 and so $(V_1 + V_2) \cap V_1^\perp = W_2$; similarly $(V_1 + V_2) \cap V_2^\perp = W_1$. It can be shown that the converse is also true: if either of these equations holds then W_1 is orthogonal to W_2.

Suppose that we have accepted the hypothesis that $\mathbb{E}(\mathbf{Y}) \in V_1 + V_2$ and want to see if we can reduce the expectation model to V_1. To do this, we examine the size of the vector $\mathbf{P}_{V_1+V_2}\mathbf{y} - \mathbf{P}_{V_1}\mathbf{y}$, by comparing the ratio $\left\|\mathbf{P}_{V_1+V_2}\mathbf{y} - \mathbf{P}_{V_1}\mathbf{y}\right\|^2/d_1$ with MS(residual), where $d_1 = \dim(V_1 + V_2) - \dim V_1 = \dim V_2 - \dim(V_1 \cap V_2) = \dim V_2 - \dim W$. Now, $\mathbf{P}_{V_1+V_2}\mathbf{y} - \mathbf{P}_{V_1}\mathbf{y}$ is just the projection of $\mathbf{y}$ onto $(V_1 + V_2) \cap V_1^\perp$, which is $\mathbf{P}_{W_2}\mathbf{y}$ if W_1 is orthogonal to W_2.

If we accept the hypothesis that $\mathbb{E}(\mathbf{Y}) \in V_1$ then we go on to test whether we can reduce the expectation model to W. We do this by examining the size of $\mathbf{P}_{V_1}\mathbf{y} - \mathbf{P}_W\mathbf{y}$, which is $\mathbf{P}_{W_1}\mathbf{y}$.

The models we are discussing are shown in Figure 5.9. Starting at $V_1 + V_2$, there are two routes down to W, and there is no good reason to choose one rather than the other. The advantage of orthogonality is that both routes give the same result, because in both cases we are examining the sizes of $\mathbf{P}_{W_1}\mathbf{y}$ and $\mathbf{P}_{W_2}\mathbf{y}$. In other words, the test for reducing model $V_1 + V_2$ to V_1 is exactly the same as the test for reducing V_2 to W, and similarly with V_1 and V_2 interchanged.

If W_1 is not orthogonal to W_2, then there are some values of the data vector **y** which give contradictory results. Inference can be difficult in these circumstances. Of course, there is unlikely to be a problem if W_1 and W_2 are 'nearly' orthogonal. In some practical circumstances we are forced to deal with non-orthogonal models, especially when the data are observational. Nonetheless, in this book we limit ourselves to collections of expectation models that conform to all three principles.

The proof of Theorem 4.1 shows that, in general, whether or not the subspaces defined by two treatment factors satisfy the Orthogonality Principle depends on the numbers of replications of the combinations of levels. Thus the Orthogonality Principle has implications for the design of experiments.

Example 5.11 (Example 3.4 continued: Drugs at different stages of development)
Figure 3.3 shows that the only pair of subspaces that we need worry about is V_A and V_B. Now, **v** is in V_A if and only if v_ω is the same for all ω which receive a dose of the new formulation, while **v** is in V_B if and only if v_ω is the same for all ω which receive a dose of the old formulation. Hence $V_A \cap V_B = V_F$. Moreover, $V_A + V_B \subset V_T$ and $\dim(V_A + V_B) = \dim(V_A) + \dim(V_B) - \dim(V_A \cap V_B) = 4 + 4 - \dim(V_F) = 8 - 2 = 6 = \dim(V_T)$: therefore $V_A + V_B = V_T$. Finally, if $\mathbf{v} \in V_A \cap V_F^\perp$ then $v_\omega = 0$ whenever $T(\omega)$ is dose of the new formulation, while if $\mathbf{w} \in V_B \cap V_F^\perp$ then $w_\omega = 0$ whenever $T(\omega)$ is dose of the old formulation, and so the spaces $V_A \cap V_F^\perp$ and $V_B \cap V_F^\perp$ are orthogonal to each other. Therefore the collection of expectation models in Figure 3.3 satisfies the three principles.

Similar arguments about intersection and orthogonality apply to all the models given in Chapter 3. To satisfy the Sum Principle, we have to explicitly ensure that sums of models are included. For example, the collection of expectation models for Example 3.5 is shown in Figure 5.10. Here V_C is the model with only two treatment parameters, one for the control treatment and one for the rest, and V_M is the model with four treatment parameters, one for each method and one for the control treatment. The space V_Q has six treatment parameters, one for the control treatment, one for each treatment of the 'quantity' type, and one for each of the other two methods. The spaces V_R and V_S are defined similarly by the 'roughage' and 'time' types. Thus the model $V_Q + V_R$ forces all treatments of the 'time type' to have the same parameter, but otherwise allows for different treatment parameters.

Given a collection of expectation models that satisfies the three principles, we test submodels by starting at the maximal model and working downwards. At each stage we test the next submodel by examining the difference between the sums of squares for the fit in the current model and for the fit in the submodel, divided by the difference between their dimensions. This mean square is always compared to the original residual mean square. If we accept the submodel, we move down the diagram to it and continue from there. If at any stage we have rejected all submodels immediately below the current model, we decide that the current one is the smallest that is supported by the data. Because of orthogonality, it does not matter in what order we test submodels when there is a choice.

Once we have decided on the smallest model, we estimate its parameters, which are usually shown in a table of means, along with their standard errors of differences.

Fortunately, the calculations needed for estimation and testing are virtually the same. For estimating the parameters of the expectation model V_m we need $\mathbf{P}_{V_m}\mathbf{y}$. If the submodel V_n is

5.4. Decomposing the treatment subspace

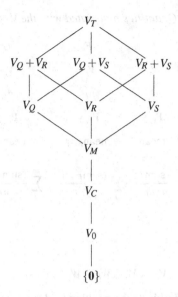

Fig. 5.10. Collection of expectation models in Example 3.5

immediately below V_m then testing for V_n needs the sum of squares $\left\|\mathbf{P}_{V_m}\mathbf{y} - \mathbf{P}_{V_n}\mathbf{y}\right\|^2$, which is equal to $\left\|\mathbf{P}_{V_m}\mathbf{y}\right\|^2 - \left\|\mathbf{P}_{V_n}\mathbf{y}\right\|^2$. These sums of squares are displayed in an analysis-of-variance table. Convention dictates the opposite order for the analysis-of-variance table to that used in the diagram of submodels, so model testing proceeds by starting at the bottom of the analysis-of-variance table and working upwards. In the next two sections this is described in detail for the case where treatments consist of all combinations of two treatment factors.

5.4 Decomposing the treatment subspace

We return to the case where treatments consist of all combinations of factors F and G. We need to show that the collection of models in Table 5.1 satisfies the Intersection Principle, the Sum Principle and the Orthogonality Principle. Figure 5.2 shows that the only pair of subspaces that we need to check is V_F and V_G. The collection of models contains the additive model $V_F + V_G$, so the Sum Principle is satisfied.

Suppose that $\mathbf{v} \in V_F \cap V_G$ and that α is a plot for which $F(\alpha) = G(\alpha) = 1$. Let β be any other plot and suppose that $G(\beta) = j$. All combinations of the levels of F and G occur, so there is a plot γ for which $F(\gamma) = 1$ and $G(\gamma) = j$. Now, $\mathbf{v} \in V_F$ so $v_\alpha = v_\gamma$. Also $\mathbf{v} \in V_G$, so $v_\gamma = v_\beta$. Hence $v_\alpha = v_\beta$ for all β and so $\mathbf{v} \in V_0$. This shows that $V_F \cap V_G = V_0$, and so the Intersection Principle is satisfied.

Theorem 5.1 shows that the Orthogonality Principle is satisfied if all combinations of levels of F and G have the same replication. That is why we insist, in most of this book, that all combinations of treatment factors must occur equally often. (The apparent exceptions in Chapter 3 will be explained in Chapter 10.) Write r for this common replication of all combinations.

Table 5.3. *Quantities associated with the V-subspaces*

symbol	0	F	G	$F \wedge G$
V-subspace	V_0	V_F	V_G	$V_{F \wedge G}$
dimension	1	n_F	n_G	$n_F n_G$
fit = vector of fitted values	$\mathbf{P}_{V_0}\mathbf{y}$	$\mathbf{P}_{V_F}\mathbf{y}$	$\mathbf{P}_{V_G}\mathbf{y}$	$\mathbf{P}_{V_{F \wedge G}}\mathbf{y}$
coordinate in fit	mean	$\text{mean}_{F=i}$	$\text{mean}_{G=j}$	$\text{mean}_{F=i,G=j}$
$\|\text{fit}\|^2 = \text{CSS}$	$\dfrac{\text{sum}^2}{n_F n_G r}$	$\sum_i \dfrac{(\text{sum}_{F=i})^2}{n_G r}$	$\sum_j \dfrac{(\text{sum}_{G=j})^2}{n_F r}$	$\sum_{i,j} \dfrac{(\text{sum}_{F=i,G=j})^2}{r}$

We now have
$$V_T = V_0 \oplus W_F \oplus W_G \oplus W_{F \wedge G}$$
orthogonally. As in Section 4.5, this decomposition leads to decompositions of the dimension of V_T and of the vector of fitted values for the full treatment model, as follows.

$$
\begin{aligned}
V_T &= V_0 \oplus W_F \oplus W_G \oplus W_{F \wedge G} \\
\text{dimension } n_F n_G &= 1 + (n_F - 1) + (n_G - 1) + (n_F - 1)(n_G - 1) \\
\text{vector } \mathbf{P}_{V_T}\mathbf{y} &= \mathbf{P}_{V_0}\mathbf{y} + \mathbf{P}_{W_F}\mathbf{y} + \mathbf{P}_{W_G}\mathbf{y} + \mathbf{P}_{W_{F \wedge G}}\mathbf{y}
\end{aligned}
$$

Here $\mathbf{P}_{V_T}\mathbf{y}$ is the fit for the full treatment model, $\mathbf{P}_{V_0}\mathbf{y}$ is the fit for the null model, $\mathbf{P}_{W_F}\mathbf{y}$ is the estimate of the main effect $\boldsymbol{\tau}_F$ of F, $\mathbf{P}_{W_G}\mathbf{y}$ is the estimate of the main effect $\boldsymbol{\tau}_G$ of G, and $\mathbf{P}_{W_{F \wedge G}}\mathbf{y}$ is the estimate of the F-by-G interaction $\boldsymbol{\tau}_{FG}$.

We need to calculate various quantities associated with the W-subspaces. The easiest way to do this is to calculate them by subtraction from the quantities associated with the V-subspaces, which are shown in Table 5.3. Here we write V_T as $V_{F \wedge G}$, because $T = F \wedge G$. Since each combination occurs on r plots, each level of F occurs on $n_G r$ plots and there are $n_F n_G r$ plots altogether.

From these we calculate the quantities associated with the W-subspaces by successively subtracting W-quantities from V-quantities, as shown in Table 5.4.

Example 5.12 (Protein in feed for chickens) Eight newly-hatched chicks took part in a feeding experiment. Four different feeds (A, B, C and D) were made available to two chicks each. The protein in feeds A and B was groundnuts, while the protein in feeds C and D was soya bean. Moreover, feeds B and D contained added fishmeal. Thus the treatments consisted of all combinations of levels of two treatments factors: P (protein) with levels g and s; and M (fishmeal) with levels $+$ and $-$.

The fourth column of Table 5.5 shows the weights of the chicks (in grams) at the end of six weeks.

Table 5.4. *Quantities associated with the W-subspaces: here 'effect' is short for 'estimated effect'*

symbol	0	F	G	F∧G
W-subspace	$W_0 = V_0$	W_F	W_G	$W_{F\wedge G}$
dimension	1	$n_F - 1$	$n_G - 1$	$(n_F - 1)(n_G - 1)$
effect = extra fit	$\mathbf{P}_{V_0}\mathbf{y} = \mathbf{P}_{W_0}\mathbf{y}$	$\mathbf{P}_{V_F}\mathbf{y} - \mathbf{P}_{V_0}\mathbf{y} = \mathbf{P}_{W_F}\mathbf{y}$	$\mathbf{P}_{V_G}\mathbf{y} - \mathbf{P}_{V_0}\mathbf{y} = \mathbf{P}_{W_G}\mathbf{y}$	$\mathbf{P}_{V_{F\wedge G}}\mathbf{y} - \mathbf{P}_{W_0}\mathbf{y} - \mathbf{P}_{W_F}\mathbf{y} - \mathbf{P}_{W_G}\mathbf{y} = \mathbf{P}_{W_{F\wedge G}}\mathbf{y}$
coordinate in effect	mean	$\text{mean}_{F=i} - \text{mean}$	$\text{mean}_{G=j} - \text{mean}$	$\text{mean}_{F=i,G=j} - \text{mean}_{F=i} - \text{mean}_{G=j} + \text{mean}$
$\|\text{effect}\|^2$ = sum of squares	CSS(mean)	CSS(F) − SS(mean)	CSS(G) − SS(mean)	CSS(F∧G) − SS(mean) − SS(F) − SS(G)

Table 5.5. *Calculating fits, effects and sums of squares in Example 5.12*

Feed (T)	Protein (P)	Fishmeal (M)	Weight (gm)	Fit for null	Fit for P	Fit for M	Fit for T	Main effect of P	Main effect of M	Fit for P + M	P-by-M interaction	Residual
A	g	−	410	441.50	417.25	421.75	401.50	−24.25	−19.75	397.50	4.00	8.50
A	g	−	393	441.50	417.25	421.75	401.50	−24.25	−19.75	397.50	4.00	−8.50
B	g	+	442	441.50	417.25	461.25	433.00	−24.25	19.75	437.00	−4.00	9.00
B	g	+	424	441.50	417.25	461.25	433.00	−24.25	19.75	437.00	−4.00	−9.00
C	s	−	443	441.50	465.75	421.75	442.00	24.25	−19.75	446.00	−4.00	1.00
C	s	−	441	441.50	465.75	421.75	442.00	24.25	−19.75	446.00	−4.00	−1.00
D	s	+	500	441.50	465.75	461.25	489.50	24.25	19.75	485.50	4.00	10.50
D	s	+	479	441.50	465.75	461.25	489.50	24.25	19.75	485.50	4.00	−10.50
squared length			1567860.0	1559378.0	1564082.5	1562498.5	1567331.0	4704.5	3120.5	1567203.0	128.0	529.0

The next four columns of Table 5.5 are the fits for four of the expectation models: V_0, V_P, V_M and V_T. All the coordinates are obtained as simple averages of entries from the 'weight' vector. The squared length of each of these vectors can be calculated either as the sum of the squares of all its entries or by using the appropriate formula for a crude sum of squares. For example, $\text{sum}_{P=g} = 1669$ and $\text{sum}_{P=s} = 1863$ so $\text{mean}_{P=g} = 417.25$ and $\text{mean}_{P=s} = 465.75$. Then $\text{CSS}(P) = 4 \times 417.25^2 + 4 \times 465.75^2 = 1669^2/4 + 1863^2/4 = 1564082.5$.

The column for the estimate of the main effect of P is obtained by subtracting the column for the fit for the null model from the column for the fit for P. The squared length of this vector can be calculated either as the sum of the squares of all its entries (that is, 8×24.25^2) or by using the appropriate formula for a sum of squares, that is, by subtracting the squared length of the fit for the null model from the squared length of the fit for P (that is, $1564082.5 - 1559378.0$). The column for the estimate of the main effect of M is similar.

The column for the fit for $P + M$ is obtained by adding the columns for the fit for the null model, the main effect of P and the main effect of M. The squared length is equal to the sum of the squares of all its entries (that is, $2 \times 397.50^2 + 2 \times 437.00^2 + 2 \times 446.00^2 + 2 \times 485.50^2$); it is also the sum of the squared lengths of three vectors which have been added to obtain it (that is, $1559378.0 + 4704.5 + 3120.5$).

The column for the estimate of the P-by-M interaction is obtained by subtracting the column for the fit for $P + M$ from the column for the fit for T. The squared length may be obtained by (a) calculating sum of the squares of all its entries (8×4.00^2), or (b) taking the difference between the squared length of the fit for T and the squared length of the fit for $P + M$ ($1567331.0 - 1567203.0$), or (c) using the formula for the sum of squares for interaction shown in Table 5.4 ($1567331.0 - 1559378.0 - 4704.5 - 3120.5$).

Finally, the column for the residual is the difference between the original 'weight' vector and the fit for T. Its squared length is equal to the sum of the squares of all its entries; it is also the difference between the total sum of squares (squared length of the 'weight' vector) and the crude sum of squares for treatments (squared length of the fit for T), as described in Section 2.8.

5.5 Analysis

If the treatments are all combinations of two treatment factors F and G then we replace the treatments line in the analysis-of-variance table by lines for the two main effects and one for the interaction. Theorem 2.5(ii) shows that, if $\text{Cov}(\mathbf{Y}) = \sigma^2 \mathbf{I}$, then the expected mean squares for these lines are those shown in Table 5.6.

First we use the ratio

$$\frac{\text{MS}(F \wedge G)}{\text{MS}(\text{residual})}$$

to test for interaction. If we cannot assume that the interaction is zero, then report that we cannot use the simpler, additive model, give a table of the treatment means and standard errors of their differences, and stop. If the interaction can be explained in a simple way, then do so. Consider whether it is more helpful to report the effect of F at each level of G, or vice versa.

If we can assume that the interaction is zero, report this clearly. Then test separately for each main effect. If either main effect is nonzero, give its table of means and the standard

5.5. Analysis

Table 5.6. *Treatment lines in the analysis-of-variance table for a factorial experiment with two treatment factors:* $d_{FG} = (n_F - 1)(n_G - 1)$

Source	Sum of squares	Degrees of freedom	Mean square	EMS	Variance ratio
F	$SS(F)$	$n_F - 1$	$\dfrac{SS(F)}{n_F - 1}$	$\dfrac{\|\tau_F\|^2}{n_F - 1} + \sigma^2$	$\dfrac{MS(F)}{MS(\text{residual})}$
G	$SS(G)$	$n_G - 1$	$\dfrac{SS(G)}{n_G - 1}$	$\dfrac{\|\tau_G\|^2}{n_G - 1} + \sigma^2$	$\dfrac{MS(G)}{MS(\text{residual})}$
F-by-G	$SS(F \wedge G)$	d_{FG}	$\dfrac{SS(F \wedge G)}{d_{FG}}$	$\dfrac{\|\tau_{FG}\|^2}{d_{FG}} + \sigma^2$	$\dfrac{MS(F \wedge G)}{MS(\text{residual})}$

Table 5.7. *Analysis-of-variance table for Example 5.12*

Source	SS	df	MS	VR
mean	1559378.0	1	1559378.00	11791.14
protein	4704.5	1	4704.50	35.57
fishmeal	3120.5	1	3120.50	23.60
protein $\wedge$ fishmeal	128.0	1	128.00	0.97
residual	529.0	4	132.25	–
Total	1567860.0	8		

errors of their differences.

Example 5.12 revisited (Protein in feed for chickens) The calculations in Table 5.5 give the analysis of variance in Table 5.7.

The variance ratio for the protein-by-fishmeal interaction is approximately 1, so it is clear that there is no interaction. (If there had been an interaction, it would probably have been better to report the effects of adding fishmeal to each type of protein.) The variance ratios for protein and for fishmeal are both greater than 21.20, which is the 99% point of the F-distribution on 1 and 4 degrees of freedom, so we can report that added fishmeal increases weight by 39.5 gm irrespective of type of protein and that replacing groundnuts by soya beans increases weight by 48.5 gm whether or not fishmeal is added. The standard errors of both of these differences is $\sqrt{132.25 \times (2/4)}$, which is 8.13.

Here is a warning about vocabulary. Many scientists say that two factors 'interact' to mean that they 'act together' in the sense that you can add their separate main effects. This is precisely what statisticians call 'zero interaction'. Thus it is always a good idea to report the presence or absence of interaction by pointing out what this means in the particular case.

The phrase 'main effect' can also be misinterpreted. In Example 5.12, a chicken breeder who thinks that groundnuts are the obvious source of protein may say that the 'main effect'

Table 5.8. *Treatment totals in Example 5.10*

Nitrogen	Percentage of vetch					Nitrogen total
	0	25	50	75	100	
Yes	12.77	14.14	14.29	12.80	10.49	64.49
No	10.04	11.86	12.77	12.03	9.59	56.29
Seed mixture total	22.81	26.00	27.06	24.83	20.08	120.78
(with N) − (without N)	2.73	2.28	1.52	0.77	0.90	8.20
linear fit	2.67	2.16	1.64	1.12	0.61	8.20
fit minus mean	1.03	0.52	0.00	−0.52	−1.03	0.00

of fishmeal is

$$\hat{\tau}_{g,+} - \hat{\tau}_{g,-} = 433.00 - 401.50 = 31.50.$$

Example 5.10 revisited (Vetch and oats) Table 5.8 shows the treatment totals. In fact, the experiment was conducted in five complete blocks, so $\text{CSS}(\text{mean}) = 120.78^2/50 = 291.7562$ and $\text{CSS}(\text{seed}) = (22.81^2 + 26.00^2 + 27.06^2 + 24.83^2 + 20.08^2)/10 = 294.8275$; furthermore, $\text{CSS}(\text{nitrogen}) = (64.49^2 + 56.29^2)/25 = 293.1010$, and the crude sum of squares for treatments is $(12.77^2 + \cdots + 9.59^2)/5 = 296.4640$. From these we calculate $\text{SS}(\text{seed}) = 294.8275 - 291.7562 = 3.0713$, $\text{SS}(\text{nitrogen}) = 293.1010 - 291.7562 = 1.3448$ and the sum of squares for interaction is $296.4640 - 291.7562 - 3.0713 - 1.3448 = 0.2917$. Hence the mean square for interaction is 0.0729. The complete set of data (not shown here) gives the residual mean square to be 0.0402. At first sight, it seems that there is no evidence of interaction.

However, Figure 5.8 shows that the effect of nitrogen decreases as the percentage of vetch increases, which is exactly what we would expect. Table 5.8 has some extra rows. The first shows the difference between the two levels of nitrogen for each seed mixture. The second gives the linear fit to this difference, as a polynomial in the percentage of vetch. The third subtracts the mean of the previous row. Since each treatment total comes from five plots, this final row corresponds to values of $+0.103$ on the five plots with nitrogen but no vetch, -0.103 on the five plots with neither nitrogen nor vetch, and so on. These values sum to zero on each level of nitrogen and on each level of the seed mixture, so the vector they represent is part of the interaction, called the 'nitrogen-by-linear-vetch' part of the interaction. Its sum of squares is $20 \times (0.103^2 + 0.052^2) = 0.2663$, which accounts for most of the sum of squares for the interaction. It has one degree of freedom (for fitting the straight line), so we obtain the analysis of variance in Table 5.9 (using information about block totals from the complete data set). This shows clearly that the effect of nitrogen changes linearly with the percentage of vetch in the seed mixture, and that there is no evidence of any further interaction.

When one of the treatment factors is quantitative, it is often instructive to decompose the interaction in this way.

5.6 Three treatment factors

Now suppose that we have three treatment factors F, G and H whose sets of levels are L_F, L_G and L_H respectively. Then $n_F = |L_F|$, $n_G = |L_G|$ and $n_H = |L_H|$. We assume that the

5.6. Three treatment factors

Table 5.9. *Analysis-of-variance table for Example 5.10*

Stratum	Source	SS	df	MS	VR
mean	mean	291.7562	1	291.7562	–
blocks	blocks	3.0409	4	0.7602	–
plots	seed mixtures	3.0713	4	0.7678	76.35
	nitrogen	1.3448	1	1.3448	33.43
	nitrogen-by-linear-vetch	0.2663	1	0.72663	6.62
	rest of interaction	0.0254	3	0.0085	0.21
	residual	1.4482	36	0.0402	–
Total		300.9531	50		

treatment set consists of all combinations of levels of F, G and H, so that $\mathcal{T} = L_F \times L_G \times L_H$ and $t = n_F n_G n_H$.

Now the factor $F \wedge G$ is no longer the same as the treatment factor T. In fact, $F \wedge G$ is the function from Ω to $L_F \times L_G$ defined by

$$(F \wedge G)(\omega) = (F(\omega), G(\omega)).$$

In other words, $F \wedge G$ tells us what combination of levels of F and G is on plot ω, ignoring all other factors.

Definition A *class* of $F \wedge G$ consists of all plots having the same level of F and the same level of G.

Thus $F \wedge G$ has $n_F n_G$ levels and $n_F n_G$ classes. The vector subspace $V_{F \wedge G}$ consists of all vectors in V which are constant on each class of $F \wedge G$, so $V_{F \wedge G}$ has dimension $n_F n_G$.

Factors $F \wedge H$ and $G \wedge H$ are defined similarly. They have $n_F n_H$ and $n_G n_H$ levels respectively.

As in Section 5.1, we put

$$W_{F \wedge G} = V_{F \wedge G} \cap (V_F + V_G)^\perp,$$

so that

$$\dim W_{F \wedge G} = (n_F - 1)(n_G - 1).$$

As in Section 5.4, the fit for the expectation model $V_{F \wedge G}$, called the fit for $F \wedge G$, is the projection of the data vector $\mathbf{y}$ onto $V_{F \wedge G}$, whose coordinate on plot ω is equal to the mean of $\mathbf{y}$ on the $F \wedge G$-class containing ω. The crude sum of squares for $F \wedge G$, written $\mathrm{CSS}(F \wedge G)$, is the sum of the squares of the entries in the fit, which is equal to

$$\sum_{F \wedge G\text{-classes}} \frac{(\text{class total})^2}{\text{class size}}.$$

Also as in Section 5.4, the effect of the F-by-G interaction is defined to be the difference between the projection of $\boldsymbol{\tau}$ onto $V_{F \wedge G}$ and the projection of $\boldsymbol{\tau}$ onto $V_F + V_G$, which is estimated by

$$(\text{fit in } V_{F \wedge G}) - (\text{fit in } (V_F + V_G)).$$

Table 5.10. *Seven of the W-subspaces for three treatment factors*

Factor	Subspace	Dimension
mean	$W_0 = V_0$	1
F	W_F	$d_F = n_F - 1$
G	W_G	$d_G = n_G - 1$
H	W_H	$d_H = n_H - 1$
$F \wedge G$	$W_{F \wedge G}$	$d_{FG} = (n_F - 1)(n_G - 1)$
$F \wedge H$	$W_{F \wedge H}$	$d_{FH} = (n_F - 1)(n_H - 1)$
$G \wedge H$	$W_{G \wedge H}$	$d_{GH} = (n_G - 1)(n_H - 1)$

The sum of squares for the F-by-G interaction, written $\mathrm{SS}(F \wedge G)$, is the sum of the squares of the entries in the estimate of the effect, so that

$$\begin{aligned}
\mathrm{SS}(F \wedge G) &= \mathrm{CSS}(F \wedge G) - \mathrm{SS}(\text{mean}) - \mathrm{SS}(F) - \mathrm{SS}(G) \\
&= \mathrm{CSS}(F \wedge G) - \mathrm{CSS}(F) - \mathrm{CSS}(G) + \mathrm{CSS}(\text{mean}).
\end{aligned}$$

The subspaces $V_{F \wedge H}$, $V_{G \wedge H}$, $W_{F \wedge H}$ and $W_{G \wedge H}$, together with the associated fits, crude sums of squares, effects and sums of squares, are defined analogously.

With three treatment factors, we have $T = F \wedge G \wedge H$, and the treatment space V_T contains the seven subspaces shown in Table 5.10. Here d_F is defined to be $\dim W_F$ and d_{FG} to be $\dim W_{F \wedge G}$; and similarly for d_G, d_H, d_{FH} and d_{GH}. If all treatments have the same replication then every pair of these spaces is orthogonal to each other. Theorem 5.1 shows that W_F is orthogonal to W_G and to W_H; by construction, W_F is orthogonal to $W_{F \wedge G}$ and to $W_{F \wedge H}$; the proof that $W_{G \wedge H}$ is orthogonal to W_F and to $W_{F \wedge G}$ will be given in Chapter 10.

Now,

$$\begin{aligned}
V_{F \wedge G} + V_{F \wedge H} + V_{G \wedge H} &= (W_0 + W_F + W_G + W_{F \wedge G}) + (W_0 + W_F + W_H + W_{F \wedge H}) \\
&\quad + (W_0 + W_G + W_H + W_{G \wedge H}) \\
&= W_0 + W_F + W_G + W_H + W_{F \wedge G} + W_{F \wedge H} + W_{G \wedge H}.
\end{aligned}$$

These W-subspaces are mutually orthogonal, so

$$\begin{aligned}
\dim(V_{F \wedge G} + V_{F \wedge H} + V_{G \wedge H}) &= 1 + d_F + d_G + d_H + d_{FG} + d_{FH} + d_{GH} \\
&= 1 + d_F + d_G + d_H + d_F d_G + d_F d_H + d_G d_H \\
&= (d_F + 1)(d_G + 1)(d_H + 1) - d_F d_G d_H \\
&= n_F n_G n_H - (n_F - 1)(n_G - 1)(n_H - 1).
\end{aligned}$$

However, $V_{F \wedge G} + V_{F \wedge H} + V_{G \wedge H} \subset V_T$, and $V_T = V_{F \wedge G \wedge H}$, so we may define a new subspace $W_{F \wedge G \wedge H}$ by

$$W_{F \wedge G \wedge H} = V_{F \wedge G \wedge H} \cap (V_{F \wedge G} + V_{F \wedge H} + V_{G \wedge H})^{\perp}.$$

Put $d_{FGH} = \dim(W_{F \wedge G \wedge H})$. Then

$$\begin{aligned}
d_{FGH} &= \dim(V_{F \wedge G \wedge H}) - \dim(V_{F \wedge G} + V_{F \wedge H} + V_{G \wedge H}) \\
&= (n_F - 1)(n_G - 1)(n_H - 1).
\end{aligned}$$

5.6. Three treatment factors

The F-by-G-by-H interaction is defined to be the difference between the projection of τ onto $V_{F\wedge G\wedge H}$ and the projection of τ onto $V_{F\wedge G} + V_{F\wedge H} + V_{G\wedge H}$. This is an example of a *three-factor interaction*; by contrast, the previous interactions were *two-factor interactions*. It is estimated by

$$(\text{fit in } V_{F\wedge G\wedge H}) - (\text{fit in } (V_{F\wedge G} + V_{F\wedge H} + V_{G\wedge H})),$$

which is the projection of $\mathbf{y}$ onto $W_{F\wedge G\wedge H}$. The crude sum of squares for $F\wedge G\wedge H$, written $\mathrm{CSS}(F\wedge G\wedge H)$, is given by

$$\mathrm{CSS}(F\wedge G\wedge H) = \sum_{F\wedge G\wedge H\text{-classes}} \frac{(\text{class total})^2}{\text{class size}}.$$

The sum of squares for $F\wedge G\wedge H$, written $\mathrm{SS}(F\wedge G\wedge H)$, which is equal to the sum of the squares of the entries in the projection of $\mathbf{y}$ onto $W_{F\wedge G\wedge H}$, is given by

$$\begin{aligned}\mathrm{SS}(F\wedge G\wedge H) &= \mathrm{CSS}(F\wedge G\wedge H) - \mathrm{SS}(\text{mean}) - \mathrm{SS}(F) - \mathrm{SS}(G) - \mathrm{SS}(H)\\ &\quad - \mathrm{SS}(F\wedge G) - \mathrm{SS}(F\wedge H) - \mathrm{SS}(G\wedge H).\end{aligned}$$

Figure 5.11 shows the collection of expectation model subspaces for three treatment factors. Each edge in this diagram corresponds to a test for simplifying a model.

In the analysis, first test for the three-factor interaction. If it is nonzero, report this, give the table of means and standard errors of differences for the factor $F\wedge G\wedge H$, and stop. If the interactions can be explained in a simple way, then do so.

If the three-factor interaction is zero, then test all of the two-factor interactions. If none of them is zero then the fitted model is $V_{F\wedge G} + V_{F\wedge H} + V_{G\wedge H}$. This means that, although the effect of G and the effect of H both depend on the level of F, the G-by-H interaction is the same for all levels of F. It is rather hard to report this concisely! Give the table of means and standard errors of differences for each of the factors $F\wedge G$, $F\wedge H$ and $G\wedge H$. It is hard to work out the overall fitted values from these tables, in spite of the additivity, so it is a good idea to also give a three-way table of fitted values and their standard errors of differences.

If precisely one of the two-factor interactions is zero, suppose that this one is the G-by-H-interaction. Then the fitted model is $V_{F\wedge G} + V_{F\wedge H}$. This means that G and H both interact with F but the G-by-H interaction is zero at all levels of F. Give the tables of means and standard errors of differences for $F\wedge G$ and for $F\wedge H$, and also the three-way table of fitted values and their standard errors of differences.

If exactly two of the two-factor interactions are zero, suppose that these are the F-by-H and the G-by-H interactions. Report that H does not interact with either of the other factors. Now the model is reduced to $V_{F\wedge G} + V_H$, and we can test for the main effect of H. If the main effect of H is not zero then report this and give the tables of means and standard errors of differences both for $F\wedge G$ and for H. If the main effect of H is zero, then report this and give the table of means and standard error of differences for $F\wedge G$ alone.

If all three of the two-factor interactions are zero, then the model reduces to $V_F + V_G + V_H$. Now test separately for all the main effects. Give the table of means and standard error of differences for each factor whose main effect is nonzero.

In summary, keep testing to see whether the current model can be simplified. When it can be simplified no further, stop. Give the tables of means and standard errors of differences for

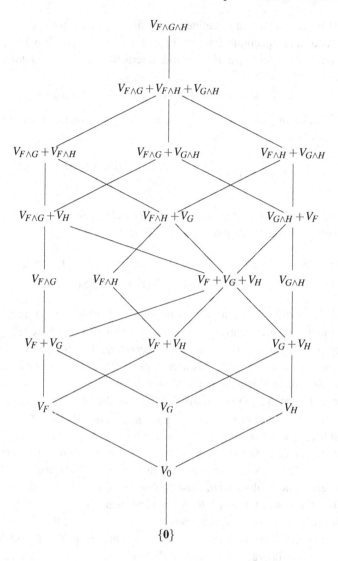

Fig. 5.11. Collection of model subspaces for three treatment factors

all maximal V-subspaces in the fitted model. If it is not easy to deduce the fitted values from these, then give the table of fitted values as well. If it is possible to interpret any interactions in the fitted model in simple terms, then do so.

Without orthogonality, we would need a different sum of squares for each of the 31 edges in Figure 5.11, excluding the bottom one. With orthogonality (which is guaranteed by equal replication of all the combinations of levels), it is sufficient for the analysis-of-variance table to show seven sums of squares, in addition to that for the grand mean, which corresponds to the bottom edge. For example, the sum of squares for the G-by-H interaction is used to test whether the model can be reduced from $V_{F \wedge G} + V_{F \wedge H} + V_{G \wedge H}$ to $V_{F \wedge G} + V_{F \wedge H}$; to test whether the model $V_{F \wedge G} + V_{G \wedge H}$ can be reduced to $V_{F \wedge G} + V_H$; to test whether the model $V_{F \wedge H} + V_{G \wedge H}$

can be reduced to $V_{F\wedge H} + V_G$; to test whether $V_{G\wedge H} + V_F$ can be reduced to $V_F + V_G + V_H$; and to test whether $V_{G\wedge H}$ can be reduced to $V_G + V_H$.

The extension of these methods to four or more treatment factors continues in the same way.

5.7 Factorial experiments

An experiment is said to be *factorial* if $\mathcal{T} = L_F \times L_G \times L_H \times \cdots$ for treatment factors, F, G, H, ...; that is, if the treatments are all combinations of levels of two or more factors.

Factorial experiments are better than experiments in which only one factor is changed at a time. This is because:

(i) they enable the best combination to be found if the interactions are nonzero;

(ii) they enable testing for the presence of interactions;

(iii) they achieve higher replication for the individual treatment factors.

Example 5.3 revisited (Cow-peas) Suppose that the three methods of cultivation are compared in an experiment in which only the variety C is sown. Figure 5.4 shows that the first method of cultivation would be chosen as the best. A second experiment to compare the five varieties, all with the first method of cultivation, would conclude that variety D is best. The best combination, which is variety E with the second method of cultivation, would not be found by this pair of experiments changing one factor at a time.

Example 5.1 revisited (Rye-grass) Compare the factorial design in Figure 1.2 with the pair of experiments that would be needed to test each factor separately. In the first year, the three cultivars are tested at zero nitrogen, on two strips each, to find the best cultivar. Two strips each is the minimum to achieve any replication, so this (first) experiment takes as much space as the one in Figure 1.2. At the end of the first experiment the data are analysed to find the best cultivar. In the second year, the best cultivar is used to test the four quantities of nitrogen. The quantities of nitrogen could be applied to six plots each, as in Figure 1.2. Three or four plots might suffice, but this is not a great saving compared to the overhead costs of the experiment.

This pair of experiments

- takes twice as long as the factorial experiment in Figure 1.2;

- costs almost twice as much;

- gives a larger variance for estimators of differences between levels of nitrogen (unless 24 plots are used in the second year);

- does not allow any detection of whether there is a nonzero cultivar-by-nitrogen interaction;

- may not find the best combination of cultivar with quantity of nitrogen.

Table 5.11. *The fourteen inorganic fertilizer treatments in the Park Grass experiment*

	Other chemicals				
	NaMg			–	
Nitrogen	PK	P	K	P	–
none	√	√		√	√
0.4 cwt/acre as sulphate of ammonia					√
0.4 cwt/acre as nitrate of soda	√				√
0.8 cwt/acre as sulphate of ammonia	√	√	√	√	
0.8 cwt/acre as nitrate of soda	√				
1.2 cwt/acre as sulphate of ammonia	√				
1.2 cwt/acre as sulphate of ammonia plus silicate of soda	√				

Example 5.13 (Park Grass) J. B. Lawes founded Rothamsted Experimental Station in 1843 to investigate the effects of fertilizer on agricultural crops. He and J. H. Gilbert established many long-term experiments that continue to this day. One of these is known as *Park Grass*.

This experiment was started in 1856 in a field that had been in continuous grass for at least one hundred years. The field was divided into twenty plots, and a certain treatment regime allocated to each plot in perpetuity. Every summer the grass is cut for hay, which is made in situ, so that seed from grass and wild flowers returns to the same plot. What is being observed is species diversity, as well as the yield of hay. Visitors to Rothamsted Experimental Station may visit Park Grass. The treatment allocated to each plot is clearly shown on a post at the end of the plot (there is no *blinding*: see Chapter 7). The plots have no fences or other physical separators between them, but their boundaries are not in doubt, because the mixture of species (clover, cow parsley, fescue, etc.) visibly changes along a sharp straight line between each adjacent pair of plots.

Initially there were twenty plots and eighteen treatments, only two of which were replicated. The importance of replication was not appreciated in 1856. Four of the treatments were organic fertilizers, such as bullocks' manure. The remaining fourteen were combinations of different quantities and types of nitrogen with various combinations of other chemicals, as shown in Table 5.11. This table shows the influence of the 'change one factor at a time' dictum (factorial designs were not advocated until the 1920s). Two of the levels of nitrogen occur with four out of the five combinations of the chemicals Na (sodium), Mg (magnesium), P (phosphorus) and K (potassium). All but one of the levels of nitrogen occur with the combination PKNaMg.

5.8 Construction and randomization of factorial designs

In simple cases, such as unstructured plots, orthogonal block designs, or row–column designs (Chapter 6), simply ignore the factorial structure on the treatments while the design is constructed and randomized.

Example 5.14 (Factorial design in unstructured plots) Suppose that there are eighteen plots, with no structure, and that there are two treatment factors: C with two levels c_1 and c_2, and D with three levels d_1, d_2 and d_3. For an equireplicate design, each treatment is assigned

5.9. Factorial treatments plus control

to three plots, giving the following systematic design.

plot	1 2 3	4 5 6	7 8 9	10 11 12	13 14 15	16 17 18
treatment	c_1d_1	c_1d_2	c_1d_3	c_2d_1	c_2d_2	c_2d_3

Suppose that the random permutation of the eighteen plots is

$$\begin{pmatrix} 1 & 2 & 3 & 4 & 5 & 6 & 7 & 8 & 9 & 10 & 11 & 12 & 13 & 14 & 15 & 16 & 17 & 18 \\ 8 & 5 & 14 & 10 & 1 & 15 & 3 & 16 & 9 & 4 & 12 & 11 & 13 & 7 & 2 & 18 & 6 & 17 \end{pmatrix}.$$

This gives the following plan.

plot	1	2	3	4	5	6	7	8	9	10	11	12	13	14	15	16	17	18
C	c_1	c_2	c_1	c_2	c_1	c_2	c_2	c_1	c_1	c_1	c_2	c_2	c_2	c_1	c_1	c_1	c_2	c_2
D	d_2	d_2	d_3	d_1	d_1	d_3	d_2	d_1	d_3	d_2	d_1	d_1	d_2	d_1	d_2	d_3	d_3	d_3

5.9 Factorial treatments plus control

Sometimes the treatments consist of a control treatment in addition to all combinations of two or more treatment factors: see Example 1.16 and Questions 3.2 and 4.3. There are no new principles involved.

As explained in Section 5.3, the non-control treatments should all have the same replication. The relative replication of the control can be decided as in Section 3.1. For an orthogonal block design, the replication of the control should be an integer multiple of the replication of the other treatments, so that each block can contain all the non-control treatments once and the control on a constant number of plots.

The design is then constructed and randomized ignoring the factorial treatment structure, as in Section 5.8.

In the analysis, the sums of squares for the main effects and interactions of the factorial treatments can be calculated by ignoring the plots with the control treatment. There is a further degree of freedom for comparing the control treatment with the mean of the rest: its sum of squares is calculated as described in Section 3.2.

Questions for discussion

5.1 A food company was interested in the precision of measuring the amount of the vitamin called niacin in bran products. Thirty-six samples of bran flakes were used. Twelve were left alone; twelve were enriched with 4 milligrams of niacin; and twelve were enriched with 8 milligrams of niacin.

Four laboratories were asked to take part in the study. The company was interested in differences between the laboratories, and also wanted to know if the laboratories were consistent in their estimation of differences in the amounts of niacin.

Three samples of each type were sent to each laboratory. Each laboratory put its nine samples into a random order, and then measured the amount of niacin in each sample according to instructions sent by the company. The measurements are shown in Table 5.12, where the data have been reordered for ease of manual calculation.

Table 5.12. *Data for Question 5.1*

	Niacin enrichment		
	+0 mg	+4 mg	+8 mg
Laboratory 1	8.03	11.50	15.10
	7.35	10.10	14.80
	7.66	11.70	15.70
Laboratory 2	8.50	11.75	16.20
	8.46	12.88	16.16
	8.53	12.64	16.48
Laboratory 3	7.31	11.11	15.00
	7.85	11.00	17.00
	7.92	11.67	15.50
Laboratory 4	8.82	12.90	17.30
	8.76	12.00	17.60
	8.52	13.50	18.40

(a) Calculate the analysis-of-variance table for these data.

(b) Are the laboratories consistent in their measurement of the differences in the amounts of niacin?

(c) Calculate the two tables of means for main effects.

(d) Give the standard error of the difference between two laboratories and the standard error of the difference between two enrichment amounts.

(e) Test the hypothesis that the method of measurement correctly gives the difference between the amount of niacin in the '+0 mg' samples and the '+4 mg' samples.

5.2 A group of people researching ways to reduce the risk of blood clotting are planning their next experiments. One says:

> We know that aspirin thins the blood. Let's experiment with the quantity of aspirin. We could enrol about 150 healthy men into the trial, give 50 of them one aspirin tablet per day for a year, another 50 one and a half aspirin tablets a day, and the final 50 will get two aspirin tablets per day.
>
> When we have decided which quantity is best, we can run another trial to find out if there is any difference between taking the aspirin after breakfast or after dinner.

How do you reply?

5.3 A completely randomized factorial experiment was conducted to find out if the recorded strength of cement depended on the person (gauger) who mixed the cement and water and worked the mixture or on the person (breaker) who tested the compressive strength of the mixture after it had set. Three gaugers and three breakers took part in the experiment, with

Questions for discussion

Table 5.13. *Data for Question 5.3*

Breaker	Gauger		
	1	2	3
1	58.0	44.2	53.6
	47.6	52.8	55.0
	52.8	55.8	61.6
	55.2	49.0	43.2
2	50.2	53.4	55.6
	43.4	62.0	57.2
	62.0	49.6	47.6
	44.0	48.8	56.2
3	53.2	41.8	56.0
	46.0	44.8	44.6
	51.8	48.0	46.8
	41.6	46.0	49.3

each breaker testing four samples of cement worked by each gauger. Thus there were 36 plots, nine treatments and the replication was four. Table 5.13 shows the data, arranged in an order to help manual calculation.

Calculate the analysis-of-variance table, the two tables of means for main effects, the standard error of the difference between two breakers, and the standard error of the difference between two gaugers. What do you conclude?

5.4 Nine types of seed of a certain plant were compared in a completely randomized design with three replicates. A known number of seeds was planted in each small plot, and the percentage which germinated was recorded. The totals are shown below.

Colour	Size		
	small	medium	large
brown	219	267	254
light red	206	273	260
dark red	66	79	146

The sum of the squares of all 27 percentages is 134896.

Analyse the data. Plot the nine treatment means on a graph like those in Figures 5.3–5.8. Is the graph consistent with the conclusion about interaction from the analysis of variance?

5.5 A completely randomized experiment was conducted with three treatment factors:

type of nitrogen fertilizer: Dutch or English;

method of fertilizer application: applied as a single dressing, or split into two halves and applied half at the normal time and half one month later;

quantity of nitrogen applied: 80 to 280 kg/ha, in increments of 40 kg/ha.

The response measured on each plot was the yield of wheat in tonnes per hectare.

The following portion of computer output shows the data, the analysis-of-variance table, the tables of means and the standard errors of differences. Interpret this analysis for the benefit of an agronomist who may not understand statistical jargon.

plot	1	2	3	4	5	6	7
method	split	split	split	single	single	single	single
type	Dutch	Dutch	Dutch	English	English	English	Dutch
nitrogen	280	120	160	160	120	280	80
yield	6.6	4.8	5.3	4.7	4.8	4.6	3.3

plot	8	9	10	11	12	13	14
method	single	single	split	split	split	single	split
type	Dutch	Dutch	English	English	English	English	Dutch
nitrogen	200	240	240	200	80	80	240
yield	3.7	4.5	5.9	6.1	4.8	3.9	5.4

plot	15	16	17	18	19	20	21
method	single	split	split	single	single	single	split
type	English	English	English	Dutch	Dutch	English	Dutch
nitrogen	240	160	280	120	280	200	200
yield	4.4	5.8	5.8	4.1	3.8	4.1	5.2

plot	22	23	24	25	26	27	28
method	split	single	split	split	split	single	split
type	Dutch	Dutch	English	Dutch	Dutch	Dutch	English
nitrogen	80	160	120	120	280	80	80
yield	4.3	3.9	4.9	4.7	5.2	3.9	3.8

plot	29	30	31	32	33	34	35
method	single	single	split	single	split	split	single
type	Dutch	English	Dutch	English	English	English	Dutch
nitrogen	200	160	160	280	200	240	240
yield	4.2	4.4	5.1	4.6	5.7	5.8	4.4

plot	36	37	38	39	40	41	42
method	single	single	split	single	split	single	split
type	English	Dutch	Dutch	Dutch	Dutch	English	English
nitrogen	120	160	240	280	80	240	120
yield	3.5	4.2	5.0	4.6	4.4	4.1	4.6

plot	43	44	45	46	47	48
method	split	single	single	split	single	split
type	English	English	Dutch	English	English	Dutch
nitrogen	160	80	120	280	200	200
yield	4.9	3.7	3.7	5.5	5.0	5.4

Questions for discussion

***** Analysis of variance *****

Variate: yield

Source of variation	d.f.	s.s.	m.s.	v.r.	F pr.
method	1	12.9169	12.9169	72.01	<.001
type	1	0.6769	0.6769	3.77	0.064
nitrogen	5	6.6760	1.3352	7.44	<.001
method.type	1	0.0352	0.0352	0.20	0.662
method.nitrogen	5	1.0044	0.2009	1.12	0.376
type.nitrogen	5	0.4094	0.0819	0.46	0.804
method.type.nitrogen	5	0.6610	0.1322	0.74	0.603
Residual	24	4.3050	0.1794		
Total	47	26.6848			

***** Tables of means *****

Variate: yield

Grand mean 4.690

method	single	split
	4.171	5.208

type	Dutch	English
	4.571	4.808

nitrogen	80	120	160	200	240	280
	4.012	4.387	4.787	4.925	4.937	5.087

method	type	Dutch	English
single		4.025	4.317
split		5.117	5.300

method	nitrogen	80	120	160	200	240	280
single		3.700	4.025	4.300	4.250	4.350	4.400
split		4.325	4.750	5.275	5.600	5.525	5.775

type	nitrogen	80	120	160	200	240	280
Dutch		3.975	4.325	4.625	4.625	4.825	5.050
English		4.050	4.450	4.950	5.225	5.050	5.125

method	type	nitrogen	80	120	160	200	240	280
single	Dutch		3.600	3.900	4.050	3.950	4.450	4.200
	English		3.800	4.150	4.550	4.550	4.250	4.600
split	Dutch		4.350	4.750	5.200	5.300	5.200	5.900
	English		4.300	4.750	5.350	5.900	5.850	5.650

*** Standard errors of differences of means ***

Table	method	type	nitrogen	method type
rep.	24	24	8	12
d.f.	24	24	24	24
s.e.d.	0.1223	0.1223	0.2118	0.1729

Table	method nitrogen	type nitrogen	method type nitrogen
rep.	4	4	2
d.f.	24	24	24
s.e.d.	0.2995	0.2995	0.4235

5.6 A factorial experiment has two treatment factors: C, which has three levels, and D, which has two levels. The design has four complete blocks. The systematic design is shown below.

Block	1	1	1	1	1	1	2	2	2	2	2	2	3	3	3	3	3	3	4	4	4	4	4	4
Plot	1	2	3	4	5	6	7	8	9	10	11	12	13	14	15	16	17	18	19	20	21	22	23	24
C	1	1	2	2	3	3	1	1	2	2	3	3	1	1	2	2	3	3	1	1	2	2	3	3
D	1	2	1	2	1	2	1	2	1	2	1	2	1	2	1	2	1	2	1	2	1	2	1	2

Which of the following methods of randomizing is correct? What is wrong with the other methods?

(a) Choose a random permutation of 24 objects and apply it to levels of both C and D at the same time.

(b) Choose a random permutation of 24 objects and apply it to levels of C; then choose another random permutation of 24 objects and apply it to levels of D.

(c) Within each block independently, choose a random permutation of six objects and apply it to levels of both C and D at the same time.

(d) Within each block independently, choose one random permutation of six objects and apply it to levels of C; then choose another random permutation of six objects and apply it to levels of D.

(e) Within each block and each level of C independently, toss a coin to decide the order of the two levels of D.

(f) Within each block and each level of D independently, choose a random permutation of three objects and apply it to the levels of C.

(g) Choose a random permutation of three objects and use it to relabel the levels of C; then do a similar thing for D.

Chapter 6

Row–column designs

6.1 Double blocking

As we noted in Section 4.1, sometimes more than one system of blocks is necessary.

Example 6.1 (Example 4.11 continued: Wine tasting) In this experiment, each of eight judges tastes each of four wines. The plan given in Figure 4.3 treats the judges as blocks and specifies the order of tasting for each judge. Perhaps the positions in tasting order should also be considered as blocks.

The plan is rewritten in Figure 6.1. Wine 4 is almost always tasted first or second. By the fourth tasting the judges may be feeling happy (and so give good marks to the other wines) or bored (and so give bad marks to the other wines). In either case, the comparison of wine 4 with the other wines could be biased.

It would be better to regard 'judges' and 'positions in tasting order' as two systems of blocks.

For the rest of this chapter we shall call one system of blocks 'rows' and the other system 'columns'. Each design will therefore be written in a rectangle, like the one in Figure 6.1. For simplicity, we assume that:

(i) each row meets each column in a single plot;

(ii) all treatments occur equally often in each row;

(iii) all treatments occur equally often in each column;

				Judge				
Tasting	1	2	3	4	5	6	7	8
1	2	4	4	2	1	2	4	4
2	1	3	1	4	4	4	2	3
3	3	2	2	3	3	1	1	1
4	4	1	3	1	2	3	3	2

Fig. 6.1. Another way of writing the plan in Figure 4.3

A	B	C	D	E	F
B	C	A	E	F	D
C	A	B	F	D	E
D	F	E	A	C	B
E	D	F	B	A	C
F	E	D	C	B	A

A	B	C	D
D	A	B	C
C	D	A	B
B	C	D	A

Fig. 6.2. Latin square of order 6   Fig. 6.3. Cyclic Latin square of order 4

(iv) there are m rows;

(v) there are n columns.

From these assumptions it follows that there are n plots per row and m plots per column; that $N = mn$; that t divides n and t divides m; and that every treatment has replication r, where $r = nm/t$.

6.2 Latin squares

The simplest way in which the above conditions can be satisfied is when $n = m = t$. Then the design is called a Latin square.

Definition A *Latin square of order t* is an arrangement of t symbols in a $t \times t$ square array in such a way that each symbol occurs once in each row and once in each column.

Example 6.2 (Latin square of order 6) Figure 6.2 shows a Latin square of order 6.

Here are some simple methods of constructing Latin squares.

Cyclic method Write the symbols in the top row in any order. In the second row, shift all the symbols to the right one place, moving the last symbol to the front. Continue like this, shifting each row one place to the right of the previous row.

A cyclic Latin square of order 4 is shown in Figure 6.3.

More generally, each row can be shifted s places to the right of the previous one, so long as s is co-prime to t.

Group method (for readers who know a little group theory) Take any group G of order t. Label its elements $g_1, g_2, \ldots, g_t$. Label the rows and columns of the square by $g_1, g_2, \ldots, g_t$. In the cell in the row labelled g_i and the column labelled g_j put the element $g_i g_j$. This gives the square shown in Figure 6.4.

For example, when $t = 4$ we can take G to be the cyclic group $\{1, g, g^2, g^3 : g^4 = 1\}$ to obtain the bordered Latin square in Figure 6.5(a). Stripping off the border of labels at the top and the left, and relabelling the symbols, gives the Latin square in Figure 6.5(b), which is a cyclic square in which each row is shifted one place to the left (equivalently, three places to the right) of the preceding row.

6.2. Latin squares

	g_1	g_2	$\cdots$	g_j	$\cdots$	g_t
g_1	g_1^2	g_1g_2	$\cdots$	g_1g_j	$\cdots$	g_1g_t
$\vdots$	$\vdots$	$\vdots$	$\ddots$	$\vdots$	$\ddots$	$\vdots$
g_i	g_ig_1	g_ig_2	$\cdots$	g_ig_j	$\cdots$	g_ig_t
$\vdots$	$\vdots$	$\vdots$	$\ddots$	$\vdots$	$\ddots$	$\vdots$
g_t	g_tg_1	g_tg_2	$\cdots$	g_tg_j	$\cdots$	g_t^2

Fig. 6.4. Latin square constructed from a group

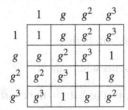

	1	g	g^2	g^3
1	1	g	g^2	g^3
g	g	g^2	g^3	1
g^2	g^2	g^3	1	g
g^3	g^3	1	g	g^2

0	1	2	3
1	2	3	0
2	3	0	1
3	0	1	2

(a) Initial construction from a cyclic group   (b) After removing border and relabelling

Fig. 6.5. Constructing a 4×4 Latin square by the group method

When $t = 6$ we can take G to be the symmetric group S_3. This gives the Latin square in Figure 6.2.

Product method Given two Latin squares S_1 of size t_1 with letters $A_1, \ldots, A_{t_1}$ and S_2 of size t_2 with letters $B_1, \ldots, B_{t_2}$, we can make a new Latin square $S_1 \otimes S_2$ with letters C_{ij} for $i = 1, \ldots, t_1$ and $j = 1, \ldots, t_2$.

Enlarge square S_2. Wherever letter B_j occurs, put the whole square S_1 but replacing letters $A_1, \ldots, A_{t_1}$ by $C_{1j}, \ldots, C_{t_1 j}$.

For example, when $t_1 = t_2 = 2$ we can take

$$S_1 = \begin{array}{|c|c|} \hline A_1 & A_2 \\ \hline A_2 & A_1 \\ \hline \end{array} \quad \text{and} \quad S_2 = \begin{array}{|c|c|} \hline B_1 & B_2 \\ \hline B_2 & B_1 \\ \hline \end{array}.$$

Then $S_1 \otimes S_2$ is the square in Figure 6.6(a), which can be rewritten as shown in Figure 6.6(b). Note that this Latin square is not cyclic. There are no permutations of rows, columns and letters which convert it to the Latin square in Figure 6.3.

C_{11}	C_{21}	C_{12}	C_{22}
C_{21}	C_{11}	C_{22}	C_{12}
C_{12}	C_{22}	C_{11}	C_{21}
C_{22}	C_{12}	C_{21}	C_{11}

C	D	E	F
D	C	F	E
E	F	C	D
F	E	D	C

(a) Initial construction as a product of two squares   (b) After removing internal lines and relabelling

Fig. 6.6. Constructing a 4×4 Latin square by the product method

6.3 Construction and randomization

We return to the more general case that both m and n are multiples of t. To construct a row–column design and randomize it, proceed as follows.

(i) Divide the $m \times n$ rectangle into $t \times t$ squares.

(ii) In each $t \times t$ square, put any Latin square of order t, using the same symbols in each square.

(iii) Randomly permute the m rows (*not* the treatments within them).

(iv) Randomly permute the n columns (*not* the treatments within them).

Note that the division into $t \times t$ squares is artificial, just to help in the construction of the design. It must be ignored at the randomization stage.

Example 6.1 revisited (Wine tasting) Here we have four wines, four rows and eight columns, so we start by dividing the 4×8 rectangle into 4×4 squares.

	1	2	3	4	:	5	6	7	8
1					:				
2					:				
3					:				
4					:				

Then we put any two 4×4 Latin squares into the spaces. Here we use the symbols A, B, C and D for the wines, to avoid confusion with the numbering of the rows and columns.

	1	2	3	4	:	5	6	7	8
1	A	B	C	D	:	C	D	A	B
2	D	A	B	C	:	D	C	B	A
3	C	D	A	B	:	A	B	C	D
4	B	C	D	A	:	B	A	D	C

6.3. Construction and randomization

	Judge							
Tasting	1	2	3	4	5	6	7	8
1	C	D	C	A	B	B	A	D
2	A	B	A	C	D	D	C	B
3	D	A	B	D	C	C	B	A
4	B	C	D	B	A	A	D	C

Fig. 6.7. Randomized plan in Example 6.1

We rub out the construction lines (the dotted vertical line in this case).

To randomize the rows, we choose a random permutation of four objects by one of the methods described in Section 2.2. If the permutation is

$$\begin{pmatrix} 1 & 2 & 3 & 4 \\ 1 & 3 & 4 & 2 \end{pmatrix}$$

then we randomize the rows of the previous rectangle to obtain

	1	2	3	4	5	6	7	8
1	A	B	C	D	C	D	A	B
3	C	D	A	B	A	B	C	D
4	B	C	D	A	B	A	D	C
2	D	A	B	C	D	C	B	A

Similarly, to randomize the columns we choose a random permutation of eight objects. The permutation

$$\begin{pmatrix} 1 & 2 & 3 & 4 & 5 & 6 & 7 & 8 \\ 3 & 6 & 5 & 7 & 8 & 2 & 1 & 4 \end{pmatrix}$$

gives the following rectangle.

	3	6	5	7	8	2	1	4
1	C	D	C	A	B	B	A	D
3	A	B	A	C	D	D	C	B
4	D	A	B	D	C	C	B	A
2	B	C	D	B	A	A	D	C

Before giving the plan to the experimenter, replace the row and column numbers by the natural orders (or the explicit names of the rows and columns, if they have any). The final plan is shown in Figure 6.7.

As we have now seen in three different situations, the construction and randomization of a design can always be broken down into the following three steps.

(1) Recognize the plot structure.

(2) Construct a systematic design (this is steps (i)–(ii) on page 108).

(3) Randomize it (steps (iii)–(iv) on page 108).

This is the same whether or not the treatments are factorial.

6.4 Orthogonal subspaces

As before, we define the treatment subspace V_T of dimension t, the one-dimensional subspace $V_0 = W_0$ consisting of constant vectors, and put $W_T = V_T \cap V_0^\perp$, which has dimension $t-1$. Spaces V_R and V_C are defined just like the block subspace V_B; that is, V_R consists of vectors which are constant on each row, and V_C consists of vectors which are constant on each column, so that $\dim V_R = m$ and $\dim V_C = n$. Then put

$$W_R = V_R \cap V_0^\perp$$

and

$$W_C = V_C \cap V_0^\perp,$$

so that $\dim W_R = m - 1$ and $\dim W_C = n - 1$.

Assumptions (i)–(iii) in Section 6.1 imply that the spaces W_T, W_R and W_C are orthogonal to each other: the proof is like the proof of Theorem 4.1. Therefore

$$V_R + V_C + V_T = W_0 \oplus W_R \oplus W_C \oplus W_T$$

orthogonally, and so

$$\dim(V_R + V_C + V_T) = 1 + (m-1) + (n-1) + (t-1).$$

Put

$$W_E = (V_R + V_C + V_T)^\perp$$

(the notation E for the subscript will be explained in Chapter 10). Then

$$\dim(W_E) = N - (1 + (m-1) + (n-1) + (t-1)) = (n-1)(m-1) - (t-1),$$

and

$$V = W_0 \oplus W_R \oplus W_C \oplus W_T \oplus W_E$$

orthogonally.

Figure 6.8 shows these orthogonal subspaces. This decomposition of V gives the analysis of variance, for both the model when rows and columns have fixed effects and the model when rows and columns have random effects.

6.5 Fixed row and column effects: model and analysis

For each plot ω, write

$$R(\omega) = \text{the row containing } \omega,$$
$$C(\omega) = \text{the column containing } \omega.$$

6.5. Fixed row and column effects: model and analysis

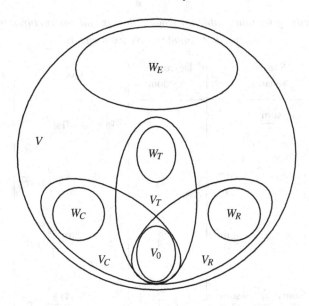

Fig. 6.8. Orthogonal subspaces when the conditions in Section 6.1 are satisfied

We call R and C the *row* factor and the *column* factor respectively.

If rows and columns have fixed effects, they both contribute to the expectation but not to the covariance. That is,

$$\mathbb{E}(Y_\omega) = \tau_{T(\omega)} + \zeta_{R(\omega)} + \eta_{C(\omega)} \tag{6.1}$$

and

$$\mathrm{Cov}(\mathbf{Y}) = \sigma^2 \mathbf{I}.$$

Here $\zeta_1, \ldots, \zeta_m$ are unknown parameters associated with the rows and $\eta_1, \ldots, \eta_n$ are unknown parameters associated with the columns. We can rewrite Equation (6.1) in vector terms as

$$\mathbb{E}(\mathbf{Y}) = \boldsymbol{\tau} + \boldsymbol{\zeta} + \boldsymbol{\eta}, \tag{6.2}$$

where $\boldsymbol{\tau} \in V_T$, $\boldsymbol{\zeta} \in V_R$, and $\boldsymbol{\eta} \in V_C$. As in Section 4.5, put $\boldsymbol{\tau}_0 = \bar{\tau} \mathbf{u}_0$, $\boldsymbol{\tau}_T = \boldsymbol{\tau} - \boldsymbol{\tau}_0$, $\boldsymbol{\zeta}_0 = \bar{\zeta} \mathbf{u}_0$, $\boldsymbol{\zeta}_R = \boldsymbol{\zeta} - \boldsymbol{\zeta}_0$, $\boldsymbol{\eta}_0 = \bar{\eta} \mathbf{u}_0$, and $\boldsymbol{\eta}_C = \boldsymbol{\eta} - \boldsymbol{\eta}_0$. Then $\boldsymbol{\tau}_0$, $\boldsymbol{\zeta}_0$ and $\boldsymbol{\eta}_0$ are all in W_0 while $\boldsymbol{\tau}_T \in W_T$, $\boldsymbol{\zeta}_R \in W_R$ and $\boldsymbol{\eta}_C \in W_C$. Then Equation (6.2) gives

$$\begin{aligned}\mathbb{E}(\mathbf{Y}) &= (\boldsymbol{\tau}_0 + \boldsymbol{\tau}_T) + (\boldsymbol{\zeta}_0 + \boldsymbol{\zeta}_R) + (\boldsymbol{\eta}_0 + \boldsymbol{\eta}_C) \\ &= (\boldsymbol{\tau}_0 + \boldsymbol{\zeta}_0 + \boldsymbol{\eta}_0) + \boldsymbol{\tau}_T + \boldsymbol{\zeta}_R + \boldsymbol{\eta}_C,\end{aligned}$$

with $\boldsymbol{\tau}_0 + \boldsymbol{\zeta}_0 + \boldsymbol{\eta}_0$ in W_0, $\boldsymbol{\tau}_T$ in W_T, $\boldsymbol{\zeta}_R$ in W_R and $\boldsymbol{\eta}_C$ in W_C. We cannot distinguish between $\boldsymbol{\tau}_0$, $\boldsymbol{\zeta}_0$ and $\boldsymbol{\eta}_0$, that is, between the overall level of treatment means, row means and column means. However, because W_T, W_R and W_C are orthogonal to each other, we can estimate treatment contrasts, row contrasts and column contrasts. In fact, Theorem 2.5 shows that

$$\mathbb{E}(\mathbf{P}_{W_T}\mathbf{Y}) = \mathbf{P}_{W_T}(\mathbb{E}\mathbf{Y}) = \mathbf{P}_{W_T}(\boldsymbol{\tau}_0 + \boldsymbol{\zeta}_0 + \boldsymbol{\eta}_0 + \boldsymbol{\tau}_T + \boldsymbol{\zeta}_R + \boldsymbol{\eta}_C) = \boldsymbol{\tau}_T$$

and

$$\mathbb{E}(\|\mathbf{P}_{W_T}\mathbf{Y}\|^2) = \|\boldsymbol{\tau}_T\|^2 + \dim(W_T)\sigma^2 = \|\boldsymbol{\tau}_T\|^2 + (t-1)\sigma^2.$$

Table 6.1. *Analysis-of-variance table for rows, columns and unstructured treatments when rows and columns have fixed effects*

Source	Sum of squares	Degrees of freedom	EMS	Variance ratio
mean	$\dfrac{\text{sum}^2}{N}$	1	$\|\tau_0 + \zeta_0 + \eta_0\|^2 + \sigma^2$	$\dfrac{\text{MS(mean)}}{\text{MS(residual)}}$
rows	$\sum_{i=1}^{m} \dfrac{(\text{sum}_{R=i})^2}{n} - \dfrac{\text{sum}^2}{N}$	$m-1$	$\dfrac{\|\zeta_R\|^2}{m-1} + \sigma^2$	$\dfrac{\text{MS(rows)}}{\text{MS(residual)}}$
columns	$\sum_{i=1}^{n} \dfrac{(\text{sum}_{C=i})^2}{m} - \dfrac{\text{sum}^2}{N}$	$n-1$	$\dfrac{\|\eta_C\|^2}{n-1} + \sigma^2$	$\dfrac{\text{MS(columns)}}{\text{MS(residual)}}$
treatments	$\sum_{i=1}^{t} \dfrac{(\text{sum}_{T=i})^2}{r} - \dfrac{\text{sum}^2}{N}$	$t-1$	$\dfrac{\|\tau_T\|^2}{t-1} + \sigma^2$	$\dfrac{\text{MS(treatments)}}{\text{MS(residual)}}$
residual	←········ by subtraction ········→		σ^2	—
Total	$\sum_\omega y_\omega^2$	N		

Similarly, $\mathbb{E}(\mathbf{P}_{W_R}\mathbf{Y}) = \zeta_R$, $\mathbb{E}(\mathbf{P}_{W_C}\mathbf{Y}) = \eta_C$,

$$\mathbb{E}(\|\mathbf{P}_{W_R}\mathbf{Y}\|^2) = \|\zeta_R\|^2 + \dim(W_R)\sigma^2 = \|\zeta_R\|^2 + (m-1)\sigma^2,$$

and

$$\mathbb{E}(\|\mathbf{P}_{W_C}\mathbf{Y}\|^2) = \|\eta_C\|^2 + \dim(W_C)\sigma^2 = \|\eta_C\|^2 + (n-1)\sigma^2.$$

This gives us the analysis-of-variance table in Table 6.1.

If the treatments are factorial, decompose the sum of squares for treatments into appropriate main effects and interactions. Test for treatment effects in the usual way.

6.6 Random row and column effects: model and analysis

If rows and columns have random effects, they make no contribution to the expectation but they do affect the pattern of covariance. Thus $\mathbb{E}(Z_\omega) = 0$ and so $\mathbb{E}(Y_\omega) = \tau_\omega$, for all plots ω, while, for a pair of plots α and β,

$$\text{cov}(Z_\alpha, Z_\beta) = \begin{cases} \sigma^2 & \text{if } \alpha = \beta \\ \rho_1 \sigma^2 & \text{if } \alpha \neq \beta \text{ but } R(\alpha) = R(\beta) \\ \rho_2 \sigma^2 & \text{if } \alpha \neq \beta \text{ but } C(\alpha) = C(\beta) \\ \rho_3 \sigma^2 & \text{otherwise.} \end{cases}$$

Typically $\rho_1 > \rho_3$ and $\rho_2 > \rho_3$. In matrix form

$$\text{Cov}(\mathbf{Y}) = \sigma^2 \mathbf{I} + \rho_1 \sigma^2 (\mathbf{J}_R - \mathbf{I}) + \rho_2 \sigma^2 (\mathbf{J}_C - \mathbf{I}) + \rho_3 \sigma^2 (\mathbf{J} - \mathbf{J}_R - \mathbf{J}_C + \mathbf{I})$$

6.6. Random row and column effects: model and analysis

$$= \sigma^2[(1-\rho_1-\rho_2+\rho_3)\mathbf{I} + (\rho_1-\rho_3)\mathbf{J}_R + (\rho_2-\rho_3)\mathbf{J}_C + \rho_3\mathbf{J}],$$

where $\mathbf{J}_R$ is the $N \times N$ matrix whose (α, β)-entry is equal to

$$\begin{cases} 1 & \text{if } R(\alpha) = R(\beta) \\ 0 & \text{otherwise,} \end{cases}$$

and $\mathbf{J}_C$ is defined similarly with respect to the column factor.

Theorem 6.1 *If*

$$\text{Cov}(\mathbf{Y}) = \sigma^2[(1-\rho_1-\rho_2+\rho_3)\mathbf{I} + (\rho_1-\rho_3)\mathbf{J}_R + (\rho_2-\rho_3)\mathbf{J}_C + \rho_3\mathbf{J}],$$

then the eigenspaces of $\text{Cov}(\mathbf{Y})$ *are* W_0, W_R, W_C *and* $(V_R + V_C)^\perp$, *with eigenvalues* ξ_0, ξ_R, ξ_C *and* ξ *respectively, where*

$$\xi_0 = \sigma^2[1 + (n-1)\rho_1 + (m-1)\rho_2 + (m-1)(n-1)\rho_3]$$
$$\xi_R = \sigma^2[1 + (n-1)\rho_1 \quad\quad - \rho_2 \quad\quad -(n-1)\rho_3]$$
$$\xi_C = \sigma^2[1 \quad\quad -\rho_1 + (m-1)\rho_2 \quad\quad -(m-1)\rho_3]$$
$$\xi = \sigma^2[1 \quad\quad -\rho_1 \quad\quad -\rho_2 \quad\quad +\rho_3].$$

Proof Put $\mathbf{C} = \text{Cov}(\mathbf{Y})$. Let $\mathbf{x}$ be a vector in V. An argument similar to the one in Section 4.6 shows that if $\mathbf{x} \in V_R$ then $\mathbf{J}_R\mathbf{x} = n\mathbf{x}$ while if $\mathbf{x} \in V_R^\perp$ then $\mathbf{J}_R\mathbf{x} = \mathbf{0}$. Likewise, if $\mathbf{x} \in V_C$ then $\mathbf{J}_C\mathbf{x} = m\mathbf{x}$ while if $\mathbf{x} \in V_C^\perp$ then $\mathbf{J}_C\mathbf{x} = \mathbf{0}$. As always, if $\mathbf{x} \in V_0$ then $\mathbf{J}\mathbf{x} = N\mathbf{x} = nm\mathbf{x}$ while if $\mathbf{x} \in V_0^\perp$ then $\mathbf{J}\mathbf{x} = \mathbf{0}$.

Now, $\mathbf{u}_0 \in V_R \cap V_C \cap V_0$ so

$$\mathbf{C}\mathbf{u}_0 = \sigma^2[(1-\rho_1-\rho_2+\rho_3) + n(\rho_1-\rho_3) + m(\rho_2-\rho_3) + nm\rho_3]\mathbf{u}_0 = \xi_0\mathbf{u}_0.$$

If $\mathbf{x} \in W_R$ then $\mathbf{x} \in V_R \cap V_C^\perp \cap V_0^\perp$ so

$$\mathbf{C}\mathbf{x} = \sigma^2[(1-\rho_1-\rho_2+\rho_3) + n(\rho_1-\rho_3)]\mathbf{x} = \xi_R\mathbf{x}.$$

Similarly, if $\mathbf{x} \in W_C$ then

$$\mathbf{C}\mathbf{x} = \sigma^2[(1-\rho_1-\rho_2+\rho_3) + m(\rho_2-\rho_3)]\mathbf{x} = \xi_C\mathbf{x}.$$

Finally, if $\mathbf{x} \in (V_R + V_C)^\perp$ then $\mathbf{x} \in V_R^\perp \cap V_C^\perp \cap V_0^\perp$ and so

$$\mathbf{C}\mathbf{x} = \sigma^2(1-\rho_1-\rho_2+\rho_3)\mathbf{x} = \xi\mathbf{x}. \blacksquare$$

Note that $\xi_C = \xi + \sigma^2 m(\rho_2 - \rho_3)$, which is bigger than ξ if $\rho_2 > \rho_3$. Similarly, $\xi_R > \xi$ if $\rho_1 > \rho_3$.

The treatment subspace W_T is contained in the eigenspace $(V_R + V_C)^\perp$, so we obtain the analysis-of-variance table shown in Table 6.2.

Estimation and testing of treatment effects is the same whether the rows and columns are fixed or random. If the rows and columns have fixed effects then we can also estimate row differences and column differences and test for these differences. All these tests are one-sided.

Table 6.2. *Analysis-of-variance table for rows, columns and unstructured treatments when rows and columns have random effects*

Stratum		Source	df	EMS	VR
V_0	mean	mean	1	$\|\tau_0\|^2 + \xi_0$	–
W_R	rows	rows	$m-1$	ξ_R	–
W_C	columns	columns	$n-1$	ξ_C	–
$(V_R+V_C)^\perp$	plots	treatments	$t-1$	$\dfrac{\|\tau_T\|^2}{t-1}+\xi$	$\dfrac{\text{MS(treatments)}}{\text{MS(residual)}}$
		residual	by subtraction	ξ	–
Total			N		

However, we do not usually test for row and column effects, because we assume that these exist when we construct the design.

If the rows and columns have random effects then we could do formal two-sided tests of the null hypotheses that $\xi_R = \xi$ and $\xi_C = \xi$. However, all that we normally do is compare MS(rows) with MS(residual) and compare MS(columns) with MS(residual) to draw broad conclusions. If MS(rows) $\gg$ MS(residual) then rows are good for blocking. Likewise, if MS(columns) $\gg$ MS(residual) then columns are good for blocking. If either MS(rows) $\ll$ MS(residual) or MS(columns) $\ll$ MS(residual) then either there are some patterns of variability or management patterns which the experimenter has not told you about or he is fiddling the data.

Example 6.3 (Straw) In the first half of the twentieth century, cereal plants had long stalks; they were used for straw (for example, to make hats) as well as for grain for food. At Rothamsted an experiment was carried out on wheat to see how the date of applying nitrogen fertilizer affected the yield of straw.

There were six treatments: no fertilizer was one, the others were the same quantity of fertilizer at different dates. These were applied to six plots of wheat each, in a 6×6 Latin square, as shown in Figure 6.9. The seed was drilled in October 1934, fertilizer applied at dates from late October to late May, and the wheat was harvested in August 1935.

Figure 6.9 also shows the yield of straw in pounds per plot. The row and column totals are also shown in Figure 6.9: the treatment totals and means are in Table 6.3. Hence we obtain the analysis of variance in Table 6.4, where the treatment sum of squares has been broken down into two parts as in Section 3.2. It is clear that rows and columns account for a lot of the variability between plots, and also that there is a large difference between the plots with and without fertilizer. However, there is no evidence that, if this quantity of fertilizer is applied, the date of application affects the yield of straw.

6.6. Random row and column effects: model and analysis

4	0	2	1	3	5	
166.0	147.9	184.9	188.4	197.4	181.8	1066.4
3	4	0	5	1	2	
190.8	193.0	168.0	198.8	191.8	197.1	1139.5
5	2	3	4	0	1	
169.2	185.6	185.8	205.2	184.8	180.8	1111.4
2	3	1	0	5	4	
188.5	191.8	174.0	172.4	189.8	168.3	1084.8
0	1	5	2	4	3	
161.1	185.0	177.8	201.7	191.1	168.0	1084.7
1	5	4	3	2	0	
168.0	170.6	170.0	190.5	188.4	134.8	1022.3
1043.6	1073.9	1060.5	1157.0	1143.3	1030.8	6509.1

sulphate of ammonia, at 0.4 cwt of nitrogen per acre, applied on these dates:

0	1	2	3	4	5
never	26 October	19 January	18 March	27 April	24 May

Fig. 6.9. Field layout in Example 6.3

Table 6.3. *Treatment totals and means in Example 6.3*

Treatment	0	1	2	3	4	5
Total	969.0	1088.0	1146.0	1124.3	1093.6	1088.0
Mean	161.5	181.3	191.0	187.4	182.3	181.3
		total 5540.1		mean 184.7		

Table 6.4. *Analysis-of-variance table for Example 6.3 if rows and columns have random effects*

Stratum	Source	SS	df	MS	VR
mean	mean	1176899.523	1	1176899.523	–
rows	rows	1324.075	5	264.815	–
columns	columns	2376.302	5	465.260	–
plots	control	2684.244	1	2684.244	37.67
	timing	455.381	4	113.845	1.60
	residual	1425.085	20	71.254	–
Total		1185114.610	36		

Questions for discussion

6.1 Look at the eelworm experiment in Question 4.3. Suppose that, in the year following the eelworm experiment, a similar experiment is to be conducted on the same plots, using four types of fertilizer at a single dose and no 'control'. How should the plots be divided into blocks?

6.2 The report-writing department of a technical research station needs to replace its old word-processing system. It plans to evaluate five new word-processors. The vendors of each word-processor have agreed to make several copies of it available to the department for a one-week trial period: the first full week in March. The head of the department has decided to abandon all normal work for that week and devote it to testing and evaluating the new word-processors. There are 15 typists in the department's typing pool. The minimum amount of time that a typist can spend with a new word-processor and sensibly be asked to evaluate it for ease of use is one day.

Design the experiment and produce a plan for the head of the department.

6.3 Derive the appropriate model and analysis when rows are fixed but columns are random.

6.4 Groups of apples were stored in a shed in the 4×4 Latin square design in Figure 6.10. There were four shelves along the side of the shed: these are the rows in Figure 6.10. Four groups of apples were stored on each shelf, so that the columns in Figure 6.10 represent distance from the door. The groups of apples were labelled A, B, C and D, where groups A and B were from one variety, groups C and D from another. Groups labelled A and C were stored for a short time, groups labelled B and D for a long time. At the end of the storage, the percentage weight loss was recorded for each group.

Analyse the data, showing the factorial treatment effects. Briefly interpret the analysis.

C	B	A	D
18.30	38.81	13.55	44.85
D	A	C	B
35.22	18.88	24.37	32.98
B	C	D	A
38.90	23.62	36.74	25.01
A	D	B	C
26.47	44.78	38.01	24.82

Fig. 6.10. Layout and data for the experiment in Question 6.4

Chapter 7
Experiments on people and animals

7.1 Introduction

In this chapter we consider experiments on people and animals. Examples are clinical trials, testing educational strategies, psychological experiments, animal nutrition experiments, and trials on ambient conditions in the workplace. There are some issues peculiar to such experiments: these are discussed in Sections 7.2 and 7.7–7.10. On the other hand, the plot structure of such experiments is typically not very complicated, usually being one of those covered in Chapters 6, 4 or 2. These plot structures are specialized to experiments on people and animals in Sections 7.3, 7.4 and 7.5 respectively.

What is an observational unit in such an experiment? Usually it is a person or animal for a set time-period. A measurement may be taken before the experiment starts. This measurement is sometimes called a *baseline* measurement. It may be used to block the people or animals. It may be used as a covariate in the analysis of the data. Thirdly, the data to be analysed may be the differences between the post-treatment measurement and the baseline measurement.

Sometimes measurements are taken at several different times without changing the treatments. This can be enforced on animals or on hospital patients, but if non-hospitalized patients are asked to come back for measurement too often they may simply drop out of the trial.

What is the experimental unit? It may be a person or animal for the duration of the trial. It may be a group of people or animals for the duration of the trial if treatments can be administered only to whole groups. For example, teaching methods in schools can normally be changed only on a whole-class basis. It may be a person or animal for a shorter time-period than the whole trial if treatments are changed during the trial.

Occasionally, several treatments can be applied simultaneously to different parts of a person or animal. Then each part is an experimental unit, and often the observational units are the same. This is discussed in Section 7.6.

What should the blocks be? This will be answered in Sections 7.3–7.6, after disposing of a specious argument for avoiding experimentation altogether.

7.2 Historical controls

When a new drug is introduced, there is often a good collection of data on the performance of the previous standard drug on a large number of people. It is, therefore, tempting to assess the efficacy of the new drug by administering it to several patients and comparing the results to those of the previous standard drug. In this situation, the patients who had received the previous drug are known as *historical controls*.

The use of historical controls is not satisfactory, for several reasons. In the first place, the new drug is given later in time than the previous one, and any observed difference between the results could conceivably be due to different conditions at different times. For example, a new drug to alleviate the symptoms of asthma might appear to be better just because it was tried out in a year when air pollution was less serious. Secondly, there is no doubt that people respond differently to treatments if they know that they are in a trial. If patients receiving the new drug know that they are taking part in a trial but the historical controls did not, then the patients on the new drug may well do better, even if there is no chemical difference between the new drug and the standard. Thirdly, if historical controls are used then the new drug is likely to be offered preferentially to those patients whom the doctors think can most benefit from having it rather than the standard drug. Again, the new drug may appear to be better without actually being so.

The only way to avoid these biases is to have a *randomized controlled* trial. Here *controlled* means that any control treatment, such as 'no treatment' or a standard drug, is included in the current trial. Patients are selected for inclusion in the trial in the knowledge that they may be allocated to any of the treatments. Randomization (and blocking) are used to ensure that differences between treatments are not confounded with time differences (such as the air pollution) or differences between patients.

Similar remarks apply to new diets, new educational methods, new training methods for athletes and new working conditions in the workplace.

7.3 Cross-over trials

A cross-over trial is suitable when each person or animal or group can be used more than once. Then we block by *person* and by *time-period*, to get a row–column design. A typical plot structure (before allocation of treatments) is shown in Figure 7.1.

The number of periods should not be so large that people are very likely to drop out of the trial before it finishes, whether through boredom, adverse reactions, death or moving away from the area where the trial is conducted. Ideally, if the number t of treatments is small, use t periods and a multiple of t people. Then the construction methods of Chapter 6 can be used.

For a cross-over trial to be practicable and useful, several constraints must be satisfied.

(i) Sufficient people or animals or groups must be available at the same time.

(ii) Those people or animals or groups must be prepared to stay in the trial until the end. Since time-periods may be as short as hours or as long as years, it is the overall length of the trial that matters, not the number of periods. There is also a difference between involuntary subjects, such as animals or groups of schoolchildren, and voluntary subjects, such as clinical patients, whether hospitalized or not.

	Person														
Period	1	2	3	4	5	6	7	8	9	10	11	12	13	14	15
1															
2															
3															

Fig. 7.1. Typical plot structure in a cross-over trial

(iii) No treatment should leave the subject in a very different state at the end of the period in which it is administered. In medicine, cross-over trials are best for chronic conditions, such as high blood pressure or asthma, and for testing drugs which may alleviate symptoms rather than curing the underlying illness. They are not good for drugs whose purpose is to cure the underlying illness; nor are they good for diseases where the patient has such a poor prognosis that he may not survive to the end of the trial.

Similarly, if the purpose of a particular teaching method is to improve pupils' understanding of Pythagoras's Theorem, then they cannot unlearn this and start afresh with another method: a cross-over trial is unsuitable. If the purpose of a particular diet is to confer long-term resistance to some disease, it should not be tested in a cross-over trial.

(iv) No treatment should have an effect which lasts into subsequent time-periods. Such effects are called *residual* effects or *carry-over* effects. Examples include drugs with delayed reactions and diets for cows which improve milk yield over several subsequent weeks.

To some extent this problem can be circumvented by separating the time-periods in the trial by neutral periods (possibly of a different length from the trial periods) in which a common non-experimental treatment is given to all subjects. Such periods are called *wash-out* periods. They must be long enough for residual effects to wear off. They increase the length of the trial, hence the cost and the probability that some subjects will drop out.

There are methods of designing and analysing cross-over trials when there are carry-over effects, but they are beyond the scope of this book.

The experiment comparing red wine and gin in Question 1.2 was a cross-over trial.

7.4 Matched pairs, matched threes, and so on

The advantage of a cross-over trial is that people are blocks, so differences between people are eliminated from differences between treatments. If a cross-over trial is not suitable, then each person can be used only once and we must block in some other way. If there are t treatments, use everything relevant that you know about the people (age, weight, state of health, educational background, ...) to group people into blocks of size t. Then do a complete-block design. If $t = 2$, this is called a *matched pairs* design. Many clinical trials have only two treatments (standard drug and new drug, or 'no treatment' and new treatment), so matched pairs designs are quite common. However, the same principles apply equally well to matched threes, matched fours, etc.

One disadvantage of having a large number of small blocks is that many degrees of freedom are assigned to within-block differences, leaving fewer for the residual, so there may be a loss of power. If there are a large number of people in the trial and few relevant blocking factors then it may be better to have block size a multiple of t rather than t itself. Design and analysis are still as in Chapter 4. In clinical trials such large blocks are sometimes called *strata*, but this use of the word is different from our use of *strata* for the eigenspaces of the covariance matrix.

For such a design to be used, the experimenter must have prior information about the people or animals involved.

7.5 Completely randomized designs

If people or animals enter the trial at different times, with no prior information about them, then no blocking may be possible. This is a common situation in clinical trials, where it may be decided that the experimental units will be the next 120 patients who arrive at the surgery satisfying a list of criteria and who agree to participate in the trial. It can also happen in a nutrition trial on young animals if animals are to enter the trial soon after birth.

Since there is no prior information about the experimental units, no blocking is possible, so a completely randomized design is used.

Decide in advance on the number N of people or animals. Produce a randomized list of treatments for the N people as in Section 2.1. As each person enters the trial, allocate them to the next treatment on the list.

Sometimes a slight modification of this procedure allows a coarse form of blocking. If several hospitals are entering patients into the trial, it can be decided in advance that each hospital will have k patients, where k is a multiple of t. Each hospital then has its own randomized list, like one of the blocks in Section 4.3. Another strategy is to block by time, so that the first k patients entering the trial form one block, the next k patients the second block, and so on.

7.6 Body parts as experimental units

Sometimes the treatments can be applied to more than one part of the body of the human or animal subject. In Example 4.1, different midge repellents can be applied to the two arms of each human volunteer. Then it is sensible to use the body parts as experimental units and the human or animal subjects as blocks. Construction and randomization are as in Chapter 4.

The within-subject variability is likely to be less than the between-subjects variability, so precision is increased. Thus, for a given precision or power, the number of human or animal subjects can be reduced in two ways. First, the smaller variability means that we can use fewer experimental units. Secondly, if k body parts are used per subject, the number of subjects can be divided by k. This is particularly important in experiments where animals have to be sacrificed.

Example 7.1 (Regenerating bone) A biomaterials scientist is interested in the properties of ceramic scaffolds. These materials have the potential to regenerate bones in humans who have lost bone matter because of disease or trauma. The regulatory authorities demand that the materials be tested for efficacy and safety in animals before being tried in humans, so he experiments on dogs, using two ceramic scaffolds and a 'do nothing' control. A portion of bone is damaged; the treatment is applied and left for several weeks; then the dog is killed so that the bone can be extracted, examined and weighed. By using three bones per dog rather than one, he is able to obtain useful information by using only one quarter of the number of dogs envisaged when it was planned to treat only one bone per dog.

If the body parts differ among themselves then the named parts can be used as a second system of blocks. This gives a row–column design, as in Chapter 6.

Example 7.2 (Example 4.1 continued: Insect repellent) Legs as well as arms could be used in this experiment, so long as the same area of skin is exposed on each limb. This doubles the number of experimental units but probably increases the within-person variability. It would be sensible to use a row–column design in which the columns are the volunteers and the rows are labelled 'right arm', 'left arm', 'right leg' and 'left leg'.

7.7 Sequential allocation to an unknown number of patients

Some experimenters would like to start a trial without specifying the number of patients in advance. If results on some patients become known before all patients have entered the trial, this presents the temptation to keep analysing intermediate results. Even if there is no effective difference between the new drug and the old, a test at the 5% significance level will show a difference one time in twenty, and the experimenter can then report the favourable result he is looking for. To avoid such spurious results, there is now an elaborate set of rules for intermediate analyses and for deciding when to stop the trial: these are beyond the scope of this book.

Even if no results are known while patients are still entering the trial, there are situations where it is unrealistic to specify the exact number of patients in advance. For example, should a trial on a rare disease have to be prolonged indefinitely while the experimenter waits for just two more patients? Some strategies have been suggested for sequential randomization of an unknown number of patients, but none is entirely satisfactory. For simplicity, I shall describe them for the case of two treatments.

The first is to simply toss a coin for each new patient. This has the disadvantage that the replications are most unlikely to be equal, so the variance of the estimator of the treatment difference will be larger than it needs be.

The second method is to toss a biased coin, so that the next patient is more likely to receive the treatment that has been under-represented so far. This has the disadvantage that a pair of successive patients are more likely to receive different treatments early on than they are later, which contradicts the randomization assumption made in Chapter 1.

The third method is to block by time. However, if blocks are large then there is still a danger of very unequal replication. If blocks are small then many degrees of freedom are lost to the between-blocks contrasts, which may lose power.

The fourth method is called *minimization*. This is a version of the biased-coin strategy that seeks to block on several factors unknown in advance (such as age, sex, smoking history). The coin is biased in such a way that replication should be almost equal within each category of each blocking factor. Unfortunately, a side effect is that replication will be very unequal in categories of combinations of two blocking factors, so the design will be poor if there is, say, an interaction between age and sex.

7.8 Safeguards against bias

Early experiments on working conditions in factories showed that the productivity of the workers in the experiment improved no matter what conditions they had. Management was taking more notice of these workers than usual, and this produced a positive effect irrespective of the experimental conditions. Thus everyone who is taking part in an experiment should be equally aware that they are doing so.

Similarly, it is now well known that patients improve if they receive a treatment which they perceive to be beneficial, whether or not it is. Thus, in a clinical trial, any null treatment should be given as a dummy treatment, which is called a *placebo*. For example, if the new drug is administered as a small blue tablet then the placebo should be a similar small blue tablet but without the active ingredient. Another reason for using placebos is to avoid confusing the effect of the drug with the effect of the regime of taking the drug.

Since people, or even animals, may respond for psychosomatic reasons, they should not know which treatment they are receiving. The trial is said to be *blind* if this condition is achieved. It is not always possible. In a comparison of methods of physiotherapy for treating lower back pain, patients will be well aware of the exercises they are doing. Similarly, in many psychological, social or educational experiments the subjects cannot avoid knowing what treatment they are getting. If two new drugs come in tablets of different colours, then it has been suggested that there should be two types of placebo, one of each colour.

Doctors, vets and other professionals involved should also not know the treatment allocated to each subject, because their own expectations or ethics may influence the result. *Assessment bias* occurs if a doctor recording a subjective rating does so more or less favourably according to whether he already thinks the treatment allocated to that patient is better or worse.

If neither patient nor doctor knows the treatment then the trial is said to be *double-blind*.

Example 7.3 (Educational psychology) An educational psychologist wanted to compare two different methods of presenting information. Her experimental units were thirty undergraduate volunteers from the Psychology department. They volunteered sequentially by arriving at the psychologist's office. In a private fifteen-minute session she presented the new information by either method A or method B and then gave the student a short test to find out how much of the information they had absorbed.

Her method of randomization was to toss a coin to decide between methods for the first volunteer, and to alternate between A and B thereafter. There are several reasons why this method is flawed. First, the students would talk among themselves and soon discover such an obvious pattern as $ABABA...$ and so could deliberately present themselves in a special order; for example, deliberately alternating people who were good at method A with people who were good at method B. Secondly, the psychologist herself was aware of the simple

7.8. Safeguards against bias

pattern and would always know the next method to use, which might unconsciously affect her decision about whether to accept the next volunteer as suitable.

The third reason is more subtle, but has already been touched on in Section 7.7. If an experiment is to be analysed as 'unstructured' by the methods of Chapter 2, then the probability that the difference $y_\alpha - y_\beta$ contributes to residual or to an estimator of a treatment difference should be the same for all pairs of distinct experimental units α and β. In other words, there must be a fixed probability p such that

$$\Pr[T(\alpha) = T(\beta)] = p$$

whenever $\alpha \neq \beta$. (For an equireplicate design, p must be equal to $(r-1)/(N-1)$, because there are $r-1$ other experimental units which receive the same treatment as α does.) The psychologist's method of randomization does not achieve this, because

$$\Pr[T(\alpha) = T(\beta)] = 1$$

if α and β are volunteers whose positions in the sequence are both even numbers or both odd numbers, and

$$\Pr[T(\alpha) = T(\beta)] = 0$$

otherwise.

Example 7.4 (Lanarkshire milk experiment) An experiment in Lanarkshire in the early twentieth century demonstrated the conflict between the ethics of the professionals involved and the statistical needs of the experiment. Extra milk was given to a random selection of pupils at some schools to see if it affected their growth. At the end of the experiment it was discovered that the teachers had altered the random allocation to ensure that extra milk was given to the most undernourished children. Their good intentions ruined the experiment, by confounding the effect of the extra milk with the initial state of health.

Example 7.5 (Doctor knows best) A consultant doctor organized a trial of drugs to cure a certain serious disease. There were three treatments: the current standard drug X, which was a very strong type of antibiotic, and two new drugs. Several general practitioners agreed to participate in the trial. They were all sent the trial protocol, and asked to telephone the consultant's secretaries when they had a patient to be entered in the trial. The secretaries had the randomization list, showing which drug to allocate to each patient in order as they entered the trial.

One day one of these general practitioners telephoned and said he had a suitable patient for the trial. The secretary asked several questions about age, weight, etc., to check whether the patient was eligible and, if so, to determine the correct dose of the allocated drug. The secretary accepted the patient, allocated the next drug on the randomization list, which was the standard drug X, worked out the dosage and informed the general practitioner that the patient should be given drug X with such-and-such a dosage regime. The general practitioner responded that his patient could not take drug X because he already knew that it was harmful to her. The secretary asked the consultant what to do, and he said that the patient should simply be allocated to the next drug on the list that was not X.

This story shows the conflict between the doctor's priorities and the statistician's. A doctor's duty is to do the best for his patients as individuals; of course, a doctor will know better than any statistician what is good for the patient. On the other hand, actions like this bias the whole trial and therefore possibly delay the introduction of a more effective treatment. It is easy to say that one patient in several hundred will introduce a negligible bias, but once you allow this sort of non-random allocation at all then it can rapidly be applied to sufficient patients to introduce a serious bias.

Nowadays, the secretaries should ask the person offering the patient a question such as 'The patient may be allocated to the standard drug X or to one of two new drugs which we believe have such-and-such properties. Can we allocate your patient to any one of these three?' If the answer is No, the patient should not be entered into the trial. Moreover, to prevent any possible cheating, the randomization list should be kept at a separate location, and the next treatment revealed only *after* the next patient has been entered in the trial.

Selection bias occurs if the experimenter or doctor consciously chooses which person to allocate to which treatment or consciously decides which person to include next in the trial on the basis of what treatment they will receive. This is what happened in Examples 7.4 and 7.5 respectively. Selection bias is not necessarily present with a randomization scheme like the one in Example 7.3, but the fact that the experimenter had the knowledge to enable her to bias her choice of subjects should make other scientists question the validity of the study. A different form of selection bias occurs if patients choose their own treatment.

Example 7.6 (AIDS tablets) An AIDS clinic in Bangkok offered its new tablets to 117 severely ill patients. Of these, 53 accepted the tablets. On average, these 53 lived for five weeks longer than the 64 patients who declined the tablets. The makers of the tablets claimed that this showed their efficacy. Other scientists pointed out that the patients who chose the tablets may have been healthier than the others, in that they still had the will to care about their future and seek cures.

In clinical trials it is rare for either the plot structure or the treatment structure to be complicated. The most important issues are usually replication and avoidance of bias.

7.9 Ethical issues

There are ethical issues in experiments on people and animals that simply do not arise in experiments on plants or on industrial processes.

An experimental treatment should not be applied to people if it is expected to cause harm. If a person under treatment appears to be suffering adverse side-effects they should probably be withdrawn from the trial. This is another reason why it is hard to achieve equal replication in clinical trials and why complicated plot structure should be avoided.

For a given illness, if there is already a standard drug which is known to be effective then it is not ethical to give no treatment in a clinical trial of a new drug for that illness. The control treatment must be the current standard drug.

Example 7.7 (Incomplete factorial) It is known that drugs A and B are both effective against a certain illness. Someone suggests a 2×2 factorial trial in which the treatments are all four combinations of (A or not) with (B or not). However, it is not ethical to give a patient no drug at all when an effective drug is available. Thus there can be only three treatments: A alone, B alone, and A and B together.

7.9. Ethical issues

Because it is unethical to give patients harmful treatments, or to force people into specified occupations or recreations or diets, information sometimes has to be gathered non-experimentally. One possibility is historical controls: for example, the long life-span of people born in the middle of the twentieth century compared to those born 100 years earlier may be associated with improved diet. Another possibility is the *observational study*, which is used for conditions which arise too rarely for any planned intervention.

A *controlled* study is better than either of these, if it is feasible. A controlled *retrospective study* (also called a *case-control* study) to investigate nystagmus might select one thousand men with the disease and one thousand without. The control group should be chosen to match the diseased men in overall age distribution and in urban/rural environment. If the aim of the study is to find out if nystagmus is associated with occupation then people should be selected for the groups without their occupation being known. If it is then found that the proportion of coal-miners among the diseased men is much higher than the proportion in the control group, then there is some evidence that coal-miners are more likely to get nystagmus.

A controlled *prospective* study (sometimes called a *cohort study*) is better than a retrospective one, but more difficult to organize. Now two groups are selected which differ in terms of the possible *cause* of the disease: for example, smokers and non-smokers. The groups are matched on other factors as closely as possible. The people are monitored for a long time, perhaps for the rest of their life, and the proportions getting lung cancer in the two groups noted. A prospective study is likely to be more accurate and less biased than a retrospective one, because it does not rely on people's memories and there is no possibility of cheating in forming the groups. However, it is much more costly. Not only does it last for several years, but more people are required initially, to allow for the fact that some of them will be *lost to follow-up* in the sense that they drop out, fail to keep in touch, or die from unrelated causes.

Example 7.8 (The British doctors' study) One famous prospective study is the 50-year investigation of British doctors. In 1951 all male British doctors were sent a questionnaire about their health and smoking habits. Over 34,000 responded. An immediate conclusion from this retrospective study was the strong association between cigarette smoking and lung cancer. The doctors were followed up, in what was now a prospective study, until 2001, with the result that very strong conclusions could be drawn about the effect of cigarette smoking (and of giving up at various ages) on mortality.

With animals, the ethical situation is less clear. Everyone agrees that animals used in experiments should not be subjected to unnecessary suffering, but there is less agreement on what suffering is necessary. Many people argue that animals may suffer, and even be killed, in the course of an experiment to find a cure for a serious human disease. Even so, the number of animals used should be kept to a minimum. Is it ethical to use animals in experiments on smoking or on cosmetics? Some people think that no animal suffering in experiments is permissible.

Example 7.9 (Frogs) Amphibian numbers declined world-wide in the 1980s and 1990s. In one experiment to find the cause, laboratory frogs were injected with one of three pesticides (DDT, malathion or dieldrin) or left in their normal state. After some days, the frogs' immune response was measured. The experiment showed that these three pesticides dramatically reduce the number of antibodies produced by the frogs. Such experiments can pave the way for banning of the harmful pesticides. Is it unethical to deliberately poison frogs in this way?

Replication is more of an ethical issue in experiments on people and animals than it is in general. Too few people (or animals) or time-periods will give insufficient power to detect genuine differences, so any suffering will be in vain. On the other hand, it is unethical to continue a clinical trial after one treatment is known to be better.

Humans in a trial should normally give their *informed consent* to participation. This means that they should be told enough details about the trial to be able to make an informed decision about whether they participate. In particular, they must be informed that there is a chance that they will be allocated an inferior treatment. In the past, such consent has not always been sought from prisoners or from people with low intelligence.

It is clear that it is unethical to conduct a clinical trial without informed consent. It is less clear whether consent is needed for educational experiments or experiments on working conditions. A particularly grey area is the clinical trial in which a whole group of people must receive the same treatment.

Example 7.10 (Educating general practitioners) A trial was conducted to test the effectiveness of implementing certain new guidelines for the treatment of diabetes. Some general practitioners were randomized to the 'intervention' treatment and asked to attend some educational sessions where the new guidelines were explained; the other general practitioners in the experiment were not invited to such sessions. However, the observational units in the experiment were the diabetic patients of the two sets of general practitioners.

Example 7.11 (Maternal dietary supplements in Gambia) In small rural communities an intervention at community level may be more effective than one at the individual level because members of the community encourage each other to take the new drug or the new diet. For this reason, a trial of maternal dietary supplements in rural Gambia used villages as experimental units. The mothers (or rather, their babies, whose birthweight was recorded) were the observational units.

7.10 Analysis by intention to treat

For minor diseases, a doctor cannot force a patient to continue to take an unpleasant medicine or continue with an intrusive therapy, even if the patient has volunteered for the experiment. So long as the results are analysed according to which treatment was *allocated* to the patient, as opposed to which treatment he actually took, this ethical principle has the effect that the results from a clinical trial are more likely to generalize to the population at large.

Example 7.12 (Mouthwash) 1200 dental students take part in a trial to see if cither of two mouthwashes (A and B) is effective in preventing gum disease. 300 students are randomized to each mouthwash, 300 to a placebo mouthwash that is actually tap water, and 300 are given no mouthwash at all (why are these last two 'treatments' included?) Suppose that A is actually more effective than B but either tastes more unpleasant or needs to be taken more often. Then it is likely that students allocated to A will be less diligent in using it than students allocated to B. Some of them will be honest about this lack of diligence but many will be ashamed to admit that they have not followed instructions.

Suppose that, on average, gum disease affects 6% of those of student age who use no mouthwash, 2% of those who use mouthwash B and 1% of those who use mouthwash A, but that one third of people who try mouthwash A give it up. Then the results of the trial on these two treatments might be as follows.

	Gum disease	No gum disease	Total
A	2	196	198
A but gave up	6	96	102
B	6	274	280
B but gave up	1	19	20

If the analysis uses only those students who claim to have persevered with their mouthwash, then the estimated disease rate for B is between 2.19% and 2.33% while that for A is between 1.02% and 2.67%, depending on how many students admit that they have given up. Thus which mouthwash appears to be better depends on the willingness of the students to admit that they have not been diligent. On the other hand, if the analysis uses all the students then A appears worse than B, and this is probably the correct conclusion if a mouthwash is to be recommended to the general public.

Questions for discussion

7.1 Describe the sort of design that would be most appropriate for each of the following situations:

  (a) investigating a new drug which purports to reduce high blood pressure;

  (b) investigating a new 'miracle' cure for the common cold;

  (c) finding out what teaching method is best for brilliant young mathematicians who go to university at age 14 or less;

  (d) comparing four diets for piglets to see which gives the greatest weight gain in the first six weeks of life;

  (e) comparing three winter feed supplements for sheep to see how they affect the quality of the wool shorn the following spring;

  (f) comparing two styles of conducting large amateur choirs;

  (g) assessing whether 'brisk' or 'motherly' policewomen obtain the most useful information from rape victims in the interview immediately after the rape;

  (h) seeing whether either of two proposed new drugs to help people give up smoking has any effect.

7.2 How would you investigate the following claims?

(a) Dentists are more likely to suffer mercury poisoning than other people.

(b) For men aged over 50, taking one aspirin a day helps to prevent stroke.

(c) Vegetarians live longer than omnivores.

(d) Children whose fathers were working at nuclear reprocessing plants at or before the time of conception are more susceptible to leukaemia than the general population.

(e) Publishing 'league tables' of schools by examination results actually decreases the amount that schoolchildren learn.

(f) Cats are healthier if given skim milk rather than full milk.

(g) People who take more exercise in their teens will suffer less from serious diseases in their fifties.

(h) The more older brothers that a man has, the more likely he is to be gay.

(i) Wild badgers spread bovine tuberculosis.

(j) People who grow up living within 500 metres of a motorway have significantly reduced lung capacity as adults.

7.3 A physician wants to compare a new treatment with the current standard treatment for a certain chronic illness. He hopes to detect a difference in response of 15 units, so he would like 90% power for detecting this difference when he does a hypothesis test at the 5% significance level. From his previous experience, he estimates that the variance σ^2 is approximately given by $\sigma = 10$ for a completely randomized design, $\sigma = 9$ for a matched-pairs design, and $\sigma = 7$ for a cross-over design with two periods. Calculate the minimum number of people needed for an experiment with each of the three types of design.

7.4 In a toxicology experiment, the toxin bromobenzene is administered to rats. The toxicologist is interested in how the body deals with the toxin over time. For example, if the toxin affects normal liver function, does the liver ever return to normal? If so, when?

Three doses of bromobenzene (low, medium and high) are each diluted in a standard quantity of corn oil. Nine rats are fed each dose. Nine further rats are fed the same quantity of corn oil, with no added bromobenzene, and nine more have no addition to their normal diet.

Six hours after administration of the corn oil (or nothing), three rats are randomly chosen from each group of nine: they are sacrificed, and their livers extracted and measured. The same thing happens at 24 hours after administration, and again at 48 hours.

(a) What are the experimental units?

(b) What are the observational units?

(c) What are the treatments?

(d) Explain why some corn oil is used without bromobenzene.

Questions for discussion

7.5 In the context of the previous question, another scientist suggests that measurements on the rats' urine can provide just as much information as measurements on their livers. Then there is no need to sacrifice rats. Moreover, the experiment can be run using only fifteen rats, three for each of the five diets. A sample of urine will be taken from each rat at six, 24 and 48 hours after administration.

(a) What are the experimental units?

(b) What are the observational units?

(c) What are the treatments?

(d) What are the advantages and disadvantages of this design compared to the design in Question 7.4?

7.6 In the issue of the *New Scientist* dated 13 January 2007, there were three news items about investigations on people. In each case, say what sort of investigation it was, comment on what you think the design and randomization were (if any), and discuss any ethical issues involved.

(a) On page 17, an opinion piece 'The curse of being different' by Robert Adler reported as follows.

> Between 2003 and 2006 Ilan Dar-Nimrod and Steven Heine at the University of British Columbia in Vancouver, Canada, tested the mathematical ability of 220 female college students (*Science*, vol 314, p 435). Before taking a maths test similar to one used by many colleagues to screen applicants to graduate programmes, some of the women read passages arguing that there are fixed gender differences in mathematical ability, while others read that differences in ability can be modified by experience. The researchers predicted that viewing ability as changeable would make it easier for women to overcome the negative stereotype that paints maths as a predominantly male pursuit.
>
> They were right: the "changeable" intervention raised women's maths scores by an astonishing 50 per cent.

(b) A short item on pages 4–5, entitled 'Perils of freedom' started as follows.

> Being released from prison may be far riskier than anyone thought. Former inmates of Washington state prison were 3.5 times more likely to die during their first two years after release than people of the same age, sex and race who had never been incarcerated.
>
> The danger was greatest during the first two weeks, when former inmates were 13 times more likely to die than their non-incarcerated peers (*The New England Journal of Medicine*, vol 356, p 157).
>
> The study tracked 30,327 former inmates released between July 1999 and December 2003.

(c) Here is the whole of the '60 seconds' item on page 5 called 'Milk spoils tea party'.

> Adding milk to tea may blunt its cardiovascular benefits. A study of 16 volunteers published in the *European Heart Journal* suggests that the effects of catechins—ingredients in tea thought to combat heart disease by dilating blood vessels—were blocked by casein proteins in milk.

7.7 An experiment was conducted to assess the effect of education about a certain disease. A large selection of volunteers was split into two groups of equal size. The splitting was done carefully, in such a way that the two groups were well matched in terms of proportion of males, age distribution and so on. A coin was tossed (in front of an independent witness!) to decide which group should receive the education. The chosen group was then split into subgroups of twenty people. Each subgroup had a one-hour session with a lecture about the disease followed by questions and discussion.

The subsequent incidence of the disease among the entire collection of volunteers was monitored.

How should this experiment have been designed and randomized? How should it have been analysed?

Chapter 8

Small units inside large units

8.1 Experimental units bigger than observational units

8.1.1 The context

Example 8.1 (Example 1.2 continued: Calf feeding) Four feed treatments are compared on 80 calves. The calves are kept in eight pens, with ten calves per pen. Each feed is applied to two whole pens, with calves feeding ad lib. Figure 8.1 shows a typical layout.

Even though every calf is weighed individually, differences between feeds should be compared with differences between pens, not with differences between calves. So should we say that the replication is 20 or 2?

Some people say that the *replication* is 2 and the *repetition* is 10.

Example 1.10 is similar: treatments are applied to whole classes but individual children are measured. So is one version of Example 1.13: pullets are fed but eggs are weighed. So are Examples 7.10 and 7.11: treatments are applied to whole doctors' practices or to villages, but it is individual people that are measured.

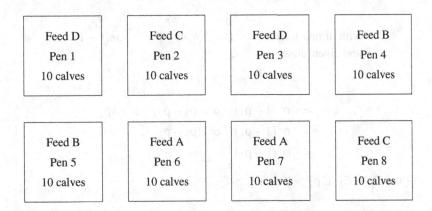

Fig. 8.1. Typical layout of the experiment in Example 8.1

To generalize this example, suppose that there are m experimental units, each of which consists of k observational units, and that there are t treatments, each of which is applied to c experimental units, so that $ct = m$.

8.1.2 Construction and randomization

This is just like the completely randomized design in Section 2.1, using the experimental units instead of the plots.

8.1.3 Model and strata

For simplicity, I use the language of Example 8.1. If pens give fixed effects then we cannot estimate anything about treatments, because any difference in response caused by different treatments could be attributed to differences between pens. Therefore we assume that pens give random effects. This gives a model like the random-effects model in Sections 4.4 and 4.6:

$$\mathbb{E}(Y_\omega) = \tau_{T(\omega)}$$

for each calf ω, and

$$\text{cov}(Y_\alpha, Y_\beta) = \begin{cases} \sigma^2 & \text{if } \alpha = \beta \\ \rho_1 \sigma^2 & \text{if } \alpha \text{ and } \beta \text{ are different calves in the same pen} \\ \rho_2 \sigma^2 & \text{if } \alpha \text{ and } \beta \text{ are in different pens.} \end{cases}$$

Usually $\rho_1 > \rho_2 > 0$ unless there is so little food that there is competition among calves in the same pen.

Put

$$V_P = \{\text{vectors in } V \text{ which take a constant value on each pen}\}$$

and

$$W_P = V_P \cap V_0^\perp.$$

From Section 4.6, we see that the eigenspaces of $\text{Cov}(\mathbf{Y})$ are

	V_0	W_P	$V_P^\perp$
with dimensions	1	$m-1$	$m(k-1)$
and eigenvalues	ξ_0	ξ_1	ξ_2,

where

$$\begin{aligned} \xi_0 &= \sigma^2(1-\rho_1) + \sigma^2 k(\rho_1 - \rho_2) + \sigma^2 mk\rho_2 \\ \xi_1 &= \sigma^2(1-\rho_1) + \sigma^2 k(\rho_1 - \rho_2) \\ \xi_2 &= \sigma^2(1-\rho_1). \end{aligned}$$

If $\rho_1 > \rho_2$ then $\xi_1 > \xi_2$; if $\rho_2 > 0$ then $\xi_0 > \xi_1$.

8.1.4 Analysis

Here I show how to build up the analysis-of-variance table in stages, so that it does not have to be memorized. The steps are shown in a diagram in Figure 8.2.

8.1. Experimental units bigger than observational units

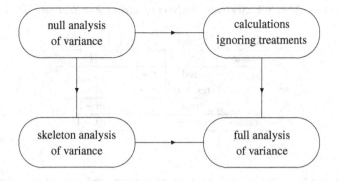

Fig. 8.2. The four steps in creating an analysis-of-variance table

The first step is to write out the list of strata and their degrees of freedom in a table, with one row for each stratum. This table is called the *null analysis of variance*. It is shown in Table 8.1.

The next two steps can be done in either order. The second step can be done before treatments are thought about. Simply take the rows of the null analysis-of-variance table, and show the calculations associated with each one. See Table 8.2. The first column contains sums of squares. As usual, the sum of squares for the mean is sum^2/N. The sum of squares for pens, SS(pens), is equal to CSS(pens) − SS(mean), where

$$\text{CSS(pens)} = \sum_{i=1}^{m} \frac{\left(\text{sum}_{\text{pen}=i}\right)^2}{k}.$$

The total sum of squares is

$$\sum_{\omega \in \Omega} y_\omega^2.$$

Then the sum of squares for calves is obtained by subtraction, so that it is equal to

$$\sum_{\omega \in \Omega} y_\omega^2 - \text{CSS(pens)}.$$

The second column contains the expected mean squares if there are no treatments. These are just the appropriate eigenvalues of $\text{Cov}(\mathbf{Y})$.

Table 8.1. *Null analysis of variance in Section 8.1*

Table 8.2. *Calculations ignoring treatments in Section 8.1*

Stratum	df	SS	EMS
mean	1	sum^2/N	ξ_0
pens	$m-1$	$\sum_i \left(\text{sum}_{\text{pen}=i}\right)^2/k - \text{SS(mean)}$	ξ_1
calves	$m(k-1)$	$\sum y_\omega^2 - \text{CSS(pens)}$	ξ_2
Total	$mk = N$	$\sum y_\omega^2$	

Table 8.3. *Skeleton analysis of variance in Section 8.1*

Stratum	Source	Degrees of freedom	
mean	mean		1
pens	feed	$t-1$	
	residual	$m-t$	
	total		$m-1$
calves	calves		$m(k-1)$
Total			N

The third step is to take the null analysis of variance and expand it to show where treatments lie. This gives the *skeleton analysis of variance*, shown in Table 8.3. An extra column for source is inserted after the stratum column, and the degrees-of-freedom column is split into two. If there is any stratum that contains no treatment subspace (except for V_0), then its name is copied into the source column and its degrees of freedom are written on the right-hand side of the 'df' column. This rule gives the rows labelled 'mean' and 'calves' in Table 8.3. The next lemma shows why there are no treatment subspaces in the calves stratum.

Lemma 8.1 *If treatments are applied to whole pens then $W_T \subseteq W_P$.*

Proof Let $\mathbf{v}$ be in V_T. If α and β are in the same pen then $T(\alpha) = T(\beta)$ so $v_\alpha = v_\beta$ because $\mathbf{v} \in V_T$. Hence $\mathbf{v} \in V_P$. Therefore $V_T \subseteq V_P$, so $W_T = V_T \cap V_0^\perp \subseteq V_P \cap V_0^\perp = W_P$. ∎

The subspaces mentioned in the proof are shown in Figure 8.3.

Since the pens stratum contains W_T, which is the subspace for differences between feeds, the row labelled 'pens' in the null analysis-of-variance table has to be split into three, with names shown in the source column. The first is W_T, labelled 'feed', whose degrees of freedom are shown on the left-hand side of the 'df' column. The third is labelled 'total'; it is really a copy of the 'pens' row in Table 8.1 so its degrees of freedom are shown on the right of the 'df' column. In between is the source labelled 'residual', which is the space $W_P \cap W_T^\perp$. Its degrees of freedom, shown on the left of the 'df' column, are calculated by subtraction as $(m-1) - (t-1) = m-t$.

The fourth step is to combine Tables 8.2 and 8.3 and insert extra information to give Table 8.4, which is the *full* analysis of variance. The sums of squares for the mean and calves strata are copied from Table 8.2 to Table 8.4. In the pens stratum, the sum of squares is copied from Table 8.2 to the 'total' line of the pens stratum in Table 8.4. The sum of squares for treatments is calculated in the usual way as

$$\text{SS(feed)} = \sum_{i=1}^{t} \frac{(\text{sum}_{T=i})^2}{ck} - \frac{\text{sum}^2}{N}.$$

The sum of squares for residual is obtained by subtraction:

$$\text{SS(residual)} = \text{SS(total)} - \text{SS(feed)}.$$

If there is space, it is useful to split the 'sum of squares' column into two just like the 'df' column.

8.1. Experimental units bigger than observational units

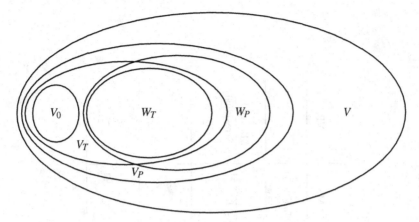

Fig. 8.3. Subspaces in Lemma 8.1

Expected mean squares are copied from Table 8.2 to Table 8.4, being put in the right-hand side of the 'EMS' column. Thus ξ_0 is written for the mean, ξ_2 for calves, and ξ_1 for both of the non-total sources in the pens stratum. To these stratum variances must be added sums of squares of expected values. As in Section 4.5, $\mathbb{E}(\mathbf{P}_{V_0}\mathbf{Y}) = \boldsymbol{\tau}_0$ and $\mathbb{E}(\mathbf{P}_{W_T}\mathbf{Y}) = \boldsymbol{\tau}_T$, so Theorem 2.11(ii) shows that $\text{EMS}(\text{mean}) = \|\boldsymbol{\tau}_0\|^2 + \xi_0$,

$$\text{EMS}(\text{feed}) = \frac{\|\boldsymbol{\tau}_T\|^2}{t-1} + \xi_1,$$

$\text{EMS}(\text{residual}) = \xi_1$ and $\text{EMS}(\text{calves}) = \xi_2$.

The final column shows what is the appropriate variance ratio, if any. As usual, there is no independent estimate of ξ_0, so there is no test for whether $\boldsymbol{\tau}_0$ is zero. If $\boldsymbol{\tau}_T = \mathbf{0}$ then $\text{EMS}(\text{feed}) = \text{EMS}(\text{residual})$ so the appropriate variance ratio to test for differences between feeds is

$$\frac{\text{MS}(\text{feed})}{\text{MS}(\text{residual})}.$$

8.1.5 Hypothesis testing

Test for differences between feeds by using the variance ratio

$$\frac{\text{MS}(\text{feed})}{\text{MS}(\text{residual})}.$$

The appropriate numbers of degrees of freedom for the F-test are $t-1$ and $m-t$.

If $c = 1$ then $m = t$ and so there is no residual in the pens stratum and so no test can be done. In this case people sometimes wrongly compare $\text{MS}(\text{feed})$ to $\text{MS}(\text{calves})$. If $\rho_1 > \rho_2$ then $\xi_1 > \xi_2$ so this test is likely to detect differences that do not really exist. This is known as *false replication* or *pseudo-replication*, because the replication of calves within the pens is falsely taken as a substitute for the replication of pens within treatments.

Probably the single most common mistake in designing experiments is to allocate treatments to large units with $c = 1$. This is the mistake in the ladybirds experiment in Example 1.1.

Table 8.4. *Full analysis of variance in Section 8.1*

Stratum	Source	df	SS	EMS	VR
mean	mean	1	$\dfrac{\text{sum}^2}{N}$	$\|\tau_0\|^2 + \xi_0$	—
pens	feed	$t-1$	$\sum \dfrac{(\text{sum}_{T=i})^2}{sk} - \dfrac{\text{sum}^2}{N}$	$\dfrac{\|\tau_T\|^2}{t-1} + \xi_1$	$\dfrac{\text{MS(feed)}}{\text{MS(residual)}}$
	residual		←············ by subtraction ············→	ξ_1	
	total	$m-1$	$\sum \dfrac{(\text{sum}_{P=i})^2}{k} - \dfrac{\text{sum}^2}{N}$		
calves	calves	$m(k-1)$	$\sum y_\omega^2 - \sum \dfrac{(\text{sum}_{P=i})^2}{k}$	ξ_2	—
Total		N	$\sum_\omega y_\omega^2$		

8.1. Experimental units bigger than observational units

Example 8.2 (Insemination of cows) An experiment on fertility in cows used about 300 cows, on two farms. All the cows on one farm were inseminated naturally; all the cows on the other farm had artificial insemination. Thus there was no way to tell whether differences in fertility were caused by the different methods of insemination or by other differences between the two farms.

Even when $c \neq 1$, some people mistakenly combine the sum of squares for residual in the pens stratum with the calves sum of squares, to give a source whose expected mean square is equal to
$$\frac{(m-t)\xi_1 + m(k-1)\xi_2}{N-t}.$$
This is smaller than ξ_1 if $\rho_1 > \rho_2$, so again there is spurious precision. There are also more degrees of freedom than there should be, which leads to spurious power.

Note that increasing k does not increase the number of degrees of freedom for residual.

8.1.6 Decreasing variance

Theorem 2.11(iv) shows that the variance of the estimator of a simple treatment difference $\tau_i - \tau_j$ is equal to
$$\frac{2}{ck}\xi_1 = \frac{2}{c}\sigma^2 \left[\frac{(1-\rho_1)}{k} + (\rho_1 - \rho_2)\right],$$
which is estimated by $(2/ck)\,\text{MS}(\text{residual})$. To decrease the variance, it is more fruitful to increase c than to increase k, even though the experimenter may find it simpler to increase k.

Example 8.1 revisited (Calf feeding) Plausible values for the two correlations are $\rho_1 = 0.3$ and $\rho_2 = 0$. Then the variance of the estimator of a difference is equal to
$$2\sigma^2 \left[\frac{0.7}{ck} + \frac{0.3}{c}\right].$$
Some combinations of values of c and k give the following coefficients of $2\sigma^2$.

	$k=7$	$k=10$	$k=15$
$c=2$	0.200	0.185	0.173
$c=3$	0.133	0.123	0.116

The original design has variance $2\sigma^2 \times 0.185$. Increasing the number of calves per pen from 10 to 15 only decreases the variance to $2\sigma^2 \times 0.173$, even though 50% more feed is used. On the other hand, replacing 8 pens of 10 calves by 12 pens of 7 calves decreases the variance substantially to $2\sigma^2 \times 0.133$ while using almost the same number of calves and almost the same amount of feed. However, the extra four pens may not be available, or they may be expensive to set up.

It is almost always easier for the experimenter to randomize treatments to a few large units than to many smaller units. Part of the statistician's job is to explain the loss of precision and power that will result from this.

Table 8.5. *Skeleton analysis of variance when types of hay and types of anti-scour are both applied to whole pens*

Stratum	Source	Degrees of freedom
mean	mean	1
pens	hay	1
	anti-scour	1
	hay $\wedge$ anti-scour	1
	residual	4
	total	7
calves	calves	72
Total		80

8.2 Treatment factors in different strata

Example 8.1 revisited (Calf feeding) Unless calves are given some kind of anti-scour treatment in their milk, they are liable to get gut infections which cause them much diarrhoea, and they then lose weight. Suppose that the four feeds consist of all combinations of two types of hay, which is put directly into the pen, with two types of anti-scour treatment, which are given to calves individually in their morning and evening pails of milk. If all calves in the same pen have the same type of anti-scour then the design and analysis are just as in Section 8.1, except that the treatments line in the analysis of variance is split into three, giving the skeleton analysis of variance in Table 8.5.

It might be better to give five calves in each pen one type of anti-scour and the other five calves the other type of anti-scour.

In general, suppose that there are m large units, each of which consists of k small units; and that there are n_H levels of factor H, each of which is applied to r_H large units (so that $n_H r_H = m$); and that there are n_A levels of treatment factor A, each of which is applied to r_A small units per large unit (so that $n_A r_A = k$). Now the number of treatments is $n_H n_A$ and they all have replication $r_H r_A$. Thus $N = mk = n_H n_A r_H r_A$.

The construction and randomization are as follows.

(i) Apply levels of H to large units just as in a completely randomized design.

(ii) Within each large unit independently, apply levels of A just as in a completely randomized design.

The model is the same as in Section 8.1. So are the strata, the null analysis of variance and the calculations ignoring treatments. However, we need to find out which strata contain the three treatment subspaces.

Let L be the factor for large units, so that $L(\omega)$ is the large unit which contains small unit ω. Define subspaces V_L and W_L like the subspaces V_P and W_P in Section 8.1.

8.2. Treatment factors in different strata

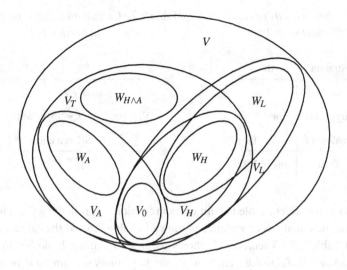

Fig. 8.4. Subspaces in Theorem 8.2

Theorem 8.2 *If levels of H are applied to large units with equal replication and levels of A are applied to small units with equal replication within each large unit, then*

(i) $W_H \subseteq W_L$;

(ii) $W_A \subseteq V_L^\perp$;

(iii) $W_{H \wedge A} \subseteq V_L^\perp$.

Proof (i) Use Lemma 8.1.

(ii) and (iii) Let Δ be any large unit. Suppose that H has level h on Δ. Let $\mathbf{v}$ be any vector in W_A or $W_{H \wedge A}$. Then $\mathbf{v} \in V_T$ and $\mathbf{v} \in V_H^\perp$. The r_H large units on which H has level h all have the same treatments, so they all have the same values of $\mathbf{v}$, although probably in different orders. Therefore

$$\sum_{\omega \in \Gamma} v_\omega = r_H \sum_{\omega \in \Delta} v_\omega, \qquad (8.1)$$

where Γ is the union of the large units on which H takes level h. Let $\mathbf{u}_h$ be the vector whose entry on plot ω is equal to

$$\begin{cases} 1 & \text{if } \omega \in \Gamma, \text{ that is, if } H(\omega) = h \\ 0 & \text{otherwise,} \end{cases}$$

and define the vector $\mathbf{w}_\Delta$ analogously. Equation (8.1) can be rewritten as

$$\mathbf{v} \cdot \mathbf{u}_h = r_H \mathbf{v} \cdot \mathbf{w}_\Delta.$$

Now, $\mathbf{u}_h$ is in V_H and $\mathbf{v}$ is orthogonal to V_H, so $\mathbf{v} \cdot \mathbf{u}_h = 0$ so $\mathbf{v} \cdot \mathbf{w}_\Delta = 0$. This is true for all large units Δ, and V_L is spanned by the vectors $\mathbf{w}_\Delta$ as Δ runs over the large units, so $\mathbf{v}$ is orthogonal to V_L. ■

These subspaces are all shown in Figure 8.4.

Table 8.6. *Null analysis of variance in Section 8.2*

Stratum	df
mean	1
large units	$m-1$
small units	$m(k-1)$
Total	$mk = N$

Table 8.7. *Calculations ignoring treatments in Section 8.2*

SS	EMS
sum^2/N	ξ_0
$\sum_i (\text{sum}_{L=i})^2/k - \text{SS(mean)}$	ξ_1
$\sum y_\omega^2 - \text{CSS(large units)}$	ξ_2
$\sum y_\omega^2$	

The analysis-of-variance table is built up as in Section 8.1, following the scheme given in Figure 8.2. The null analysis of variance is shown in Table 8.6 and the calculations ignoring treatments in Table 8.7. Theorem 8.2 shows us how to expand Table 8.6 to the skeleton analysis of variance in Table 8.8. The row for the large units stratum must be split to give a line for H, a residual line and a total line. The total line is the same as in the null analysis of variance, and the residual is the difference between total and H. The row for the small units stratum must be split even further: it has to show two treatment lines, the main effect of A and the H-by-A interaction; the residual is obtained by subtracting both of these from the total. Note that we now have two separate lines called 'residual', one in the large units stratum and one in the small units stratum. Where necessary, I distinguish these by writing, for example, 'large units residual' in full.

Tables 8.7 and 8.8 are combined to give the full analysis of variance in Table 8.9. Here $\tau_0 = \mathbf{P}_{V_0}\tau$, $\tau_H = \mathbf{P}_{W_H}\tau$, $\tau_A = \mathbf{P}_{W_A}\tau$, and $\tau_{HA} = \mathbf{P}_{W_{H\wedge A}}\tau$. Moreover,

$$\text{CSS}(L) = \sum_{l=1}^{m} \frac{(\text{sum}_{L=l})^2}{k},$$

$$\text{SS}(H) = \sum_{i=1}^{n_H} \frac{(\text{sum}_{H=i})^2}{r_H k} - \frac{\text{sum}^2}{N},$$

$$\text{SS}(A) = \sum_{j=1}^{n_A} \frac{(\text{sum}_{A=j})^2}{r_A m} - \frac{\text{sum}^2}{N}$$

and

$$\text{SS}(H \wedge A) = \sum_{i=1}^{n_H} \sum_{j=1}^{n_A} \frac{(\text{sum}_{H=i,A=j})^2}{r_H r_A} - \frac{\text{sum}^2}{N} - \text{SS}(H) - \text{SS}(A).$$

Now differences between levels of H are assessed against the variability of large units. Testing uses

$$\frac{\text{MS}(H)}{\text{MS(large units residual)}}.$$

The standard error of a difference between levels of H is

$$\sqrt{\frac{2}{r_H k} \text{MS(large units residual)}},$$

8.2. Treatment factors in different strata

Table 8.8. *Skeleton analysis of variance in Section 8.2*

Stratum	Source	Degrees of freedom
mean	mean	1
large units	H	$n_H - 1$
	residual	$m - n_H$
	total	$m - 1$
small units	A	$n_A - 1$
	$H \wedge A$	$(n_H - 1)(n_A - 1)$
	residual	$m(k-1) - n_H(n_A - 1)$
	total	$m(k-1)$
Total		N

and the variance of the estimator of such a difference is $2\xi_1/r_H k$.

The main effect of A and the H-by-A interaction are assessed against the variability of small units within large units. The tests use the variance ratios

$$\frac{\text{MS}(A)}{\text{MS(small units residual)}}$$

and

$$\frac{\text{MS}(H \wedge A)}{\text{MS(small units residual)}}$$

respectively.

The variance of the estimator of the difference between two levels of A is $2\xi_2/r_A m$, so the standard error of such a difference is

$$\sqrt{\frac{2}{r_A m} \text{MS(small units residual)}}.$$

Standard errors of differences between individual treatments require a little more care. For $i = 1, \ldots, n_H$ and $j = 1, \ldots, n_A$, let $\mathbf{u}_{ij}$ be the vector in V_T whose coordinate on plot ω is equal to

$$\begin{cases} 1 & \text{if } H(\omega) = i \text{ and } A(\omega) = j \\ 0 & \text{otherwise.} \end{cases}$$

Similarly, let $\mathbf{v}_i$ be the vector whose coordinate on ω is equal to

$$\begin{cases} 1 & \text{if } H(\omega) = i \\ 0 & \text{otherwise} \end{cases}$$

and $\mathbf{w}_j$ the vector whose coordinate on ω is equal to

$$\begin{cases} 1 & \text{if } A(\omega) = j \\ 0 & \text{otherwise.} \end{cases}$$

Table 8.9. *Full analysis of variance in Section 8.2*

Stratum	Source	df	SS	EMS	VR
mean	mean	1	$\dfrac{\text{sum}^2}{N}$	$\|\boldsymbol{\tau}_0\|^2 + \xi_0$	—
large units	H	$n_H - 1$	$SS(H)$	$\dfrac{\|\boldsymbol{\tau}_H\|^2}{n_H - 1} + \xi_1$	$\dfrac{MS(H)}{MS(\text{large units residual})}$
large units	residual	←⋯⋯⋯⋯ by subtraction ⋯⋯⋯⋯→		ξ_1	—
large units	total	$m - 1$	$CSS(L) - \dfrac{\text{sum}^2}{N}$		
small units	A	$n_A - 1$	$SS(A)$	$\dfrac{\|\boldsymbol{\tau}_A\|^2}{n_A - 1} + \xi_2$	$\dfrac{MS(A)}{MS(\text{small units residual})}$
small units	$H \wedge A$	$(n_H - 1)(n_A - 1)$	$SS(H \wedge A)$	$\dfrac{\|\boldsymbol{\tau}_{HA}\|^2}{(n_H - 1)(n_A - 1)} + \xi_2$	$\dfrac{MS(H \wedge A)}{MS(\text{small units residual})}$
small units	residual	←⋯⋯⋯⋯ by subtraction ⋯⋯⋯⋯→		ξ_2	—
small units	total	$m(k - 1)$	$\sum y_\omega^2 - CSS(L)$		
Total		N	$\sum_\omega y_\omega^2$		

8.2. Treatment factors in different strata

Fix levels i, i' of H and j, j' of A, with (i, j) different from (i', j'). Put $\mathbf{u} = (\mathbf{u}_{ij} - \mathbf{u}_{i'j'})/r$, where $r = r_H r_A$. Then $\mathbf{u} \in V_0^\perp$. Let $\mathbf{v}$ be the projection of $\mathbf{u}$ onto V_H. Theorem 2.4(vii) shows that $\mathbf{v} = (\mathbf{v}_i - \mathbf{v}_{i'})/rn_A$, which is in W_H, so $\mathbf{v} = \mathbf{P}_{W_H}\mathbf{u}$. Similarly, if $\mathbf{w} = \mathbf{P}_{W_A}\mathbf{u}$ then $\mathbf{w} = (\mathbf{w}_j - \mathbf{w}_{j'})/rn_H$. Therefore

$$\mathbf{u} = \mathbf{v} + \mathbf{w} + \mathbf{x}, \tag{8.2}$$

where $\mathbf{x} \in W_{H \wedge A}$.

The vectors on the right-hand side of Equation (8.2) are mutually orthogonal, and so

$$\|\mathbf{x}\|^2 = \|\mathbf{u}\|^2 - \|\mathbf{v}\|^2 - \|\mathbf{w}\|^2.$$

Now, $\|\mathbf{u}\|^2 = 2/r$; $\|\mathbf{v}\|^2$ is equal to $2/rn_A$ if $i \neq i'$ and to zero otherwise; and $\|\mathbf{w}\|^2$ is equal to $2/rn_H$ if $j \neq j'$ and to zero otherwise. Therefore

$$\|\mathbf{x}\|^2 = \begin{cases} 2(n_H n_A - n_H - n_A)/N & \text{if } i \neq i' \text{ and } j \neq j' \\ 2n_A(n_H - 1)/N & \text{if } i = i' \text{ and } j \neq j' \\ 2n_H(n_A - 1)/N & \text{if } i \neq i' \text{ and } j = j'. \end{cases}$$

Parts (iii) and (iv) of Theorem 2.11 show that $\mathbf{v} \cdot \mathbf{Y}$ is the best linear unbiased estimator of $\mathbf{v} \cdot \boldsymbol{\tau}$ and that it has variance $\|\mathbf{v}\|^2 \xi_1$. Similarly, $\mathbf{w} \cdot \mathbf{Y}$ is the best linear unbiased estimator of $\mathbf{w} \cdot \boldsymbol{\tau}$ and it has variance $\|\mathbf{w}\|^2 \xi_2$, and $\mathbf{x} \cdot \mathbf{Y}$ is the best linear unbiased estimator of $\mathbf{x} \cdot \boldsymbol{\tau}$ and it has variance $\|\mathbf{x}\|^2 \xi_2$.

Hence $\mathbf{u} \cdot \mathbf{Y}$ is the best linear unbiased estimator of $\mathbf{u} \cdot \boldsymbol{\tau}$, which is equal to $\tau_{ij} - \tau_{i'j'}$. Theorem 2.11(v) shows that three estimators are uncorrelated, and so

$$\text{Var}(\mathbf{u} \cdot \mathbf{Y}) = \|\mathbf{v}\|^2 \xi_1 + \|\mathbf{w}\|^2 \xi_2 + \|\mathbf{x}\|^2 \xi_2.$$

If $i = i'$ while $j \neq j'$ then

$$\text{Var}(\mathbf{u} \cdot \mathbf{Y}) = \left[\frac{2}{rn_H} + \frac{2n_A(n_H - 1)}{N}\right] \xi_2 = \frac{2n_A}{N}[1 + (n_H - 1)]\xi_2 = \frac{2n_A n_H}{N}\xi_2 = \frac{2}{r}\xi_2.$$

If $i \neq i'$ but $j = j'$ then

$$\text{Var}(\mathbf{u} \cdot \mathbf{Y}) = \frac{2\xi_1}{rn_A} + \frac{2n_H(n_A - 1)\xi_2}{N} = \frac{2}{N}[n_H \xi_1 + n_H(n_A - 1)\xi_2] = \frac{2}{rn_A}[\xi_1 + (n_A - 1)\xi_2].$$

If $i \neq i'$ and $j \neq j'$ then

$$\text{Var}(\mathbf{u} \cdot \mathbf{Y}) = \frac{2\xi_1}{rn_A} + \frac{2\xi_2}{rn_H} + \frac{2(n_H n_A - n_H - n_A)\xi_2}{N} = \frac{2}{N}[n_H \xi_1 + n_H(n_A - 1)\xi_2]$$

$$= \frac{2}{rn_A}[\xi_1 + (n_A - 1)\xi_2].$$

Thus the standard error of a difference is equal to

$$\sqrt{\frac{2}{r}\text{MS(small units residual)}}$$

if the levels of H are the same; otherwise it is equal to

$$\sqrt{\frac{2}{rn_A}[\text{MS(large units residual)} + (n_A - 1)\text{MS(small units residual)}]}.$$

Table 8.10. *Skeleton analysis of variance when types of hay are applied to whole pens and each type of anti-scour is given to five calves per pen*

Stratum	Source	Degrees of freedom
mean	mean	1
pens	hay	1
	residual	6
	total	7
calves	anti-scour	1
	hay ∧ anti-scour	1
	residual	70
	total	72
Total		80

Since we expect ξ_2 to be smaller than ξ_1, this design gives more precise estimates of differences between levels of A than the design in Section 8.1. It also gives more precise estimates of differences between any two levels of $H \wedge A$. There is increased power for testing the main effect of A and the H-by-A interaction, for not only is ξ_2 smaller than ξ_1 but the number of residual degrees of freedom in the small units stratum is substantially more than the number of residual degrees of freedom in the large units stratum in the design in Section 8.1. There is even a slight increase in power for testing the main effect of H, because the number of residual degrees of freedom in the large units stratum increases from $m - n_H n_A$ to $m - n_H$.

Example 8.1 revisited (Calf feeding) If each type of hay is applied to four whole pens and each type of anti-scour is given to five calves per pen, then we obtain the skeleton analysis of variance in Table 8.10. Compare this with Table 8.5. The improved precision for anti-scour and for hay ∧ anti-scour is immediately evident.

To see the improved power for testing for the main effect of hay, we argue as we did in Sections 2.13 and 4.8. There is one degree of freedom for hay, and an F-test with one degree of freedom for the numerator is equivalent to a two-sided t-test on the square root of the statistic. The vector τ_H has just two distinct entries, one for each type of hay. Let δ be the modulus of their difference.

The design summarized in Table 8.5 has 4 residual degrees of freedom in the pens stratum. If we test at the 5% significance level then we need the 0.975 point of the t-distribution on 4 degrees of freedom, which is 2.776. The 0.90 point is 1.553. Thus the argument of Section 2.13 shows that to have probability at least 0.9 of detecting that the main effect of hay is nonzero we need

$$\delta > (2.776 + 1.533) \times \sqrt{\frac{2}{40}\xi_1} = 4.309\sqrt{\frac{\xi_1}{20}}.$$

On the other hand, the design summarized in Table 8.10 has 6 residual degrees of freedom in the pens stratum. The 0.975 and 0.90 points of the t-distribution on 6 degrees of freedom are 2.447 and 1.440 respectively. In order to have probability at least 0.9 of detecting that the main effect of hay is nonzero, we need

$$\delta > (2.447 + 1.440) \times \sqrt{\frac{2}{40}\xi_1} = 3.887\sqrt{\frac{\xi_1}{20}}.$$

8.2. Treatment factors in different strata

Table 8.11. *Data for Example 8.3*

Block	Spray	Pruning method					Block total	Spray total
		1	2	3	4	5		
1	a	46.38	30.66	21.62	47.24	28.41	174.31	
2	a	71.80	40.26	48.08	49.68	39.71	249.53	423.84
3	b	54.05	26.64	28.60	40.38	22.68	172.35	
4	b	29.47	50.00	24.10	40.23	21.51	165.31	337.66
5	c	56.57	48.65	43.51	45.23	30.08	224.04	
6	c	62.50	46.70	50.37	57.65	45.23	262.45	486.49
Pruning total		320.77	242.91	216.28	280.41	187.62	1247.99	

Thus the second design can detect a difference between the types of hay which is 10% smaller than the difference detectable by the first design.

Example 8.3 (Cider apples) In the 1920s and 1930s, Long Ashton Research Station conducted a long-term factorial experiment on cider-apple trees in twelve complete blocks. The treatment factors were five methods of pruning and several varieties of apple. In one year, they decided to do a limited experiment to see if spraying the crop with a very weak solution of naphthalene would help to prevent the apples dropping to the ground before being picked. They limited the new experiment to half of the orchard (six blocks) and the trees of a single variety (King Edward VII). There were three spray treatments:

   a   spray at 5 parts per million
   b   spray at 10 parts per million
   c   no spray.

Each spray treatment was applied in September to all the trees of variety King Edward VII in each of two whole blocks.

In October there was an unusually strong gale, during which many of the apples fell. The next day, the experimenters counted how many apples had fallen from each tree, and how many remained. Table 8.11 shows the number fallen as a percentage of the total for the tree. The data are presented in treatment order; rows are blocks.

The calculations ignoring treatments give the following sums of squares:

$$\begin{aligned}
\text{mean} &\quad\quad 51915.97 \\
\text{blocks} &\quad 53751.01 - 51915.97 = 1835.04 \\
\text{trees} &\quad 56758.71 - 53751.01 = 3007.70
\end{aligned}$$

These lead to the analysis of variance in Table 8.12.

Table 8.12. *Analysis of variance in Example 8.3*

Stratum	Source	SS	df	MS	VR
mean	mean	51915.97	1	51915.97	–
blocks	sprays	1116.75	2	558.37	2.33
	residual	718.29	3	239.43	–
	total	1835.04	5		
trees	pruning	1835.15	4	458.79	6.19
	sprays $\wedge$ pruning	284.13	8	35.52	0.48
	residual	888.42	12	74.04	–
	total	3007.70	24		
Total		56758.71	30		

The 90% point of F on 2 and 3 degrees of freedom is 5.46, so we have insufficient evidence to reject the null hypothesis of no difference between the sprays. This is not really surprising, given the design employed. On the other hand, we can conclude that there is no interaction between sprays and pruning, and that there are differences between the methods of pruning. Their means are

pruning method	1	2	3	4	5
mean	53.46	40.48	36.05	46.74	31.27

with standard error of a difference 4.97.

8.3 Split-plot designs

8.3.1 Blocking the large units

In the situation described in either of Sections 8.1–8.2, the large units may themselves be grouped into blocks, as in Chapter 4. Then it is convenient to think of the large units as *small blocks* and the blocks as *large blocks*. Thus we have b large blocks each of which contains s small blocks each of which contains k plots.

Such large and small blocks often occur naturally.

Example 8.4 (Example 1.7 continued: Rye-grass) Here the large blocks are the fields and the small blocks are the strips. Thus $b = 2$, $s = 3$ and $k = 4$.

Example 8.5 (Animal breeding) In animal breeding it is typical to mate each *sire* (that is, male parent) with several *dams* (female parents), each of which may then produce several offspring. Then the large blocks are the sires, the small blocks are the dams, and the plots are the offspring.

Confusingly, the small blocks are sometimes called *whole plots* or *main plots* while the plots are called *subplots*. This is because they are sometimes used when the plots in a long-running experiment are subdivided for the application of the levels of a new treatment factor.

8.3. Split-plot designs

For simplicity, I shall describe only the classic *split-plot* design. This is like the second design in Section 8.2 except that

- the large units (small blocks) are grouped into b large blocks of size s;
- each level of H is applied to one small block per large block (so $n_H = s$ and $r_H = b$);
- each level of A is applied to one plot per small block (so $n_A = k$ and $r_A = 1$).

Example 8.4 is a classic split-plot design.

Example 8.3 revisited (Cider apples) Since the modified experiment used just the trees of one variety in each block, it is likely that the original experiment was a split-plot design, with variety as the factor H and pruning method as the factor A.

Example 8.6 (Concrete) In Question 1.4, there are two recipes for making concrete. Each of five operators mixes two batches of concrete, one of each type, casts three cylinders from each batch, then leaves them to harden for three different lengths of time before evaluating their breaking strength. It is clear that cylinders within operatives should vary less than cylinders of different operators, so we can regard operators as (large) blocks. It is less clear that we should regard batches as small blocks. To see this, imagine that each operator mixes two batches of each type of concrete. We should expect cylinders in each batch to be more alike than cylinders in different batches, even those made by the same operator with the same type of concrete.

Now the levels of the H treatment factor are the types of concrete, and the levels of the A treatment factor are the lengths of time. This also needs some thought, because time is sometimes a plot factor and sometimes a treatment factor. In a cross-over experiment, time is a plot factor because we cannot change the time of an observational unit. Here time is a treatment factor because we randomly choose which cylinder to break at which time.

(Time plays yet another role in experiments like the one in Question 7.5. Here rats are given different amounts of toxin, then each is measured at several time-points, to see how the relevant body function changes over time. The rats are the experimental units and the observational units, because each response is a vector of measurements, indexed by the time-points. The experiment in Question 1.3 is similar. Such experiments are called *repeated measurements* experiments or *time-course* experiments. Their analysis is outside the scope of this book.)

8.3.2 Construction and randomization

(i) Apply levels of H to small blocks just as in a complete-block design.

(ii) Within each small block independently, apply levels of A just as in a completely randomized design.

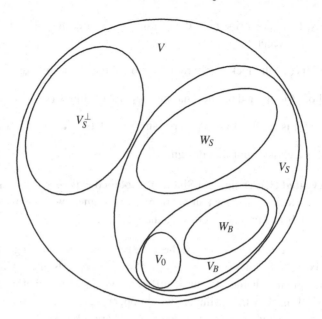

Fig. 8.5. Subspaces defined by small blocks and large blocks

8.3.3 Model and strata

We assume that large blocks and small blocks both give random effects. Thus

$$\mathbb{E}(Y_\omega) = \tau_{H(\omega)A(\omega)}$$

for each plot ω, and

$$\text{cov}(Y_\alpha, Y_\beta) = \begin{cases} \sigma^2 & \text{if } \alpha = \beta \\ \rho_1 \sigma^2 & \text{if } \alpha \text{ and } \beta \text{ are different plots in the same small block} \\ \rho_2 \sigma^2 & \text{if } \alpha \text{ and } \beta \text{ are in different small blocks in the same} \\ & \text{large block} \\ \rho_3 \sigma^2 & \text{if } \alpha \text{ and } \beta \text{ are in different large blocks.} \end{cases} \quad (8.3)$$

Usually $\rho_1 > \rho_2 > \rho_3$.

Define V_B as in Chapter 4 and put $W_B = V_B \cap V_0^\perp$, so that $\dim V_B = b$ and $\dim W_B = b - 1$. Further, put

$$V_S = \{\text{vectors in } V \text{ which take a constant value on each small block}\}.$$

Then $\dim V_S = bs$. The argument in the proof of Lemma 8.1 shows that $V_B \subseteq V_S$, so it is natural to put $W_S = V_S \cap V_B^\perp$. Then $\dim W_S = \dim V_S - \dim V_B = bs - b = b(s-1)$. Finally, $\dim V_S^\perp = N - bs = bsk - bs = bs(k-1)$. These subspaces are shown in Figure 8.5.

The proof of the following proposition is an extended version of the argument that finds the eigenspaces and eigenvectors of $\text{Cov}(\mathbf{Y})$ in Section 4.6.

8.3. Split-plot designs

Table 8.13. *Null analysis of variance for the classic split-plot design*

Table 8.14. *Calculations ignoring treatments for the classic split-plot design*

Stratum		df	SS	EMS
V_0	mean	1	sum$^2/N$	ξ_0
W_B	large blocks	$b-1$	$\sum_i (\text{sum}_{B=i})^2/sk - \text{SS(mean)}$	ξ_B
W_S	small blocks	$b(s-1)$	$\sum_j (\text{sum}_{S=j})^2/k - \text{CSS}(B)$	ξ_S
$V_S^\perp$	plots	$bs(k-1)$	$\sum y_\omega^2 - \text{CSS}(S)$	ξ
	Total	bsk	$\sum y_\omega^2$	

Proposition 8.3 *Under assumption (8.3), the eigenspaces of* $\text{Cov}(\mathbf{Y})$ *are* V_0, W_B, W_S *and* $V_S^\perp$, *with eigenvalues* ξ_0, ξ_B, ξ_S *and* ξ *respectively, where*

$$\begin{aligned}
\xi_0 &= \sigma^2[(1-\rho_1) + k(\rho_1 - \rho_2) + ks(\rho_2 - \rho_3) + bks\rho_3] \\
\xi_B &= \sigma^2[(1-\rho_1) + k(\rho_1 - \rho_2) + ks(\rho_2 - \rho_3)] \\
\xi_S &= \sigma^2[(1-\rho_1) + k(\rho_1 - \rho_2)] \\
\xi &= \sigma^2(1-\rho_1).
\end{aligned}$$

If $\rho_1 > \rho_2 > \rho_3$ *then* $\xi_B > \xi_S > \xi$.

8.3.4 Analysis

Once more we follow the scheme given in Figure 8.2. Proposition 8.3 shows that the null analysis of variance is as shown in Table 8.13. The calculations ignoring treatments are shown in Table 8.14. Here S is the factor for small blocks,

$$\text{CSS}(B) = \text{CSS}(\text{large blocks}) = \sum_{i=1}^{b} \frac{(\text{sum}_{B=i})^2}{sk},$$

and

$$\text{CSS}(S) = \text{CSS}(\text{small blocks}) = \sum_{j=1}^{bs} \frac{(\text{sum}_{S=j})^2}{k}.$$

To work out the skeleton analysis of variance, we have to decide which stratum contains each treatment subspace. Every treatment occurs once in each block, so Theorem 4.1 shows that W_T is orthogonal to V_B; that is, $W_T \subseteq V_B^\perp$. The argument of Lemma 8.1 shows that $W_H \subseteq V_S$. Since $W_H \subset W_T \subseteq V_B^\perp$, we obtain $W_H \subseteq V_S \cap V_B^\perp = W_S$. Theorem 8.2 shows that W_A and $W_{H \wedge A}$ are both contained in $V_S^\perp$. See Figure 8.6. Hence we obtain the skeleton analysis of variance in Table 8.15.

Putting all the parts together gives the full analysis of variance in Table 8.16.

Table 8.15. *Skeleton analysis of variance for the classic split-plot design*

Stratum	Source	Degrees of freedom
mean	mean	1
large blocks	large blocks	$b-1$
small blocks	H	$s-1$
	residual	$(b-1)(s-1)$
	total	$b(s-1)$
plots	A	$k-1$
	$H \wedge A$	$(s-1)(k-1)$
	residual	$(b-1)s(k-1)$
	total	$bs(k-1)$
Total		bsk

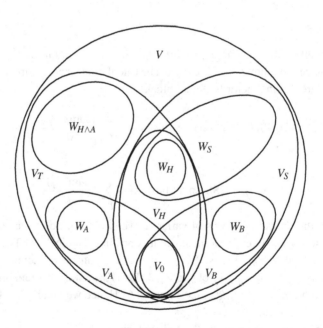

Fig. 8.6. Subspaces for a split-plot design

Table 8.16. *Full analysis of variance for the classic split-plot design*

Stratum	Source	df	SS	EMS	VR
mean	mean	1	$\dfrac{\text{sum}^2}{bsk}$	$\|\tau_0\|^2 + \xi_0$	—
large blocks	large blocks	$b-1$	$\text{CSS}(B) - \text{SS}(\text{mean})$	ξ_B	—
small blocks	H	$s-1$	$\text{SS}(H)$	$\dfrac{\|\tau_H\|^2}{s-1} + \xi_S$	$\dfrac{\text{MS}(H)}{\text{MS(small blocks residual)}}$
small blocks	residual	$(b-1)(s-1)$	by subtraction	ξ_S	—
small blocks	total	$b(s-1)$	$\text{CSS}(S) - \text{CSS}(B)$		
plots	A	$k-1$	$\text{SS}(A)$	$\dfrac{\|\tau_A\|^2}{k-1} + \xi$	$\dfrac{\text{MS}(A)}{\text{MS(plots residual)}}$
plots	$H \wedge A$	$(s-1)(k-1)$	$\text{SS}(H \wedge A)$	$\dfrac{\|\tau_{HA}\|^2}{(s-1)(k-1)} + \xi$	$\dfrac{\text{MS}(H \wedge A)}{\text{MS(plots residual)}}$
plots	residual	$(b-1)s(k-1)$	by subtraction	ξ	—
plots	total	$bs(k-1)$	$\sum_\omega y_\omega^2 - \text{CSS}(S)$		
Total		bsk	$\sum_\omega y_\omega^2$		

8.3.5 Evaluation

Just as in the second design in Section 8.2, A and $H \wedge A$ are estimated more precisely than H, and with more power. So why would anyone deliberately choose a split-plot design? There are three good reasons.

(i) Treatment factor H is already applied to small blocks in a long-term experiment. At a later stage, it is decided to include treatment factor A, and it is practicable to apply levels of A to smaller units than the small blocks.

(ii) It is not feasible to apply levels of H to such small units as is feasible for levels of A. This is the case in Example 8.4. Unless sowing is done by hand, varieties typically must be sown on relatively large areas while levels of other agricultural treatment factors, such as fertilizer, fungicide or pesticide, can be applied to smaller areas.

(iii) The differences between levels of H are already known and the main purpose of the experiment is to investigate A and the H-by-A interaction.

If none of those three conditions applies, think twice before recommending a split-plot design.

8.4 The split-plot principle

The split-plot idea is so simple to describe and to implement that there is a temptation to overuse it. Suppose that there are treatment factors F, G, H, ... Some scientists like to design experiments with repeated splitting as follows.

(i) Construct a complete-block design for F, or an equireplicate completely randomized design for F.

(ii) Split each plot into n_G subplots and apply levels of G randomly to subplots in each plot, independently within each plot.

(iii) Split each subplot into n_H sub-subplots and apply levels of H randomly to sub-subplots within each subplot, independently within each subplot.

(iv) And so on.

This is simple to understand, to construct and to randomize. It makes the experiment easy to conduct. There may be practical reasons why levels of F have to be applied to larger units than levels of G, and so on. If not, this is a bad design, because F has higher variance and fewer residual degrees of freedom than G and $F \wedge G$, which in turn have higher variance and fewer residual degrees of freedom than H, $F \wedge H$, $G \wedge H$ and $F \wedge G \wedge H$, and so on. This matters less if the main purpose of the experiment is to test for interactions.

Example 8.7 (Insecticides on grasshoppers) An experiment was conducted on a prairie in Western Canada to find out if insecticides used to control grasshoppers affected the weight of young chicks of ring-necked pheasants, either by affecting the grass around the chicks or by affecting the grasshoppers eaten by the chicks. Three insecticides were used, at low and high

8.4. The split-plot principle

Table 8.17. *Skeleton analysis of variance for the grasshopper experiment in Example 8.7*

Stratum	Source	Degrees of freedom
mean	mean	1
weeks	weeks	2
strips	insecticide	2
	residual	4
	total	6
swathes	dose	1
	insecticide ∧ dose	2
	residual	6
	total	9
pens	food	1
	insecticide ∧ food	2
	dose ∧ food	1
	insecticide ∧ dose ∧ food	2
	residual	12
	total	18
chicks	chicks	180
Total		216

doses. The low dose was the highest dose recommended by the department of agriculture; the high dose was four times as much as the recommended dose, to assess the effects of mistakes.

The experimental procedure took place in each of three consecutive weeks. On the first day of each week a number of newly-hatched female pheasant chicks were placed in a brooder pen. On the third day, the chicks were randomly divided into twelve groups of six chicks each. Each chick was given an identification tape and weighed.

On the fourth day, a portion of the field was divided into three strips, each of which was divided into two swathes. The two swathes within each strip were sprayed with the two doses of the same insecticide. Two pens were erected on each swathe, and one group of pheasant chicks was put into each pen.

For the next 48 hours, the chicks were fed with grasshoppers which had been collected locally. Half the grasshoppers were anaesthetized and sprayed with insecticide; the other half were also anaesthetized and handled in every way like the first half except that they were not sprayed. All grasshoppers were frozen. The experimenters maintained a supply of frozen grasshoppers to each pen, putting them on small platforms so that they would not absorb further insecticide from the grass. In each swathe, one pen had unsprayed grasshoppers while the other had grasshoppers sprayed by the insecticide which had been applied to that swathe. At the end of the 48 hours, the chicks were weighed again individually.

Table 8.17 shows the skeleton analysis of variance.

Questions for discussion

8.1 For the experiment in Example 8.1, Alice wants to do the analysis of variance shown in Table 8.4, and use it to calculate the standard errors of treatment differences. Bob wants to start by calculating the mean value for each pen, then analysing those means by the methods of Chapter 2. Who is right?

8.2 On Monday 3 March 1997 a surgeon sent a message to *Allstat*, the email list for British statisticians, asking their advice. He said that he was going to use nine animals, of the same species, in an experiment to compare three methods of grafting skin. He intended to use three animals for each method. After the graft was complete he would take a sample of new skin from each animal. He would cut each sample into 20 (tiny!) pieces and use a precision instrument to measure the thickness of each piece. He wanted to know how to analyse the data.

(a) What are the observational units, and how many are there?

(b) What are the experimental units, and how many are there?

(c) What are the treatments, and how many are there?

(d) Construct the skeleton analysis-of-variance table.

8.3 A cattle breeder wants to find a way of protecting his cattle against a particular stomach disease. He wants to compare the effects of

        S: spraying the grass in the paddock with a special chemical
        N: no spray.

He also wants to compare the effects of

            $+$:   injecting each animal with a special vaccine
            $-$:   no injection.

He has several paddocks. Each paddock contains 20 animals, labelled 1, ..., 20 with ear-tags. He wants to apply the treatments once. A month later he will assess the amount of stomach disease in each animal by counting the number of a certain type of bacterium in a sample from the stomach contents. The logarithm of this number will be analysed.

He wants to use two paddocks to find out the effect of the spray, using S on one paddock and N on the other. He wants to use another two paddocks to find out the effect of the injection, using $+$ on one paddock and $-$ on the other. He asks you:

Since N and $-$ are both 'no treatment', can I just use three paddocks, one for S, one for $+$ and one for 'no treatment'?

How do you answer?

Table 8.18. *Data for Question 8.4*

Variety	Date of third cutting	Block					
		I	II	III	IV	V	VI
Ladak	A	2.17	1.88	1.62	2.34	1.58	1.66
	S1	1.58	1.26	1.22	1.59	1.25	0.94
	S20	2.29	1.60	1.67	1.91	1.39	1.12
	O7	2.23	2.01	1.82	2.10	1.66	1.10
Cossack	A	2.33	2.01	1.70	1.78	1.42	1.35
	S1	1.38	1.30	1.85	1.09	1.13	1.06
	S20	1.86	1.70	1.81	1.54	1.67	0.88
	O7	2.27	1.81	2.01	1.40	1.31	1.06
Ranger	A	1.75	1.95	2.13	1.78	1.31	1.30
	S1	1.52	1.47	1.80	1.37	1.01	1.31
	S20	1.55	1.61	1.82	1.56	1.23	1.13
	O7	1.56	1.72	1.99	1.55	1.51	1.33

8.4 A split-plot experiment used three varieties of alfalfa (lucerne) on the whole plots in six blocks. Each whole plot was divided into four subplots, to which four cutting schemes were applied. In the summer, the alfalfa on all subplots was cut twice, the second cut taking place on 27 July. The cutting schemes were: no further cut (A), or a third cut on 1 September (S1), 20 September (S20) or 7 October (O7). Table 8.18 gives the yields for the following year in tons per acre. They have been rearranged from field order to make the arithmetic easier.

(a) Calculate the analysis-of-variance table and interpret it.

(b) Calculate the means for the two main effects.

(c) Give the standard error of a difference between two varieties.

(d) Give the standard error of the difference between two cutting schemes.

8.5 In a biotechnology experiment, plants were grown in 90 pots. Five different quantities of potassium were randomized to the pots, so that each quantity was applied to the soil in 18 pots. After each of 15, 30 and 45 days, one plant was randomly chosen from each pot and removed; nine small pieces were cut from it to be used for tissue culture. Three different levels of nutrition were each applied to three of the pieces of each plant removed at each time. After a certain further length of time, the number of plantlets growing in each piece of tissue was counted.

(a) What are the observational units, and how many are there?

(b) What are the experimental units, and how many are there?

(c) What are the treatments, and how many are there?

(d) Construct the skeleton analysis-of-variance table.

8.6 Read the first five pages of the paper 'Effects of Rodeo® and Garlon® 3A on nontarget wetland species in central Washington', written by Susan C. Gardner and Christian E. Grue and published in the journal *Environmental Toxicology and Chemistry* in 1996 (Volume 15, pages 441–451).

(a) Describe the experimental design used, and comment on it.

(b) Suggest a better design.

(c) Comment on the statistical analysis used.

Chapter 9

More about Latin squares

9.1 Uses of Latin squares

Let S be an $n \times n$ Latin square. Figure 9.1 shows a possible square S when $n = 4$, using the symbols 1, 2, 3, 4 for the 'letters'. Such a Latin square can be used to construct a design for an experiment in surprisingly many ways.

9.1.1 One treatment factor in a square

The square S is used to allocate n treatments to a row–column array with n rows and n columns, as in Chapter 6. Then randomize rows and randomize columns. This design can be summarized by the skeleton analysis of variance in Table 9.1, assuming that rows and columns have random effects.

Example 9.1 (Common cold) Four friends (Adam, Beth, Carl and Daisy) want to compare some suggested methods of protection against the common cold during the British winter: taking a tablet of echinacea every morning; taking a vitamin C pill every morning; drinking a hot honey-and-lemon drink each evening. They decide to include the control treatment, consisting of adding nothing at all to their normal diet. They think that a month is long enough for each method to show its effects, so they decide to use the months November–February, with each person using a different method each month. Each month, each person will record the number of days on which they show any symptoms of the common cold.

The Latin square in Figure 9.1 can be used to give the initial design in Figure 9.2. One possibility after randomization is shown in Figure 9.3.

1	2	3	4
2	4	1	3
3	1	4	2
4	3	2	1

Fig. 9.1. A Latin square of order 4, used to construct four types of design

Table 9.1. *Skeleton analysis of variance in Section 9.1.1*

Stratum	Source	Degrees of freedom
mean	mean	1
rows	rows	$n-1$
columns	columns	$n-1$
plots	treatments	$n-1$
	residual	$(n-1)(n-2)$
	total	$(n-1)^2$
Total		n^2

	Adam	Beth	Carl	Daisy
November	echinacea	vitamin C	honey & lemon	nothing
December	vitamin C	nothing	echinacea	honey & lemon
January	honey & lemon	echinacea	nothing	vitamin C
February	nothing	honey & lemon	vitamin C	echinacea

Fig. 9.2. *Initial design in Example 9.1*

9.1.2 More general row–column designs

The square S is used to allocate n treatments to a row–column array with nm rows and nl columns, where m and l are positive integers, as described in Chapter 6. The rectangular array is split into ml squares of size $n \times n$, and a copy of S is put into each one. The skeleton analysis of variance is shown in Table 9.2.

Example 6.1 has this form, with $n = 4$, $m = 1$ and $l = 2$.

Of course, just as in Chapter 6, the rows and columns need not be physical rows and columns. In Example 6.1, the rows are tasting positions and the columns are people. In an experiment on plots of land, a previous experiment may have used b complete blocks of t treatments each. If there are n treatments in the current experiment, and n divides both b and t, then it is sensible to use the previous blocks and the previous treatments as rows and columns for the current design.

	Adam	Beth	Carl	Daisy
November	echinacea	nothing	vitamin C	honey & lemon
December	nothing	echinacea	honey & lemon	vitamin C
January	honey & lemon	vitamin C	echinacea	nothing
February	vitamin C	honey & lemon	nothing	echinacea

Fig. 9.3. *One possible randomized plan in Example 9.1*

9.1. Uses of Latin squares

Table 9.2. *Skeleton analysis of variance in Section 9.1.2*

Stratum	Source	Degrees of freedom
mean	mean	1
rows	rows	$nm - 1$
columns	columns	$nl - 1$
plots	treatments	$n - 1$
	residual	$n^2ml - nm - nl - n + 2$
	total	$(nm-1)(nl-1)$
Total		n^2ml

9.1.3 Two treatment factors in a block design

Suppose that there are n blocks of size n and that there are two n-level treatment factors F and G. If we can assume that the F-by-G interaction is zero then we need estimate only the main effects of F and G. We can use S to allocate levels of F and G to the plots in such a way that all combinations of F and G occur and that each of F and G is orthogonal to blocks. This design depends on the interaction being zero, so it is called a *main-effects-only* design. It is also a *single-replicate* design, because each treatment occurs exactly once.

Construction Rows are blocks. Columns are levels of F, say $f_1, f_2, \ldots, f_n$. 'Letters' are levels of G, say $g_1, g_2, \ldots, g_n$. If the letter in row i and column j is l then treatment combination $f_j g_l$ comes in block i.

The square in Figure 9.1 gives the following block design.

Block 1			
$f_1 g_1$	$f_2 g_2$	$f_3 g_3$	$f_4 g_4$

Block 2			
$f_1 g_2$	$f_2 g_4$	$f_3 g_1$	$f_4 g_3$

Block 3			
$f_1 g_3$	$f_2 g_1$	$f_3 g_4$	$f_4 g_2$

Block 4			
$f_1 g_4$	$f_2 g_3$	$f_3 g_2$	$f_4 g_1$

Randomization Now blocks do not all contain the same treatments, so we need to randomize whole blocks among themselves as well as randomizing plots within blocks.

(i) Randomize blocks (that is, randomly permute the order or the names of the four whole blocks);

(ii) randomize plots within blocks (that is, within each block independently, randomly permute plots).

Be careful not to think of this as randomizing levels of either F or G, because we do not want to destroy the sets of n treatment combinations that come together in a block.

Table 9.3. *Skeleton analysis of variance in Section 9.1.3*

Stratum	Source	Degrees of freedom
mean	mean	1
blocks	blocks	$n-1$
plots	F	$n-1$
	G	$n-1$
	residual	$(n-1)(n-2)$
	total	$n(n-1)$
Total		n^2

Block	1	1	1	1	2	2	2	2	3	3	3	3	4	4	4	4
Plot	1	2	3	4	5	6	7	8	9	10	11	12	13	14	15	16
Spray	a	b	c	d	a	b	c	d	a	b	c	d	a	b	c	d
Pruning method	2	3	4	5	3	5	2	4	4	2	5	3	5	4	3	2

Fig. 9.4. Initial design in Example 9.2

Skeleton analysis of variance The null analysis of variance has strata for the mean, blocks and plots, just like any block design. By construction, blocks and F and G are all orthogonal to each other (use Theorems 4.1 and 5.1), so the skeleton analysis of variance is as shown in Table 9.3.

Example 9.2 (Example 8.3 continued: Cider apples) In Example 8.3 we found no interaction between sprays and pruning methods. Suppose that we decide to continue the experiment in the following year, but applying sprays to individual plots so that we can obtain more information about them. We include an extra spray, so that the sprays are a, b, c and d, and exclude pruning method 1, which is actually no pruning. We use the trees of a single variety in four plots of each of four blocks of the original experiment.

Figure 9.4 shows the initial design obtained from the Latin square S, and Figure 9.5 shows one of the possible randomized layouts.

Block	1	1	1	1	2	2	2	2	3	3	3	3	4	4	4	4
Plot	1	2	3	4	5	6	7	8	9	10	11	12	13	14	15	16
Spray	d	a	c	b	a	d	c	b	a	d	b	c	d	a	b	c
Pruning method	5	2	4	3	5	2	3	4	4	3	2	5	4	3	5	2

Fig. 9.5. One possible randomized layout in Example 9.2

Table 9.4. *Skeleton analysis of variance in Section 9.1.4*

Stratum	Source	Degrees of freedom
mean	mean	1
plots	F	$n-1$
	G	$n-1$
	H	$n-1$
	residual	$(n-1)(n-2)$
	total	n^2-1
Total		n^2

9.1.4 Three treatment factors in an unblocked design

Suppose that F, G and H are three n-level treatment factors. If we can assume that all interactions are zero then we can construct a main-effects-only design in n^2 plots with no blocking. Not all the treatment combinations occur, so this is called a *fractional replicate*.

Construction Rows are levels of F. Columns are levels of G. Letters are levels of H. If the letter in row i and column j is l then the treatment $f_i g_j h_l$ occurs in the design.

Thus the Latin square in Figure 9.1 gives the following 16 treatments out of the total possible 64. The design is called a *quarter replicate*.

$f_1g_1h_1$, $f_1g_2h_2$, $f_1g_3h_3$, $f_1g_4h_4$, $f_2g_1h_2$, $f_2g_2h_4$, $f_2g_3h_1$, $f_2g_4h_3$, $f_3g_1h_3$, $f_3g_2h_1$, $f_3g_3h_4$, $f_3g_4h_2$, $f_4g_1h_4$, $f_4g_2h_3$, $f_4g_3h_2$, $f_4g_4h_1$

Randomization Completely randomize all the plots. Once again, do not think of this as randomizing levels of any of the treatment factors, or the careful choice of fractional replicate may be ruined.

Skeleton analysis of variance There are no blocks, so Section 2.14 shows that the strata are V_0 and $V_0^\perp$. By construction, F, G and H are all orthogonal to each other, so we obtain the skeleton analysis of variance in Table 9.4.

Example 9.2 revisited (Cider apples) If we believe that there are also no interactions between sprays and varieties of apple or between pruning methods and varieties of apple, or among all three factors, then we can do a completely randomized experiment using four varieties of apple (J, K, L and M) in sixteen plots of a single block of the original experiment. A quarter-replicate obtained from Latin square S is in Figure 9.6. One possible randomized layout is shown in Figure 9.7.

Plot	1	2	3	4	5	6	7	8	9	10	11	12	13	14	15	16
Spray	a	a	a	a	b	b	b	b	c	c	c	c	d	d	d	d
Pruning method	2	3	4	5	2	3	4	5	2	3	4	5	2	3	4	5
Variety of apple	J	K	L	M	K	M	J	L	L	J	M	K	M	L	K	J

Fig. 9.6. Initial design in the quarter-replicate version of Example 9.2

Plot	1	2	3	4	5	6	7	8	9	10	11	12	13	14	15	16
Spray	c	d	b	b	a	d	a	d	c	a	c	a	c	b	b	d
Pruning method	3	5	2	4	5	2	3	3	5	4	4	2	2	3	5	4
Variety of apple	J	J	K	J	M	M	K	L	K	L	M	J	L	M	L	K

Fig. 9.7. One possible randomized layout in the quarter-replicate version of Example 9.2

9.2 Graeco-Latin squares

Definition Two $n \times n$ Latin squares are *orthogonal* to each other if each letter of the first square occurs in the same position as each letter of the second square exactly once.

Such a pair is often called a *Graeco-Latin* square, because traditionally Latin letters are used for the first square and Greek letters for the second square.

Example 9.3 (Mutually orthogonal Latin squares of order three) The following two 3×3 Latin squares are orthogonal to each other.

A	B	C
C	A	B
B	C	A

α	β	γ
β	γ	α
γ	α	β

When the two squares are superimposed, we obtain the following Graeco-Latin square.

A α	B β	C γ
C β	A γ	B α
B γ	C α	A β

As usual, let V_R, V_C, V_L and V_G be the subspaces of V consisting of vectors which are constant on each row, column, Latin letter, Greek letter respectively. Then put $W_R = V_R \cap V_0^\perp$, $W_C = V_C \cap V_0^\perp$, $W_L = V_L \cap V_0^\perp$ and $W_G = V_G \cap V_0^\perp$. Then $W_R \perp W_C$. Further, because each square is a Latin square, W_L and W_G are both orthogonal to W_R and W_C. The extra condition for a Graeco-Latin square ensures that $W_L \perp W_G$: that is why the two squares are said to be *orthogonal* to each other.

Note that not every Latin square has another orthogonal to it. For example, there is no Latin square orthogonal to the one in Figure 6.3 (if you try to construct one by trial and error you will soon find that it is impossible).

9.2. Graeco-Latin squares

However, there are some simple methods for constructing pairs of mutually orthogonal Latin squares.

Prime numbers If $n = p$, a prime number, label the rows and columns of the square by the integers modulo p. Choose a to be an integer modulo p which is not equal to 0 or 1. In the first square, the letter in row x and column y should be $x+y$; in the second it should be $ax+y$.

	in the first square			in the second square	
	y			y	
x	$x+y$		x	$ax+y$	

Once the first square has been constructed it is easy to construct the second: for each value of x, take the row labelled ax in the first square and put it in the row labelled x in the second square.

For example, taking $n = p = 5$ and $a = 2$ gives the following pair of mutually orthogonal Latin squares of order 5.

	0	1	2	3	4
0	0	1	2	3	4
1	1	2	3	4	0
2	2	3	4	0	1
3	3	4	0	1	2
4	4	0	1	2	3

	0	1	2	3	4
0	0	1	2	3	4
1	2	3	4	0	1
2	4	0	1	2	3
3	1	2	3	4	0
4	3	4	0	1	2

In general, different values of a give squares orthogonal to each other.

Finite fields If n is a power of a prime but not itself prime, use a similar construction using the *finite field* $GF(n)$ with n elements. Do not worry if you know nothing about finite fields. For practical purposes it is enough to know about $GF(4)$, $GF(8)$ and $GF(9)$.

The elements of the field $GF(4)$ can be written as 0, 1, u and $u+1$, where all operations are modulo 2 and $u^2 = u+1$. For example, $u^3 = u(u^2) = u(u+1) = u^2 + u = (u+1) + u = 1$.

The elements of the field $GF(8)$ can be written in the form $m_1 + m_2 v + m_3 v^2$, where m_1, m_2 and m_3 are integers modulo 2. All operations are modulo 2, and $v^3 = v+1$.

The elements of the field $GF(9)$ can be written in the form $m_1 + m_2 i$, where m_1 and m_2 are integers modulo 3. All operations are modulo 3, and $i^2 = -1 = 2$. For example, if we take $a = i$ then we obtain the pair of mutually orthogonal 9×9 Latin squares in Figure 9.8.

	0	1	2	i	$1+i$	$2+i$	$2i$	$1+2i$	$2+2i$
0	0	1	2	i	$1+i$	$2+i$	$2i$	$1+2i$	$2+2i$
1	1	2	0	$1+i$	$2+i$	i	$1+2i$	$2+2i$	$2i$
2	2	0	1	$2+i$	i	$1+i$	$2+2i$	$2i$	$1+2i$
i	i	$1+i$	$2+i$	$2i$	$1+2i$	$2+2i$	0	1	2
$1+i$	$1+i$	$2+i$	i	$1+2i$	$2+2i$	$2i$	1	2	0
$2+i$	$2+i$	i	$1+i$	$2+2i$	$2i$	$1+2i$	2	0	1
$2i$	$2i$	$1+2i$	$2+2i$	0	1	2	i	$1+i$	$2+i$
$1+2i$	$1+2i$	$2+2i$	$2i$	1	2	0	$1+i$	$2+i$	i
$2+2i$	$2+2i$	$2i$	$1+2i$	2	0	1	$2+i$	i	$1+i$

	0	1	2	i	$1+i$	$2+i$	$2i$	$1+2i$	$2+2i$
0	0	1	2	i	$1+i$	$2+i$	$2i$	$1+2i$	$2+2i$
1	i	$1+i$	$2+i$	$2i$	$1+2i$	$2+2i$	0	1	2
2	$2i$	$1+2i$	$2+2i$	0	1	2	i	$1+i$	$2+i$
i	2	0	1	$2+i$	i	$1+i$	$2+2i$	$2i$	$1+2i$
$1+i$	$2+i$	i	$1+i$	$2+2i$	$2i$	$1+2i$	2	0	1
$2+i$	$2+2i$	$2i$	$1+2i$	2	0	1	$2+i$	i	$1+i$
$2i$	1	2	0	$1+i$	$2+i$	i	$1+2i$	$2+2i$	$2i$
$1+2i$	$1+i$	$2+i$	i	$1+2i$	$2+2i$	$2i$	1	2	0
$2+2i$	$1+2i$	$2+2i$	$2i$	1	2	0	$1+i$	$2+i$	i

Fig. 9.8. Pair of mutually orthogonal Latin squares of order 9, constructed from GF(9)

9.2. Graeco-Latin squares

A	B	C	D	E	F	G	H	I
B	C	A	E	F	D	H	I	G
C	A	B	F	D	E	I	G	H
D	E	F	G	H	I	A	B	C
E	F	D	H	I	G	B	C	A
F	D	E	I	G	H	C	A	B
G	H	I	A	B	C	D	E	F
H	I	G	B	C	A	E	F	D
I	G	H	C	A	B	F	D	E

A	B	C	D	E	F	G	H	I
D	E	F	G	H	I	A	B	C
G	H	I	A	B	C	D	E	F
C	A	B	F	D	E	I	G	H
F	D	E	I	G	H	C	A	B
I	G	H	C	A	B	F	D	E
B	C	A	E	F	D	H	I	G
E	F	D	H	I	G	B	C	A
H	I	G	B	C	A	E	F	D

Fig. 9.9. Pair of mutually orthogonal Latin squares of order 9, rewritten using letters

It may be less confusing to rewrite these two squares as in Figure 9.9, where the elements of GF(9) have been replaced by letters.

Product method If S_1 and T_1 are mutually orthogonal Latin squares of order n_1 and S_2 and T_2 are mutually orthogonal Latin squares of order n_2 then the product squares $S_1 \otimes S_2$ and $T_1 \otimes T_2$ are orthogonal to each other and have order $n_1 n_2$.

There is no Graeco-Latin square of order 1, 2 or 6. However, Graeco-Latin squares exist for all other orders. The above methods can be combined to give a pair of orthogonal Latin squares of order n whenever n is odd or divisible by 4. If n is even but not divisible by 4 then the construction is more complicated. A pair of mutually orthogonal Latin squares of order 10 is shown in Figure 9.10.

A	H	I	D	J	F	G	B	C	E
H	I	A	J	C	D	E	F	G	B
I	E	J	G	A	B	H	C	D	F
B	J	D	E	F	H	I	G	A	C
J	A	B	C	H	I	F	D	E	G
E	F	G	H	I	C	J	A	B	D
C	D	H	I	G	J	B	E	F	A
F	B	E	A	D	G	C	H	I	J
D	G	C	F	B	E	A	J	H	I
G	C	F	B	E	A	D	I	J	H

α	β	γ	λ	ε	μ	ν	η	ζ	δ
γ	δ	λ	ζ	μ	ν	β	α	η	ε
ε	λ	η	μ	ν	γ	δ	β	α	ζ
λ	α	μ	ν	δ	ε	ζ	γ	β	η
β	μ	ν	ε	ζ	η	λ	δ	γ	α
μ	ν	ζ	η	α	λ	γ	ε	δ	β
ν	η	α	β	λ	δ	μ	ζ	ε	γ
η	ζ	ε	δ	γ	β	α	λ	μ	ν
ζ	ε	δ	γ	β	α	η	μ	ν	λ
δ	γ	β	α	η	ζ	ε	ν	λ	μ

Fig. 9.10. Pair of mutually orthogonal Latin squares of order 10

9.3 Uses of Graeco-Latin squares

There are many ways of using a Graeco-Latin square of order n to construct an experiment for n^2 plots. All the designs described here have $(n-1)(n-3)$ residual degrees of freedom.

9.3.1 Superimposed design in a square

If last year's experiment was a Latin square, and there are the same number of treatments this year, on the same plots, then this year's experiment should be a Latin square orthogonal to last year's. If you know before you do the first experiment that there will be a second one, then choose any Graeco-Latin square, randomize its rows and columns, use the Latin letters for Year 1 and the Greek letters for Year 2.

If the experimental units are expensive and long-lived, such as trees, then an experiment next year is quite likely, so it is safer to always use a Latin square that has another Latin square orthogonal to it.

9.3.2 Two treatment factors in a square

If the experimental units form an $n \times n$ square and there are two n-level treatment factors, F and G, whose interaction is assumed to be zero, then a Graeco-Latin square can be used to construct a main-effects-only single-replicate design. The Latin letters give the levels of F, the Greek letters the levels of G. Randomize rows and columns independently.

Example 9.4 (Oceanography) A team of oceanographers wanted to investigate the effects of two factors on the number of sea-animals living on the sea-bottom in a fjord. One factor was the frequency of disturbance: this had seven levels, ranging from 'never' to 'once per day'. The other factor was the quantity of organic enrichment, which also had seven levels.

The oceanographers filled 49 plastic buckets with muddy sand from the bed of the fjord. They arranged the buckets in a 7×7 square in a concrete tank next to the fjord. Sea-water was pumped from the bottom of the fjord and allowed to flow through the tank. This was done by having a tap at the North end of each column letting sea-water into that column, and a tap at the South end of each column letting it flow out again. Thus it was expected that there might be differences between columns. Also, there might be differences between rows, because these corresponded to distance from the inlet taps.

The oceanographers assumed that there was no interaction between the frequency of disturbance and the quantity of organic enrichment, so they used the Graeco-Latin square in Figure 9.11. They ran the experiment for twelve weeks, then counted the number of capitellid worms in each bucket. These numbers are also shown in Figure 9.11.

9.3.3 Three treatment factors in a block design

Suppose that there are n blocks of size n and three n-level treatment factors F, G and H. If all interactions among the treatment factors can be assumed to be zero then we can use a Graeco-Latin square to construct a main-effects-only fractional replicate. Rows are blocks; columns are levels of F; Latin and Greek letters are levels of G and H respectively. If the Latin and Greek letters in row i and column j are l and m respectively then the treatment combination $f_j g_l h_m$ occurs in block i.

C η	G δ	A ε	F γ	E β	D α	B φ
151	170	190	336	127	41	106
E φ	B γ	C δ	A β	G α	F η	D ε
247	231	66	99	27	132	382
B δ	F α	G β	E η	D φ	C ε	A γ
158	545	167	360	219	169	214
F β	C φ	D η	B ε	A δ	G γ	E α
162	87	328	122	101	681	29
A α	E ε	F φ	D δ	C γ	B β	G η
74	275	473	339	136	75	437
D γ	A η	B α	G φ	F ε	E δ	C β
476	603	41	429	348	255	102
G ε	D β	E γ	C α	B η	A φ	F δ
487	131	178	66	87	69	249

	Level of disturbance		Organic enrichment
A	daily	α	400 g carbon/m^2
B	thrice per week	β	200 g carbon/m^2
C	twice per week	γ	100 g carbon/m^2
D	weekly	δ	50 g carbon/m^2
E	fortnightly	ε	25 g carbon/m^2
F	once per four weeks	φ	12.5 g carbon/m^2
G	never	η	none

Fig. 9.11. Design and data in Example 9.4

Randomization is as in Section 9.1.3.

9.3.4 Four treatment factors in an unblocked design

If there are no interactions among the four n-level treatment factors F, G, H and K then a Graeco-Latin square can be used to construct a main-effects-only fractional replicate in n^2 plots. Rows are levels of F; columns are levels of G; Latin and Greek letters are levels of H and K respectively. If the Latin and Greek letters in row i and column j are l and m respectively then the treatment combination $f_i g_j h_l k_m$ occurs in the design.

Randomization is as in Section 9.1.4.

Questions for discussion

9.1 Construct a 4×4 Graeco-Latin square and an 11×11 Graeco-Latin square.

9.2 Construct the skeleton analysis-of-variance table for a general experiment in an $n \times n$ square in Section 9.3.2.

9.3 Analyse the data in Example 9.4.

9.4 Construct a 12×12 Graeco-Latin square.

9.5 Last year, five varieties of potato were compared in an experiment using five complete blocks. The layout is shown below, with the varieties coded A, B, C, D and E.

Block 1
D	A	E	C	B

Block 2
B	A	D	E	C

Block 3
A	B	C	E	D

Block 4
A	D	C	E	B

Block 5
E	D	B	C	A

The potatoes were dug up for harvest, but small traces may remain in the soil. This year the intention is to compare five chemicals for their ability to kill the remaining traces of potato plant. The chemicals are coded α, β, γ, δ and ε. It is believed that there is no interaction between variety of potato and type of chemical.

Construct and randomize the design for this year's experiment, presenting the plan in a format consistent with last year's.

9.6 A horticultural research station intends to investigate the effects of two treatment factors on the total weight of apples produced from apple trees. One treatment factor is the method of thinning; that is, removing fruitlets at an early stage of development so that those remaining will be able to grow larger. There are five methods of thinning, coded as A, B, C, D, E. The second treatment factor is the type of grass to grow around the base of the tree to prevent the growth of weeds. There are five types of grass, coded as a, b, c, d, e. It is assumed that there is no interaction between method of thinning and type of grass.

There are 25 trees available for the experiment. They are arranged in a 5×5 rectangle. Construct a suitable design.

9.7 A road safety organization wishes to compare four makes of car tyre. The organization has four test cars and four test drivers. One working day is needed to fit new tyres to a car, take it for an exhaustive test-drive, take relevant measurements on tyre treads, record all details of the test-drive, and prepare the car for the next session. The organization has only one week in which to perform its tests. To keep each car properly balanced, the organization has decided that all four tyres on a car at any one time should be of the same make.

Construct a suitable design for this trial.

Chapter 10

The calculus of factors

10.1 Introduction

In the preceding chapters we have met all the principles of orthogonal designs (which will be defined in Section 10.12). The plots may be unstructured, as in Chapter 2. There may be a single system of blocks, as in Chapter 4. If there are two systems of blocks then we may have each block of one system meeting each block of the other system, as in Chapter 6 and Section 7.3, or each block of one system may contain several blocks of the other system, as in Section 8.3. The experimental units may be the same as the observational units, or each experimental unit may contain several observational units, as in Section 8.1.

The treatments also may be unstructured, as in Chapter 2. They may be divided into different types, such as control and new treatments, as in Chapter 3. They may consist of all combinations of two or more treatment factors, as in Chapter 5. The different treatment effects may all be estimated in the same stratum, or they may be estimated in two or more strata, as in Sections 8.2–8.4. We may even assume that some interactions are zero, as in Chapter 9.

The purpose of the present chapter is to give a unifying framework that not only encompasses all the designs we have met so far but also permits the construction and analysis of infinitely many more. Once this framework is understood there is absolutely no need to memorize the structure of any individual named design.

The reader may wish to omit proofs on a first reading of this chapter.

10.2 Relations on factors

10.2.1 Factors and their classes

Consider factors F, G, ... on a set, which might be either the set Ω of observational units or the set $\mathcal{T}$ of treatments. If the set is Ω then we write $F(\omega)$ for the level of F which occurs on plot ω. The *class* of F containing plot α is defined to be $\{\omega \in \Omega : F(\omega) = F(\alpha)\}$. Let us write this as

$$F[[\alpha]] = \{\omega \in \Omega : F(\omega) = F(\alpha)\}.$$

Likewise, if G is a factor on $\mathcal{T}$ and i is any treatment then $G(i)$ denotes the level of G on treatment i, and $G[[i]]$ is the G-class containing i, that is, the set of all treatments which have the same level of G as i does.

10.2.2 Aliasing

Definition Let F and G be factors on the same set. Then F is *equivalent* to G, or *aliased* with G, if every F-class is also a G-class; that is, F and G are the same apart from the names of their levels.

Notation Write $F \equiv G$ if F is equivalent to G.

If the set is Ω, then $F \equiv G$ if and only if $F[[\omega]] = G[[\omega]]$ for all ω in Ω.

Example 10.1 (Example 1.7 continued: Rye-grass) There are two fields, each consisting of three strips, each of which contains four plots. Thus field and strip are relevant factors on Ω even before we apply treatments, so we call them *plot factors*. The treatments consist of all combinations of the factors cultivar and nitrogen. The levels of cultivar are Cropper, Melle and Melba, each applied to one whole strip per field. The levels of nitrogen are 0, 80, 160 and 240 kg/ha, each applied to one plot per strip. Thus cultivar and nitrogen are *treatment factors*, as is T itself.

On the data sheet, we may well code the cultivars Cropper, Melle and Melba as 1, 2 and 3 respectively, and abbreviate cultivar to C. Thus $C(\omega) = 1$ if and only if the cultivar on ω is Cropper. Thus C is merely a renaming of cultivar and so $C \equiv$ cultivar.

Aliasing has more uses than simple renaming. If there are two nuisance factors on the plots, it may be beneficial to alias them.

Example 10.2 (Car tyres) There are three nuisance factors in Question 9.7: car, driver and day. If we use a Graeco-Latin square for the design then each of these nuisance factors uses up three degrees of freedom, so there are only three residual degrees of freedom. However, the experimenter is not interested in comparing cars or drivers or days, so this is wasteful. If we allocate each driver to the same car every day then we can use a Latin square for the design and there are six residual degrees of freedom. Now driver $\equiv$ car even though drivers and cars are not inherently the same.

Example 10.3 (Example 8.6 continued: Concrete) In Question 1.4 there are two types of concrete, with different quantities of cement, water and aggregate per cubic metre. We could regard cement, water and aggregate as treatment factors, but that is rather pointless in this example, as they are all aliased with each other.

If a treatment factor is aliased with a plot factor, it may indicate false replication.

Example 10.4 (Example 1.1 continued: Ladybirds) The field was divided into three areas, from each of which three samples were taken. Thus the plots were the nine samples and area was a plot factor. The treatment factor pesticide had three levels, each one being applied to a single area. Thus pesticide $\equiv$ area, which immediately shows that there was no proper replication.

10.2.3 One factor finer than another

Definition Let F and G be factors on the same set. Then F is *finer* than G, or G is *coarser* than F, if every F-class is contained in a G-class but $F \not\equiv G$.

Notation Write $F \prec G$ if F is finer than G. Write $F \preccurlyeq G$ if $F \prec G$ or $F \equiv G$. Also write $G \succ F$ for $F \prec G$ and $G \succcurlyeq F$ for $F \preccurlyeq G$.

If the set is Ω then $F \preccurlyeq G$ means that $F[[\omega]] \subseteq G[[\omega]]$ for all ω in Ω. Hence, if $\alpha \in F[[\omega]]$ then $\alpha \in G[[\omega]]$.

In Example 10.1, strip $\prec$ field and strip $\prec$ cultivar. Also $T \prec$ cultivar and $T \prec$ nitrogen.

10.2.4 Two special factors

There are two special factors on Ω (indeed, on any set).

The universal factor U This is defined by $U(\omega) = 1$ for all ω in Ω. Thus U has a single class, which is the whole of Ω, or the *universe*. It is the **u**ncaring factor, because it makes no distinctions between units.

The equality factor E This is defined by $E(\omega) = \omega$ for all ω in Ω. Thus $E[[\omega]] = \{\omega\}$ for all ω in Ω, and so E has as many classes as there are plots. There is a class for **e**ach and **e**very plot. If α and β are in the same E-class then $\alpha = \beta$: they are **e**qual.

In Example 10.1, $E =$ plot and plot $\prec$ strip $\prec$ field. In Example 10.2, $E = $ car $\wedge$ day. In Example 10.4, $E = $ sample and sample $\prec$ area. In Example 8.1, $E = $ calf and calf $\prec$ pen. In Example 6.1, $E = $ judge $\wedge$ tasting position.

Similarly, the special factors U and E are defined on $\mathcal{T}$, the set of treatments. On this set, $E = T$.

For every factor F, we have

$$E \preccurlyeq F \preccurlyeq U. \tag{10.1}$$

10.3 Operations on factors

10.3.1 The infimum of two factors

We have already met the idea of $F \wedge G$, where F and G are factors on the same set. The levels of $F \wedge G$ (which may be pronounced 'F down G') are the combinations of levels of F and G. If F and G are factors on $\mathcal{T}$ then

$$(F \wedge G)(i) = (F(i), G(i)).$$

The class of $F \wedge G$ containing treatment i is

$$(F \wedge G)[[i]] = \{j \in \mathcal{T} : F(j) = F(i) \text{ and } G(j) = G(i)\} = F[[i]] \cap G[[i]].$$

Similarly, if F and G are plot factors then the $(F \wedge G)$-class containing plot ω is

$$(F \wedge G)[[\omega]] = \{\alpha \in \Omega : F(\alpha) = F(\omega) \text{ and } G(\alpha) = G(\omega)\} = F[[\omega]] \cap G[[\omega]].$$

Formally we make the following definition.

Definition The *infimum* of factors F and G on the same set is the factor $F \wedge G$ whose classes are the non-empty intersections of F-classes with G-classes.

Example 10.1 revisited (Rye-grass) Within each field, each cultivar occurs on one whole strip. Therefore field $\wedge$ cultivar = strip.

Notice that the notation $\wedge$ does not imply that all combinations of the two factors occur. In Example 1.16, T = amount $\wedge$ time but Figure 1.5(b) shows that not all combinations of amount and time occur.

The factor $F \wedge G$ is called the infimum of F and G because it satisfies the following two conditions:

(i) $F \wedge G \preccurlyeq F$ and $F \wedge G \preccurlyeq G$;

(ii) if H is a factor such that $H \preccurlyeq F$ and $H \preccurlyeq G$ then $H \preccurlyeq F \wedge G$.

Property (i) is true because $(F \wedge G)[[\omega]] = F[[\omega]] \cap G[[\omega]]$, which is a subset of both $F[[\omega]]$ and $G[[\omega]]$. As for property (ii), if $H \preccurlyeq F$ and $H \preccurlyeq G$ then, for all ω, we have $H[[\omega]] \subseteq F[[\omega]]$ and $H[[\omega]] \subseteq G[[\omega]]$ so $H[[\omega]] \subseteq F[[\omega]] \cap G[[\omega]] = (F \wedge G)[[\omega]]$.

The concept of infimum should already be familiar from the natural numbers, with 'is finer than or equivalent to' replaced by 'divides'. The *highest common factor* h of two natural numbers n and m satisfies

(i) h divides n and h divides m;

(ii) if k is a natural number such that k divides n and k divides m then k divides h.

There is a dual concept, that of the *least common multiple*. Now, l is the least common multiple of two natural numbers n and m if

(i) n divides l and m divides l;

(ii) if k is a natural number such that n divides k and m divides k then l divides k.

10.3.2 The supremum of two factors

The dual concept also occurs for factors on a set, defined very like the least common multiple but with 'divides' replaced by 'is finer than or equivalent to'.

Definition The *supremum* of factors F and G on the same set is the unique factor (up to equivalence) $F \vee G$ which satisfies

(i) $F \preccurlyeq F \vee G$ and $G \preccurlyeq F \vee G$;

(ii) if H is a factor such that $F \preccurlyeq H$ and $G \preccurlyeq H$ then $F \vee G \preccurlyeq H$.

The supremum $F \vee G$ may be pronounced 'F up G'.

Unfortunately, the recipe for writing down the class of $F \vee G$ containing ω is not so simple as it was for $F \wedge G$. Starting with ω, write down all the plots in the same F-class as ω. Then write down all the plots which are in the same G-class as any plot written down so far.

10.3. Operations on factors

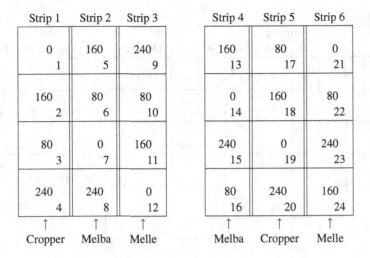

Fig. 10.1. Layout of the experiment in Example 10.1

Then write down all the plots which are in the same F-class as any plot written down so far. Continue alternating between F and G until no new plots are added. Then the set of plots which have been written down forms the $(F \vee G)$-class containing ω.

In other words, α and β are in the same class of $F \vee G$ if there is a finite sequence of elements (plots if we are dealing with Ω, treatments if we are dealing with $\mathcal{T}$)

$$\alpha = \alpha_1, \beta_1, \alpha_2, \beta_2, \ldots, \alpha_n, \beta_n = \beta$$

such that

α_i and β_i have the same level of F for $i = 1, 2, \ldots, n$

and

β_j and α_{j+1} have the same level of G for $j = 1, \ldots, n-1$.

In a complete-block design, every block contains a plot with each treatment. If B is the block factor and T the treatment factor, then $T[[\alpha]]$ contains a plot in every block so the first two steps of this process show that $T \vee B = U$.

Similarly, in Example 6.1, judge $\vee$ tasting position $= U$.

In general, if all combinations of levels of F and levels of G occur together then $F \vee G = U$.

Example 10.1 revisited (Rye-grass) The plan for this factorial experiment is shown again in Figure 10.1, this time with the plots numbered. We shall calculate strip $\vee T$.

Start at plot number 16. The plots in the same strip are plots 13, 14, 15 and 16. The plots with the same treatment as plot 13 are plots 5 and 13; the plots with the same treatment as plot 14 are plots 7 and 14; the plots with the same treatment as plot 15 are plots 8 and 15; the plots with the same treatment as plot 16 are plots 6 and 16. So far we have written down $\{5, 6, 7, 8, 13, 14, 15, 16\}$. These form two whole strips, and so the process stops. These plots are precisely those that have the cultivar Melba: in other words,

$$(\text{strip} \vee T)(16) = \text{cultivar}(16).$$

Strip	Treatment												
	Cropper				Melle				Melba			
	0	80	160	240	0	80	160	240	0	80	160	240
1	√ 1	√ 3	√ 2	√ 4								
2									√ 7	√ 6	√ 5	√ 8
3					√ 12	√ 10	√ 11	√ 9				
4									√ 14	√ 16	√ 13	√ 15
5	√ 19	√ 17	√ 18	√ 20								
6					√ 21	√ 22	√ 24	√ 23				

Fig. 10.2. A different presentation of the information in Figure 10.1

Fig. 10.3. A hypothetical example where the calculation of $F \vee G$ requires more than two steps

A similar argument applies no matter what plot we start with, and therefore strip $\vee$ T = cultivar.

Another way to calculate $F \vee G$ is to draw a rectangular diagram like those in Figures 1.5(b) and 1.6. The rows and columns represent the levels of F and G (in either order), and a tick in a cell indicates that there is at least one element which has the given levels of F and G. Here an 'element' is a plot if the set is Ω and a treatment if the set is $\mathcal{T}$. Thus the cells with ticks are the classes of $F \wedge G$. Starting at any element, you may move to any other cell in the same row that has a tick in it; then you may move within the column to any other cell with a tick in it; and so on. All the elements which can be reached in this way make up the class of $F \vee G$ which contains the starting element.

Example 10.1 revisited (Rye-grass) The information in Figure 10.1 is presented again in rectangular form in Figure 10.2. This shows that strip $\vee$ T = cultivar and strip $\wedge$ T = E.

The process of finding all the elements in a class of $F \vee G$ does not always stop after two steps. Figure 10.3 shows a hypothetical example in which $F \wedge G$ has 19 classes and $F \vee G$ has two classes.

Consider a factorial experiment with treatment factors F, G and H. Inside each F-class, all combinations of levels of G and H occur. This shows that $(F \wedge G) \vee (F \wedge H) = F$.

10.4. Hasse diagrams

Table 10.1. *Treatment factors in Example 10.5*

treatment	1	2	3	4	5	6	7	8	9
dose	0	1	2	1	2	1	2	1	2
type	Z	S	S	K	K	M	M	N	N
fumigant	1	2	2	2	2	2	2	2	2

dose	type				
	Z	S	K	M	N
none	√				
single		√	√	√	√
double		√	√	√	√

Fig. 10.4. *The nine treatments in Example 10.5*

Example 10.5 (Nematodes) The experiment in Question 4.3 has nine treatments. One is a control: no fumigant is applied. Each of the others consists of one of four types of fumigant (coded S, K, M and N) in either a single or a double dose. The best way to think about this is to consider the factor type to have five levels—S, K, M, N and none, which was coded Z in Question 4.3—and the factor dose to have the three levels none, single and double, which were coded 0, 1 and 2. Thus the nine treatments have the structure shown in Figure 10.4.

This figure shows that dose $\vee$ type has two classes. One contains just the control treatment (no fumigant) while the other contains all the other treatments. It is convenient to name this factor fumigant, as in Table 10.1 and Question 4.3. Thus we have dose $\vee$ type = fumigant.

Note that if $F \preccurlyeq G$ then $F \wedge G = F$ and $F \vee G = G$.

10.3.3 Uniform factors

Definition A factor is *uniform* if all of its classes have the same size.

Notation If factor F is uniform then k_F denotes the size of all its classes.

Let us now extend the notation n_F so that if F is *any* factor then n_F denotes its number of levels. Then if F is a uniform factor on Ω then $n_F k_F = N$, while if F is a uniform factor on $\mathcal{T}$ then $n_F k_F = t$.

The factors U and E are always uniform, with $k_E = 1$ and k_U being the size of the set. On the other hand, $n_U = 1$ and n_E is the size of the set.

In Example 10.1, field, strip, plot, cultivar and nitrogen are all uniform. In Example 10.5, none of type, dose and fumigant is uniform on $\mathcal{T}$.

10.4 Hasse diagrams

It is convenient to show the relationships between factors by drawing Hasse diagrams, which are named after the German mathematician Hasse. We have already used Hasse diagrams to show the relationships between expectation models: see Figures 2.4, 3.1–3.3, 5.2 and 5.9–5.11. Now we use the same idea for collections of factors.

Draw a dot for each factor. If $F \prec G$ then draw the dot for G (roughly) above the dot for F and join F to G with an upwards line (which may bend or may go through other dots).

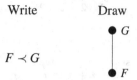

The dot for U is always at the top. If E is included, it is at the bottom.

If neither of F and G is finer than the other, make sure that the diagram contains a dot for $F \vee G$, as follows.

If F and G are plot factors and neither is finer than the other, make sure that the diagram contains a dot for $F \wedge G$, as follows.

If F and G are treatment factors and neither is finer than the other, then we include the dot for $F \wedge G$ unless we believe that the interaction between F and G is zero.

It is usually best to draw two separate Hasse diagrams, one for the plot factors and one for the treatment factors. I distinguish them by using a filled dot for a plot factor and an open dot for a treatment factor. Beside the dot for each factor F, write its number of levels, n_F.

If the set is unstructured then the Hasse diagram is very simple: it contains just U and E and the line joining them. For example, the Hasse diagram for the plot factors in Example 5.12 is in Figure 10.5(a) and the Hasse diagram for the treatment factors in Example 8.1 is in Figure 10.5(b).

If there is only one factor apart from U and E then the three factors form a chain. The plot factors for Examples 1.11 and 8.1 are shown in Figures 10.6(a) and (b). Figure 10.6(c)

Fig. 10.5. Hasse diagrams for two unstructured sets

10.4. Hasse diagrams

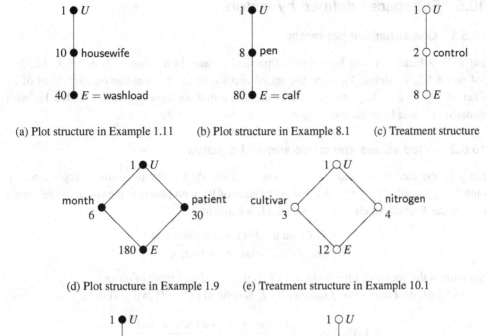

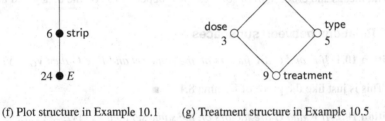

Fig. 10.6. Hasse diagrams for several examples

shows the Hasse diagram for the treatment factors if there are eight treatments, one of which is a control. Note that there is nothing in the Hasse diagram to indicate whether a factor is uniform or not. If there were eight treatments divided into two types, with four of each type, then the Hasse diagram would still look like Figure 10.6(c).

If there are two factors other than U and E and neither is finer than the other then we obtain the diamond shape in Figures 10.6(d) and (e). If one is finer than the other then we have a chain of four factors, as in Figure 10.6(f). The treatment structure in Example 10.5 is more complicated, and is shown in Figure 10.6(g). We shall meet further types of Hasse diagram later in the chapter.

10.5 Subspaces defined by factors

10.5.1 One subspace per factor

Let F be a factor. As we have done in particular cases in Sections 2.3, 3.2–3.3, 4.2, 5.1, 6.4 and 8.1.3, we define V_F to be the set of vectors which are constant on each level of F. Then $\dim V_F = n_F$. In particular, V_U is the space that we have previously called V_0, with dimension 1, and V_E is the whole space.

10.5.2 Fitted values and crude sums of squares

Let $\mathbf{y}$ be the data vector and F a factor on Ω. Then $\mathbf{P}_{V_F}\mathbf{y}$ is the orthogonal projection of $\mathbf{y}$ onto V_F: it is called the *fit* for F. The coordinate of $\mathbf{P}_{V_F}\mathbf{y}$ on plot ω is the mean of the values of $\mathbf{y}$ on the F-class which contains ω, which is equal to

$$\frac{\text{total of } \mathbf{y} \text{ on the } F\text{-class containing } \omega}{\text{size of the } F\text{-class containing } \omega}.$$

Sometimes the vector of fitted values for F is displayed as a *table of means*. As usual, the *crude sum of squares* for F is defined to be $\left\|\mathbf{P}_{V_F}\mathbf{y}\right\|^2$. Thus

$$\text{CSS}(F) = \sum_{F\text{-classes}} \frac{(\text{total of } \mathbf{y} \text{ on the } F\text{-class})^2}{\text{size of the } F\text{-class}}.$$

The fit, means and crude sum of squares for F depend only on the data $\mathbf{y}$ and the factor F.

10.5.3 Relations between subspaces

Proposition 10.1 *If F and G are factors on the same set and $F \preccurlyeq G$ then $V_G \subseteq V_F$.*

Proof This is just like the proof of Lemma 8.1. ∎

Proposition 10.2 *If F and G are factors on the same set then $V_F \cap V_G = V_{F \vee G}$.*

Proof We have $F \preccurlyeq F \vee G$, so Proposition 10.1 shows that $V_{F \vee G} \subseteq V_F$. Similarly, $V_{F \vee G} \subseteq V_G$. Therefore $V_{F \vee G} \subseteq V_F \cap V_G$.

Conversely, we need to show that $V_F \cap V_G \subseteq V_{F \vee G}$. Suppose that $\mathbf{v}$ is a vector in $V_F \cap V_G$. We can turn $\mathbf{v}$ into a factor H by putting $H(\omega) = v_\omega$. Since $\mathbf{v} \in V_F$, we know that $\mathbf{v}$ is constant on each class of F, which implies that each class of F is contained in a single class of H. In other words, $F \preccurlyeq H$. Similarly, $G \preccurlyeq H$. By the definition of supremum, $F \vee G \preccurlyeq H$. By Proposition 10.1, $V_H \subseteq V_{F \vee G}$. However, it is clear that $\mathbf{v} \in V_H$, and so $\mathbf{v} \in V_{F \vee G}$. Thus every vector in $V_F \cap V_G$ is in $V_{F \vee G}$, and so $V_F \cap V_G \subseteq V_{F \vee G}$.

Hence $V_F \cap V_G = V_{F \vee G}$. ∎

10.6 Orthogonal factors

10.6.1 Definition of orthogonality

If F and G are treatment factors then we want to consider both V_F and V_G to be expectation models. By the Intersection Principle, we should also consider $V_F \cap V_G$ to be an expectation

10.6. Orthogonal factors

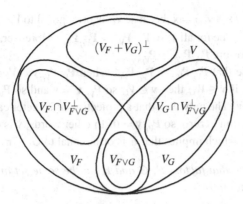

Fig. 10.7. Subspaces defined by the two factors F and G

model. Proposition 10.2 shows that this intersection is just $V_{F \vee G}$. Therefore we insist that $F \vee G$ must be a treatment factor if F and G are.

The Orthogonality Principle says that if V_F and V_G are both expectation models then the subspace $V_F \cap (V_F \cap V_G)^\perp$ should be orthogonal to the subspace $V_G \cap (V_F \cap V_G)^\perp$. Since Proposition 10.2 shows that $V_F \cap V_G = V_{F \vee G}$, this motivates the following definition.

Definition Factors F and G on the same set are *orthogonal* to each other if the subspace $V_F \cap V_{F \vee G}^\perp$ is orthogonal to the subspace $V_G \cap V_{F \vee G}^\perp$.

These subspaces are shown in Figure 10.7. Equivalent definitions of orthogonality are that $V_F \cap V_{F \vee G}^\perp$ is orthogonal to V_G, or that $V_G \cap V_{F \vee G}^\perp$ is orthogonal to V_F.

Note that if $F \preccurlyeq G$ then $F \vee G = G$ and so

$$V_G \cap V_{F \vee G}^\perp = V_G \cap V_G^\perp = \{\mathbf{0}\},$$

which is orthogonal to all vectors, so F is orthogonal to G according to our definition.

This definition is consistent with the definition of orthogonal block design in Section 4.2. So long as $B \vee T = U$, a block design is orthogonal precisely when the treatment factor T is orthogonal to the block factor B.

Our definition of orthogonality is convenient for verifying the Orthogonality Principle but fairly useless for checking whether two factors are orthogonal to each other. We now show that orthogonality is equivalent to two more tractable conditions.

10.6.2 Projection matrices commute

Theorem 10.3 *Factors F and G on the same set are orthogonal to each other if and only if $\mathbf{P}_{V_F}\mathbf{P}_{V_G} = \mathbf{P}_{V_G}\mathbf{P}_{V_F}$. If they are orthogonal to each other then $\mathbf{P}_{V_F}\mathbf{P}_{V_G} = \mathbf{P}_{V_{F \vee G}}$.*

Proof Put $W_F = V_F \cap V_{F \vee G}^\perp$, $W_G = V_G \cap V_{F \vee G}^\perp$ and $W = (V_F + V_G)^\perp$. If F is orthogonal to G then the whole space is the orthogonal direct sum $V_{F \vee G} \oplus W_F \oplus W_G \oplus W$. Thus any vector $\mathbf{v}$ has a unique expression as $\mathbf{v} = \mathbf{x} + \mathbf{v}_F + \mathbf{v}_G + \mathbf{w}$ where $\mathbf{x} \in V_{F \vee G}$, $\mathbf{v}_F \in W_F$, $\mathbf{v}_G \in W_G$ and $\mathbf{w} \in W$. Now $\mathbf{P}_{V_G}\mathbf{v} = \mathbf{x} + \mathbf{v}_G$, because $\mathbf{x}$ and $\mathbf{v}_G$ are in V_G while $\mathbf{v}_F$ and $\mathbf{w}$ are orthogonal to V_G.

Hence $\mathbf{P}_{V_F}(\mathbf{P}_{V_G}\mathbf{v}) = \mathbf{P}_{V_F}(\mathbf{x}+\mathbf{v}_G) = \mathbf{x}$, because $\mathbf{v}_G$ is orthogonal to V_F. Similarly, $\mathbf{P}_{V_G}(\mathbf{P}_{V_F}\mathbf{v}) = \mathbf{P}_{V_G}(\mathbf{x}+\mathbf{v}_F) = \mathbf{x}$. This is true for all $\mathbf{v}$, so $\mathbf{P}_{V_F}\mathbf{P}_{V_G} = \mathbf{P}_{V_G}\mathbf{P}_{V_F}$. Moreover, $V_{F\vee G}^{\perp} = W_F \oplus W_G \oplus W$, and so $\mathbf{P}_{V_{F\vee G}}\mathbf{v} = \mathbf{x}$: therefore $\mathbf{P}_{V_F}\mathbf{P}_{V_G} = \mathbf{P}_{V_{F\vee G}}$.

Conversely, suppose that $\mathbf{P}_{V_F}\mathbf{P}_{V_G} = \mathbf{P}_{V_G}\mathbf{P}_{V_F}$. Then $\mathbf{P}_{V_F}\mathbf{P}_{V_G}\mathbf{v} \in V_F \cap V_G = V_{F\vee G}$ for every vector $\mathbf{v}$. In particular, if $\mathbf{v} \in W_G$ then $\mathbf{v} \in V_G$ so $\mathbf{P}_{V_G}\mathbf{v} = \mathbf{v}$ and so $\mathbf{P}_{V_F}\mathbf{v} = \mathbf{P}_{V_F}\mathbf{P}_{V_G}\mathbf{v} \in V_{F\vee G}$, which is orthogonal to $\mathbf{v}$. The only way that a projection of a vector can be orthogonal to that vector is that the projection is zero, so $\mathbf{P}_{V_F}\mathbf{v} = \mathbf{0}$. In other words, $\mathbf{v}$ is orthogonal to V_F. Thus W_G is orthogonal to V_F, which implies that F is orthogonal to G. ∎

Corollary 10.4 *Suppose that factors F, G and H on the same set are pairwise orthogonal. Then $F \vee G$ is orthogonal to H.*

Proof Because F is orthogonal to G, $\mathbf{P}_{V_{F\vee G}} = \mathbf{P}_{V_F}\mathbf{P}_{V_G}$. Because H is orthogonal to both F and G, $\mathbf{P}_{V_H}$ commutes with both $\mathbf{P}_{V_F}$ and $\mathbf{P}_{V_G}$. Therefore

$$\mathbf{P}_{V_{F\vee G}}\mathbf{P}_{V_H} = \mathbf{P}_{V_F}\mathbf{P}_{V_G}\mathbf{P}_{V_H} = \mathbf{P}_{V_F}\mathbf{P}_{V_H}\mathbf{P}_{V_G} = \mathbf{P}_{V_H}\mathbf{P}_{V_F}\mathbf{P}_{V_G} = \mathbf{P}_{V_H}\mathbf{P}_{V_{F\vee G}}$$

and so $F \vee G$ is orthogonal to H. ∎

10.6.3 Proportional meeting

Theorem 10.5 *Factors F and G on the same set are orthogonal to each other if and only if the following two conditions are satisfied within each class of $F \vee G$ separately:*

(i) every F-class meets every G-class;

(ii) all these intersections have size proportional to the product of the sizes of the relevant F-class and G-class.

Proof Let p_i be the size of the i-th class of F, q_j the size of the j-th class of G, and s_{ij} the size of their intersection. Then s_{ij} must be zero whenever the i-th class of F and the j-th class of G are in different classes of $F \vee G$. Hence conditions (i) and (ii) together are equivalent to the existence of constants c_Δ, for each class Δ of $F \vee G$, such that

$$s_{ij} = c_\Delta p_i q_j \tag{10.2}$$

whenever the i-th class of F and the j-th class of G are in Δ. (A counting argument shows that c_Δ must be the reciprocal of the size of Δ, but we do not need to use that.)

Let $\mathbf{y}$ be the data vector. We showed in Section 2.6 that the coordinate of $\mathbf{P}_{V_F}\mathbf{y}$ is equal to $\text{sum}_{F=i}/p_i$ on every element in the i-th class of F. Hence the coordinate of $\mathbf{P}_{V_G}\mathbf{P}_{V_F}\mathbf{y}$ is equal to

$$\left(\sum_i s_{ij} \frac{\text{sum}_{F=i}}{p_i}\right) \Big/ q_j \tag{10.3}$$

on every element in the j-th class of G. If F is orthogonal to G then $\mathbf{P}_{V_G}\mathbf{P}_{V_F} = \mathbf{P}_{V_{F\vee G}}$ and so expression (10.3) has the same value for all j such that the j-th class of G is in a given class Δ of $F \vee G$. This is true no matter what the data $\mathbf{y}$ are, so $s_{ij}/q_j = s_{ij'}/q_{j'}$ whenever the j-th and j'-th classes of G are contained in Δ. Similarly, $s_{ij}/p_i = s_{i'j}/p_{i'}$ for all i and i' such that the i-th and i'-th classes of F are in Δ. Therefore Equation (10.2) is true.

10.6. Orthogonal factors

(a)

dose	type				
	Z	S	K	M	N
none	16				
single		4	4	4	4
double		4	4	4	4

(b)

dose	type				
	Z	S	K	M	N
none	8				
single		4	5	3	4
double		4	5	3	4

(c)

dose	type				
	Z	S	K	M	N
none	20				
single		5	5	5	5
double		3	3	3	3

(d)

dose	type				
	Z	S	K	M	N
none	12				
single		3	6	9	6
double		1	2	3	2

Fig. 10.8. Several different patterns of replication in Example 10.5 that make dose orthogonal to type

Conversely, let Δ be the class of $F \vee G$ containing the j-th class of G. If Equation (10.2) holds then $s_{ij}\text{sum}_{F=i}$ is equal to $c_\Delta p_i q_j \text{sum}_{F=i}$ if the i-th class of F is in Δ and to zero otherwise. Hence expression (10.3) is equal to $c_\Delta \text{sum}_{F \vee G = \Delta}$, so this is the coordinate of $\mathbf{P}_{V_G}\mathbf{P}_{V_F}\mathbf{y}$ on every element of Δ. Similarly, the coordinate of $\mathbf{P}_{V_F}\mathbf{P}_{V_G}\mathbf{y}$ is equal to $c_\Delta \text{sum}_{F \vee G = \Delta}$ on every element of Δ. Hence $\mathbf{P}_{V_F}\mathbf{P}_{V_G} = \mathbf{P}_{V_G}\mathbf{P}_{V_F}$ and so F is orthogonal to G. ∎

Note that condition (i) of Theorem 10.5 implies that the process of calculating $F \vee G$ is complete in two steps. Condition (ii) is most often achieved by having all classes of $F \wedge G$ of the same size within each class of $F \vee G$, but possibly different sizes in different classes of $F \vee G$.

10.6.4 How replication can affect orthogonality

For treatment factors, the property of orthogonality may be different on the treatment set $\mathcal{T}$ from what it is on the set Ω of plots. This is an important point that we shall return to in Section 10.12.

Example 10.5 revisited (Nematodes) On $\mathcal{T}$, the factors dose and type are orthogonal to each other.

On Ω, if the control treatment is replicated r_1 times and all other treatments are replicated r_2 times, then dose is orthogonal to type. Figure 10.8(a) shows the replication in the experiment in Question 4.3: this makes dose orthogonal to type. Some other patterns of replication that also lead to orthogonality are shown in Figures 10.8(b)–(d).

10.6.5 A chain of factors

We have already noted that F is orthogonal to G if $F \prec G$. In that case, $F \vee G = G$ and so $\mathbf{P}_{V_F}\mathbf{P}_{V_G} = \mathbf{P}_{V_G}\mathbf{P}_{V_F} = \mathbf{P}_{V_G}$.

Table 10.2. *Treatment factors in Example 10.6*

Treatment (T)	1	2	3	4	5
control	1	2	2	2	2
fumigant	1	2	2	2	3
state	1	2	2	3	4

A sequence of factors $F_1, F_2, \ldots, F_n$ on the same set is a *chain* if

$$F_1 \prec F_2 \prec \cdots \prec F_n.$$

Thus the collection of factors in a chain are mutually orthogonal.

Figures 10.6(a), (b), (c) and (f) show chains of factors.

Example 10.6 (Soil fungicide) In an experiment on eucalypts, the treatments are fungicides applied to the soil. Treatment 1 is 'no fungicide'. Treatments 2, 3 and 4 are fumigants, which leave little residue in the soil, whereas treatment 5 is not a fumigant, and so leaves a large residue in the soil. Treatment 2 is in liquid form; treatment 3 is liquid under pressure; and treatment 4 is granular.

Table 10.2 shows relevant factors on these treatments. Differences between levels of control show whether there is any difference between no fungicide and some fungicide. Differences between levels of fumigant within level 2 of control show whether there is any difference between fumigants and non-fumigants. Differences between levels of state within level 2 of fumigant show whether there is any difference between liquid fumigants and solid fumigants. Finally, differences between levels of T within level 2 of state show whether there is any difference between the two liquid fungicides.

Thus we have the chain

$$T \prec \text{state} \prec \text{fumigant} \prec \text{control} \prec U,$$

and these five factors are mutually orthogonal.

10.7 Orthogonal decomposition

10.7.1 A second subspace for each factor

As we have previously done in particular cases, we now want to define a W-subspace associated with each factor in such a way that W_F is orthogonal to W_G if F and G are different factors. If $F \prec G$ then $V_G \subset V_F$ so we want W_F to be contained in V_F but orthogonal to V_G. This is the case for all G coarser than F, so we want W_F to be orthogonal to the space $\sum_{G \succ F} V_G$. It turns out that the Intersection Principle and the Orthogonality Principle give us just the right conditions to make the W-subspaces an orthogonal decomposition of the whole space.

Theorem 10.6 *Let $\mathcal{F}$ be a set of mutually inequivalent factors on the same set. Suppose that $\mathcal{F}$ satisfies*

(a) *if $F \in \mathcal{F}$ and $G \in \mathcal{F}$ then $F \vee G \in \mathcal{F}$;*

10.7. Orthogonal decomposition

(b) if $F \in \mathcal{F}$ and $G \in \mathcal{F}$ then F is orthogonal to G.

Define subspaces W_F for F in $\mathcal{F}$ by

$$W_F = V_F \cap \left(\sum_{G \succ F} V_G \right)^\perp$$

and put $d_F = \dim W_F$. Then

(i) if F and G are different factors in $\mathcal{F}$ then W_F is orthogonal to W_G;

(ii) if $F \in \mathcal{F}$ then V_F is the orthogonal direct sum of those W_G for which $G \succcurlyeq F$: in particular, if $E \in \mathcal{F}$ then the whole space is equal to

$$\bigoplus_{F \in \mathcal{F}} W_F;$$

(iii) if $F \in \mathcal{F}$ then

$$d_F = n_F - \sum_{G \succ F} d_G. \qquad (10.4)$$

Proof (i) If $F \neq G$ then at least one of F and G is different from $F \vee G$. Suppose that $F \neq F \vee G$. Then $F \prec F \vee G$ so $W_F \subseteq V_F \cap V_{F \vee G}^\perp$, while $W_G \subseteq V_G$. Since the factors F and G are orthogonal to each other, the subspaces $V_F \cap V_{F \vee G}^\perp$ and V_G are orthogonal to each other. Therefore W_F is orthogonal to W_G.

(ii) We use induction on F, starting with the coarsest factors in $\mathcal{F}$. If there is no G in $\mathcal{F}$ such that $G \succ F$ then $W_F = V_F$ and the result is true.

Suppose that the result is true for all factors G in $\mathcal{F}$ for which $G \succ F$. Then

$$\sum_{G \succ F} V_G = \sum_{G \succ F} \left(\bigoplus_{H \succcurlyeq G} W_H \right) = \bigoplus_{G \succ F} W_G.$$

The definition of W_F shows that

$$V_F = W_F \oplus \left(\sum_{G \succ F} V_G \right) = W_F \oplus \left(\bigoplus_{G \succ F} W_G \right).$$

Hence

$$V_F = \bigoplus_{G \succcurlyeq F} W_G. \qquad (10.5)$$

(iii) The dimension of the right-hand side of Equation (10.5) is equal to $\sum_{G \succcurlyeq F} d_F$, which is equal to $d_F + \sum_{G \succ F} d_G$. The dimension of the left-hand side is equal to n_F. Therefore $n_F = d_F + \sum_{G \succ F} d_G$, and Equation (10.4) follows. ∎

Note that

$$\left(\sum_{G \succ F} V_G \right)^\perp = \bigcap_{G \succ F} V_G^\perp,$$

and so
$$W_F = V_F \cap \bigcap_{G \succ F} V_G^\perp.$$

The number d_F is called the number of *degrees of freedom for F*. It is important to realize that, although n_F depends only on F, d_F depends also on what other factors are included in $\mathcal{F}$.

10.7.2 Effects and sums of squares

Effects, and sums of squares, can be defined unambiguously only when the hypotheses of Theorem 10.6 are satisfied.

Theorem 10.7 *If $\mathcal{F}$ is a set of mutually inequivalent factors which satisfies the hypotheses of Theorem 10.6, then*

(i) *if F and G are different factors in $\mathcal{F}$ then $\mathbf{P}_{W_F}\mathbf{y}$ is orthogonal to $\mathbf{P}_{W_G}\mathbf{y}$;*

(ii) *if $F \in \mathcal{F}$ then $\mathbf{P}_{V_F}\mathbf{y} = \sum_{G \succcurlyeq F} \mathbf{P}_{W_G}\mathbf{y}$: in particular, if $E \in \mathcal{F}$ then*
$$\mathbf{y} = \mathbf{P}_V \mathbf{y} = \mathbf{P}_{V_E}\mathbf{y} = \sum_{F \in \mathcal{F}} \mathbf{P}_{W_F}\mathbf{y};$$

(iii) *if $F \in \mathcal{F}$ then $\left\|\mathbf{P}_{V_F}\mathbf{y}\right\|^2 = \sum_{G \succcurlyeq F} \left\|\mathbf{P}_{W_G}\mathbf{y}\right\|^2.$*

Proof (i) Theorem 10.6(i) shows that W_F is orthogonal to W_G, and so every vector in W_F is orthogonal to every vector in W_G.

(ii) Theorem 10.6(ii) shows that V_F is the orthogonal direct sum of the spaces W_G for which $G \succcurlyeq F$. Hence the orthogonal projection of $\mathbf{y}$ onto V_F is the sum of the orthogonal projections of $\mathbf{y}$ onto those W_G.

(iii) Since the vectors $\mathbf{P}_{W_G}\mathbf{y}$ with $G \succcurlyeq F$ are mutually orthogonal and sum to $\mathbf{P}_{V_F}\mathbf{y}$, the result follows from Pythagoras's Theorem. ∎

When the conditions of Theorem 10.6 are satisfied then we call $\mathbf{P}_{W_F}\mathbf{y}$ the *effect* of F and $\left\|\mathbf{P}_{W_F}\mathbf{y}\right\|^2$ the *sum of squares* for F. Theorem 10.7(ii) states that

$$\text{fit for } F = \mathbf{P}_{V_F}\mathbf{y} = \text{effect of } F + \sum_{G \succ F} (\text{effect of } G),$$

and so
$$\text{effect of } F = \text{fit for } F - \sum_{G \succ F} (\text{effect of } G). \tag{10.6}$$

Similarly, Theorem 10.7(iii) states that
$$\text{CSS}(F) = \left\|\mathbf{P}_{V_F}\mathbf{y}\right\|^2 = \text{SS}(F) + \sum_{G \succ F} \text{SS}(G),$$

and so
$$\text{SS}(F) = \text{CSS}(F) - \sum_{G \succ F} \text{SS}(G). \tag{10.7}$$

Thus the effect of F and the sum of squares for F depend on the data $\mathbf{y}$, the factor F, and also on what other factors are included in $\mathcal{F}$.

All the analyses which we have looked at so far use special cases of Theorems 10.6 and 10.7.

10.8. Calculations on the Hasse diagram

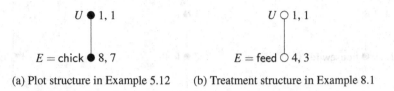

(a) Plot structure in Example 5.12   (b) Treatment structure in Example 8.1

Fig. 10.9. Hasse diagrams for two unstructured sets, showing numbers of levels and degrees of freedom

10.8 Calculations on the Hasse diagram

10.8.1 Degrees of freedom

It is very convenient to calculate degrees of freedom by using the Hasse diagram. Start at the top. There is no factor coarser than U, so Equation (10.4) shows that $d_U = n_U = 1$. Then work down the Hasse diagram, using Equation (10.4) at each dot F to calculate d_F from n_F and the degrees of freedom for those dots above F. At each dot, write d_F beside n_F.

When doing this process by hand, you can use one colour for the numbers of levels n_F and another for the degrees of freedom d_F. In this book I adopt the convention that d_F is always shown either immediately to the right of n_F or immediately below n_F.

Figures 10.5 and 10.6 are redisplayed as Figures 10.9 and 10.10 with the degrees of freedom included.

Example 10.7 (Example 3.4 continued: Drugs at different stages of development) The treatments consist of three doses of the old formulation A and three doses (not comparable with the first three) of a new formulation B. Table 10.3 repeats the information that was given in Chapter 3, showing a factor F which distinguishes between the two formulations, and factors A and B which are designed for testing for differences among the different doses of each formulation. These give the Hasse diagram in Figure 10.11(a). Notice that $d_E = 0$; that is, the space W_E consists of the zero vector only. Thus there are only three treatment sums of squares in addition to the mean. This agrees with the finding in Chapter 3.

Example 10.8 (Rats) Six different diets were fed to 60 rats, ten rats per diet. The rats were weighed at the beginning and the end of the experiment, and their weight gain was recorded. The six diets consisted of three sources of protein, each at a high or low amount. The sources of protein were beef, pork and cereal, so there is a relevant factor animal which distinguishes between the two animal sources and the cereal. The treatment factors are displayed in Table 10.4 and Figure 10.11(b).

Table 10.3. *Treatment factors in Example 10.7*

	Old formulation			New formulation		
treatment	1	2	3	4	5	6
F	1	1	1	2	2	2
A	1	2	3	0	0	0
B	0	0	0	1	2	3

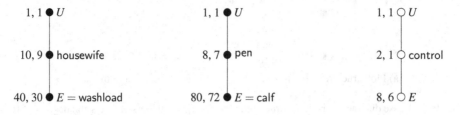

(a) Plot structure in Example 1.11   (b) Plot structure in Example 8.1   (c) Treatment structure

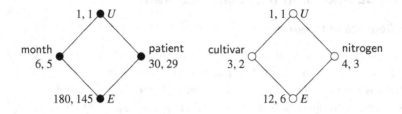

(d) Plot structure in Example 1.9   (e) Treatment structure in Example 10.1

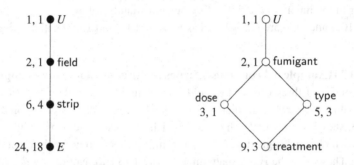

(f) Plot structure in Example 10.1   (g) Treatment structure in Example 10.5

Fig. 10.10. Hasse diagrams for several examples, showing numbers of levels and degrees of freedom

If the treatments consist of all combinations of the levels of three treatment factors F, G and H then we obtain the Hasse diagram in Figure 10.12.

Example 10.9 (Example 1.15 continued: Fungicide factorial) Fungicide was sprayed early, mid-season or late, or at any combination of those times (including none at all). This might

Table 10.4. *Treatment factors in Example 10.8*

treatment	1	2	3	4	5	6
source	beef		pork		cereal	
amount	low	high	low	high	low	high
animal	1	1	1	1	2	2
animal ∧ amount	1	2	1	2	3	4

10.8. Calculations on the Hasse diagram

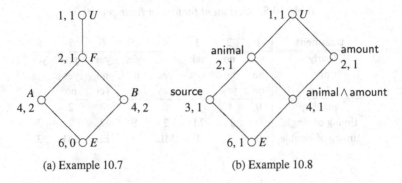

(a) Example 10.7

(b) Example 10.8

Fig. 10.11. Hasse diagrams for two treatment structures

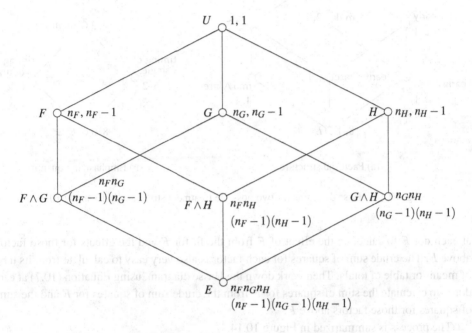

Fig. 10.12. Hasse diagram for a factorial treatment structure with three factors F, G and H

be considered as a $2 \times 2 \times 2$ factorial experiment, giving the first three rows of Table 10.5 and the Hasse diagram in Figure 10.13(a). However, the total quantity of the spray may be more important than the timing, so it may be more relevant to use the bottom three rows of Table 10.5, which give the Hasse diagram in Figure 10.13(b).

10.8.2 Sums of squares

Although it is impractical to write effects or sums of squares on the Hasse diagram, the same method of calculation gives both of these. The fit for each factor is easy to calculate: it is little more than its table of means. Then work down the Hasse diagram, using Equation (10.6)

Table 10.5. *Treatment factors in Example 10.9*

treatment	1	2	3	4	5	6	7	8
early	no	no	no	no	yes	yes	yes	yes
mid	no	no	yes	yes	no	no	yes	yes
late	no	yes	no	yes	no	yes	no	yes
quantity	0	1	1	2	1	2	2	3
timing of single	0	L	M	2	E	2	2	3
timing of double	0	1	1	ML	1	EL	EM	3

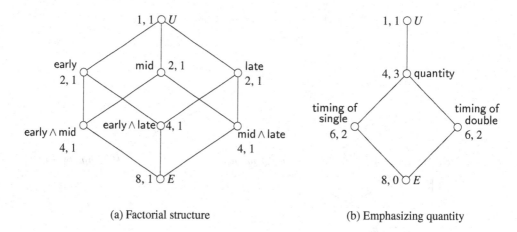

(a) Factorial structure (b) Emphasizing quantity

Fig. 10.13. Hasse diagrams for two possible treatment structures in Example 10.9

at each dot F to calculate the effect of F from the fit for F and the effects for those factors above F. The crude sum of squares for each factor is also very easy to calculate from its table of means or table of totals. Then work down the Hasse diagram, using Equation (10.7) at each dot F to calculate the sum of squares for F from the crude sum of squares for F and the sums of squares for those factors above F.

The process is summarized in Figure 10.14.

Example 10.10 (Example 8.3 continued: Cider apples) Pruning method 1 was no pruning at all. The other four methods were all combinations of two times of pruning with two parts of the tree to be pruned. These methods are shown in Figure 10.15(a), together with the data

easy calculation	n_F	fit for F	CSS(F)	for each F individually
use the Hasse diagram	↓	↓	↓	
	d_F	effect of F	SS(F)	using all G with $G \succcurlyeq F$

Fig. 10.14. Obtaining degrees of freedom, effects, and sums of squares

10.9. Orthogonal treatment structures

time	part			total
	none	open centre	modified leader	
never	1 320.77			320.77
winter only		2 242.91	4 280.41	523.32
winter and summer		3 216.28	5 187.62	403.90
total	320.77	459.19	468.03	1247.99
		927.22		

(a) Pruning methods and data totals

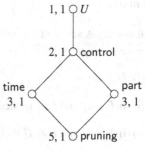

(b) Factors for pruning methods

Fig. 10.15. *Diagrams for Example 10.10*

total for each method. Figure 10.15(b) gives the Hasse diagram for these treatment factors.

The totals in Figure 10.15 give the crude sums of squares which are shown in the second column of Table 10.6. For example, CSS(control) = $320.77^2/6 + 927.22^2/24 = 52971.27$ and CSS(part) = $320.77^2/6 + 459.19^2/12 + 468.03^2/12 = 52974.53$. Then successive subtractions give the sums of squares in the third column of Table 10.6. Each source has a single degree of freedom, so the mean squares are the same as the sums of squares. From Table 8.12, the relevant residual mean square is 74.04 on 12 degrees of freedom, which is used to give the variance ratios in the final column of Table 10.6. We conclude that the differences between the five pruning methods can be accounted for by the difference between no pruning and some pruning, and by the difference between pruning in the winter only and pruning in both winter and summer.

10.9 Orthogonal treatment structures

10.9.1 Conditions on treatment factors

In Chapter 4 we saw that if the plots are structured then V_0 needs to be removed from both the treatment subspace and the block subspace. In other words, the universal factor U must be considered both a treatment factor and a plot factor.

Table 10.6. *Calculations in Example 10.10*

Factor	CSS	SS	df	VR
U	51915.97	51915.97	1	–
control	52971.27	1055.30	1	14.25
part	52974.53	3.26	1	0.04
time	53565.48	594.21	1	8.03
part ∧ time	53751.12	182.38	1	2.46

This consideration and the discussion at the start of Section 10.7.1 motivate the following definition.

Definition A set $\mathcal{G}$ of mutually inequivalent factors on the treatment set $\mathcal{T}$ is an *orthogonal treatment structure* if

(i) $U \in \mathcal{G}$;

(ii) if $F \in \mathcal{G}$ and $G \in \mathcal{G}$ then $F \vee G \in \mathcal{G}$;

(iii) if $F \in \mathcal{G}$ and $G \in \mathcal{G}$ then F is orthogonal to G.

Condition (i) lets us remove the grand mean from all treatment subspaces. If we did not have condition (ii), the degrees of freedom for $F \vee G$ would be included in whichever of F and G was fitted first, but not in the other. Condition (iii) is needed for the Orthogonality Principle.

For the rest of this section we assume that $\mathcal{G}$ is an orthogonal treatment structure.

10.9.2 Collections of expectation models

For each F in $\mathcal{G}$, we want to consider V_F as an expectation model. By the Sum Principle, if F and G are both in $\mathcal{G}$ then $V_F + V_G$ should also be an expectation model. Thus the expectation models should be all sums of zero, one, two or more of the spaces V_F for F in $\mathcal{G}$. These are subspaces of the form $\sum_{K \in \mathcal{K}} V_K$ for subsets $\mathcal{K}$ of $\mathcal{G}$. Theorem 10.6 shows that each such subspace is the direct sum of some of the spaces W_F with F in $\mathcal{G}$.

Theorem 10.8 *Let $\mathcal{H}$ be a subset of $\mathcal{G}$. Then $\bigoplus_{H \in \mathcal{H}} W_H$ is an expectation model if and only if $\mathcal{H}$ satisfies the following condition:*

$$\text{if } F \in \mathcal{H} \text{ and } G \in \mathcal{G} \text{ and } F \preccurlyeq G \text{ then } G \in \mathcal{H}. \tag{10.8}$$

Proof Put $M = \bigoplus_{H \in \mathcal{H}} W_H$. First suppose that M is an expectation model. Then there is some subset $\mathcal{K}$ of $\mathcal{G}$ such that $M = \sum_{K \in \mathcal{K}} V_K$. If $F \in \mathcal{H}$ then $W_F \subseteq M$. Theorem 10.6(ii) shows that if $K \in \mathcal{G}$ then either $W_F \subseteq V_K$ or W_F is orthogonal to V_K. Hence there is some K in $\mathcal{K}$ such that $W_F \subseteq V_K \subseteq M$. Theorem 10.6(ii) shows that $K \preccurlyeq F$. If $G \in \mathcal{G}$ and $F \preccurlyeq G$ then $K \preccurlyeq G$. Hence $W_G \subseteq V_G \subseteq V_K \subseteq M$ and so $G \in \mathcal{H}$.

Conversely, suppose that $\mathcal{H}$ satisfies Condition (10.8). If $F \in \mathcal{H}$ then $G \in \mathcal{H}$ for all G with G in $\mathcal{G}$ and $F \preccurlyeq G$: hence $M \supseteq \bigoplus_{G \succcurlyeq F} W_G = V_F$. Therefore $M = \sum_{F \in \mathcal{H}} V_F$, which is an expectation model. ∎

We shall now show that this collection of expectation models satisfies the three principles discussed in Section 5.3. Let $M_1 = \bigoplus_{H \in \mathcal{H}_1} W_H$ and $M_2 = \bigoplus_{H \in \mathcal{H}_2} W_H$, where $\mathcal{H}_1$ and $\mathcal{H}_2$ both satisfy Condition (10.8). Then

$$M_1 + M_2 = \bigoplus_{H \in \mathcal{H}_1 \cup \mathcal{H}_2} W_H.$$

If $F \in \mathcal{H}_1 \cup \mathcal{H}_2$ then $F \in \mathcal{H}_i$ for $i = 1$ or 2. If $F \preccurlyeq G$ then G is also in $\mathcal{H}_i$ and so $G \in \mathcal{H}_1 \cup \mathcal{H}_2$. Hence $\mathcal{H}_1 \cup \mathcal{H}_2$ satisfies Condition (10.8) and so the Sum Principle is satisfied.

10.9. Orthogonal treatment structures

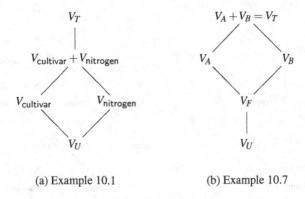

Fig. 10.16. Collections of expectation models

Because the W-spaces are mutually orthogonal,

$$M_1 \cap M_2 = \bigoplus_{H \in \mathcal{H}_1 \cap \mathcal{H}_2} W_H.$$

If $F \in \mathcal{H}_1 \cap \mathcal{H}_2$ then $F \in \mathcal{H}_1$ and $F \in \mathcal{H}_2$. If $F \preccurlyeq G$ also, then $G \in \mathcal{H}_1$ and $G \in \mathcal{H}_2$ and so $G \in \mathcal{H}_1 \cap \mathcal{H}_2$. Hence $\mathcal{H}_1 \cap \mathcal{H}_2$ satisfies Condition (10.8) and so the Intersection Principle is satisfied.

Finally, put $\mathcal{H}'_1 = \{H \in \mathcal{H}_1 : H \notin \mathcal{H}_2\}$ and $\mathcal{H}'_2 = \{H \in \mathcal{H}_2 : H \notin \mathcal{H}_1\}$. Then

$$M_1 \cap (M_1 \cap M_2)^\perp = \bigoplus_{H \in \mathcal{H}'_1} W_H \quad \text{and} \quad M_2 \cap (M_1 \cap M_2)^\perp = \bigoplus_{H \in \mathcal{H}'_2} W_H.$$

Because $\mathcal{H}'_1$ and $\mathcal{H}'_2$ have no factors in common, these two spaces are orthogonal to each other, and so the Orthogonality Principle is satisfied.

The diagram showing the relationships between the expectation models is the opposite way up to the Hasse diagram for the factors. Of course, we could draw the Hasse diagram for the factors the other way up too, but then it would no longer correspond to the conventional order in the analysis-of-variance table. In general, the diagram for the expectation models has more points than the Hasse diagram for the factors because a model such as $V_F + V_G$ may not be a factor subspace if neither of F and G is finer than the other.

Each line in the model diagram corresponds to a single W-space, which is used to test whether we can simplify down that line.

Example 10.1 revisited (Rye-grass) Here the treatments are all combinations of the factors cultivar and nitrogen. The Hasse diagram for the treatment factors is in Figure 10.10(e). The expectation models (apart from the zero model) are shown in Figure 10.16(a).

Example 10.7 revisited (Drugs at different stages of development) The treatment structure is shown in Figure 10.11(a). Although this has more treatment factors than the previous example, it has the same number of expectation models, because $V_A + V_B = V_T$. Figure 10.16(b) repeats the diagram of the expectation models from Figure 3.3, omitting the zero model.

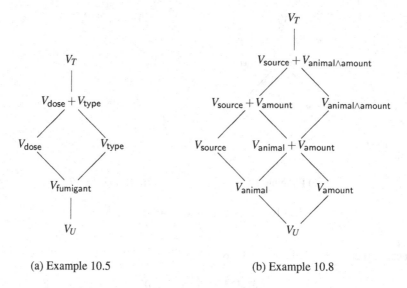

(a) Example 10.5   (b) Example 10.8

Fig. 10.17. More collections of expectation models

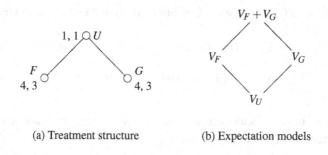

(a) Treatment structure   (b) Expectation models

Fig. 10.18. Diagrams for Example 10.11

Example 10.5 revisited (Nematodes) Figure 10.10(g) gives the Hasse diagram for the treatment factors. The nonzero expectation models are shown in Figure 10.17(a).

Example 10.8 revisited (Rats) The Hasse diagram for the treatment factors is shown in Figure 10.11(b). The expectation models, other than the zero model, are in Figure 10.17(b).

If the treatments consist of all levels of three treatment factors then we have the Hasse diagram in Figure 10.12 and the nonzero expectation models in Figure 5.11.

Example 10.11 (Main-effects only design in blocks) The design in Section 9.1.3 is for two treatment factors F and G, each with four levels, on the assumption that their interaction is zero. Thus the treatment structure is simply the one shown in Figure 10.18(a), with the collection of expectation models in Figure 10.18(b), omitting the zero model.

10.10. Orthogonal plot structures

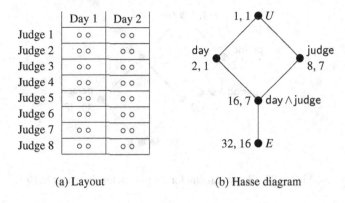

(a) Layout (b) Hasse diagram

Fig. 10.19. Plot structure in Example 10.12

10.10 Orthogonal plot structures

10.10.1 Conditions on plot factors

When the plot structure is any more complicated than those in Chapters 2, 4 and 6, it is usual to assume that the plot structure gives random effects rather than fixed effects; that is, the plot structure makes no difference to the expectation model but does determine the covariance matrix. Thus we hope for a plot structure which defines a covariance matrix whose strata can be determined.

Since the strata should give an orthogonal decomposition of the whole space, and one of the strata should be V_0, a reasonable plot structure must satisfy at least the three conditions for an orthogonal treatment structure. In addition, it must satisfy three more conditions.

First, E must be included, so that we obtain a decomposition of the *whole space* V_E. As Example 10.11 shows, this condition is not necessary for an orthogonal treatment structure.

Secondly, all factors must be uniform. If there are blocks of different sizes then we cannot relabel them by randomization. Furthermore, it may not be reasonable to assume that the covariance between a pair of plots in a larger block is the same as the covariance between a pair of plots in a smaller block. Again this contrasts with treatment structure, where we have seen several examples of treatment factors that are not uniform.

Thirdly, if F and G are both plot factors then so must $F \wedge G$ be. Two further examples should make the reason for this clear.

Example 10.12 (Soap pads) A manufacturer of soap pads tries to improve them by varying the amount of detergent, the solubility of the detergent and the coarseness of the pad. Each of these treatment factors has just two levels, high and low, and so there are eight treatments.

When a quantity of each type of soap pad has been made, the manufacturer compares them in a two-day trial during which each of eight judges tests two soap pads per day and scores them on a subjective scale from 1 to 5. Thus the plot structure is schematically as shown in Figure 10.19(a), while the Hasse diagram for the plot factors is in Figure 10.19(b).

Consider the pattern of correlations in this plot structure. We might expect a correlation ρ_U between plots which are in different judges and different days, another correlation ρ_J between plots which are in the same judge but different days, and a third correlation ρ_D between plots

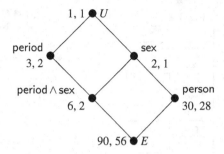

Fig. 10.20. Hasse diagram for the plot factors in Example 10.13

which are in the same day but different judges. However, the correlation between a pair of plots which are in the same day and the same judge should probably be different from all of the previous three. In other words, we need to take account of the factor day ∧ judge.

Thinking about randomization leads us to a similar conclusion. Any method of randomization must preserve the grouping of the plots into judges, and it must preserve the grouping of the plots into days. Thus it cannot help but preserve the classes of day ∧ judge.

Example 10.13 (Cross-over with blocks) Suppose that a cross-over trial for three treatments uses 30 people, of whom half are men and half are women. If sex is considered a relevant plot factor then the correlation between different plots in the same time-period should depend on whether or not they are in the same sex. Then it is logical that the correlation between plots in different time-periods and different subjects should also depend on whether or not they are in the same sex. In other words, the plot factor period ∧ sex is relevant.

The Hasse diagram is in Figure 10.20.

Definition A set $\mathcal{F}$ of mutually inequivalent factors on Ω is an *orthogonal plot structure* if

(i) every factor in $\mathcal{F}$ is uniform;

(ii) $U \in \mathcal{F}$;

(iii) $E \in \mathcal{F}$;

(iv) if $F \in \mathcal{F}$ and $G \in \mathcal{F}$ then $F \vee G \in \mathcal{F}$;

(v) if $F \in \mathcal{F}$ and $G \in \mathcal{F}$ then $F \wedge G \in \mathcal{F}$;

(vi) if $F \in \mathcal{F}$ and $G \in \mathcal{F}$ then F is orthogonal to G.

For the rest of this section we assume that $\mathcal{F}$ is an orthogonal plot structure.

10.10.2 Variance and covariance

Suppose that α and β are in Ω. Then $U(\alpha) = U(\beta)$, so there is at least one factor in $\mathcal{F}$ for which α and β are in the same class. If F and G are two such factors then $\beta \in F[[\alpha]] \cap G[[\alpha]] =$

10.10. Orthogonal plot structures

$(F \wedge G)[[\alpha]]$, so $F \wedge G$ is another, by condition (v). Consequently there is a factor F in $\mathcal{F}$ with this property that it is *finest* in the sense that any other factor in $\mathcal{F}$ which has this property must be coarser than F. In fact, if $F_1, F_2, \ldots, F_n$ are the factors in $\mathcal{F}$ which take the same level on α and β, then $F = F_1 \wedge F_2 \wedge \cdots \wedge F_n$.

Now, it is reasonable to suppose that the covariance between Y_α and Y_β just depends on this finest factor F. When $\alpha = \beta$ then this finest factor is E, and so the assumption says that all the responses have the same variance, say σ^2. More generally, there are constant correlations ρ_F, for F in $\mathcal{F}$, such that the covariance between Y_α and Y_β is $\rho_F \sigma^2$ when F is the finest factor for which α and β are in the same class.

10.10.3 Matrix formulation

As we did in Chapter 4 for the block factor B, and in Chapter 6 for the row factor R and the column factor C, for each factor F in $\mathcal{F}$ we define an $N \times N$ matrix $\mathbf{J}_F$ whose (α, β)-entry is equal to

$$\begin{cases} 1 & \text{if } F(\alpha) = F(\beta) \\ 0 & \text{otherwise.} \end{cases}$$

For an orthogonal plot structure we can also define another $N \times N$ matrix $\mathbf{A}_F$ by putting the (α, β)-entry of $\mathbf{A}_F$ equal to

$$\begin{cases} 1 & \text{if } F \text{ is the finest factor in } \mathcal{F} \text{ such that } F(\alpha) = F(\beta) \\ 0 & \text{otherwise.} \end{cases}$$

Thus our assumption is that

$$\text{Cov}(\mathbf{Y}) = \sigma^2 \left(\sum_{F \in \mathcal{F}} \rho_F \mathbf{A}_F \right), \tag{10.9}$$

with $\rho_E = 1$.

Now, if $F(\alpha) = F(\beta)$ then the finest factor G which has α and β in the same class must satisfy $G \preccurlyeq F$. This shows that

$$\mathbf{J}_F = \sum_{G \preccurlyeq F} \mathbf{A}_G$$

for all F in $\mathcal{F}$. Thus

$$\mathbf{J}_F = \mathbf{A}_F + \sum_{G \prec F} \mathbf{A}_G,$$

which can be rewritten as

$$\mathbf{A}_F = \mathbf{J}_F - \sum_{G \prec F} \mathbf{A}_G. \tag{10.10}$$

This implies that the $\mathbf{A}$-matrices can be calculated from the $\mathbf{J}$-matrices by using the Hasse diagram, but this time starting at the bottom and working up. At the bottom, Equation (10.10) gives

$$\mathbf{A}_E = \mathbf{J}_E = \mathbf{I}.$$

When the $\mathbf{A}$-matrices for all points below F have been calculated, use Equation (10.10) to calculate $\mathbf{A}_F$ from $\mathbf{J}_F$. In fact, we do not actually need to do this calculation, but we need to know that it can be done in principle, because it shows that every $\mathbf{A}$-matrix is a linear combination of the $\mathbf{J}$-matrices.

10.10.4 Strata

Theorem 10.9 *Let $\mathcal{F}$ be an orthogonal plot structure on Ω. Suppose that*

$$\mathrm{Cov}(\mathbf{Y}) = \sigma^2 \sum_{F \in \mathcal{F}} \rho_F \mathbf{A}_F,$$

for some (unknown) variance σ^2 and (unknown) correlations ρ_F for F in $\mathcal{F}$ (with $\rho_E = 1$). Then the W-subspaces defined in Theorem 10.6 are the eigenspaces of $\mathrm{Cov}(\mathbf{Y})$; that is, they are the strata.

Proof First we note that, because $\mathcal{F}$ is an orthogonal plot structure, it does satisfy the hypotheses of Theorem 10.6. Hence the definition of W-subspaces given there makes sense. Moreover, these spaces are orthogonal to each other and the whole space V is the direct sum of the W_F for F in $\mathcal{F}$ (excluding any W_F that happens to be zero).

Put $\mathbf{C} = \mathrm{Cov}(\mathbf{Y})$. Then $\mathbf{C} = \sigma^2 \sum_{F \in \mathcal{F}} \rho_F \mathbf{A}_F$. Since each $\mathbf{A}$-matrix is a linear combination of the $\mathbf{J}$-matrices, there must be constants θ_F, for F in $\mathcal{F}$, such that $\mathbf{C} = \sum_{F \in \mathcal{F}} \theta_F \mathbf{J}_F$.

Let $\mathbf{x}$ be a vector in V and let F be a factor in $\mathcal{F}$. Then F is uniform. An argument like the one in Section 4.6 shows that if $\mathbf{x} \in V_F$ then $\mathbf{J}_F \mathbf{x} = k_F \mathbf{x}$ while if $\mathbf{x} \in V_F^\perp$ then $\mathbf{J}_F \mathbf{x} = \mathbf{0}$. If $\mathbf{x} \in W_G$ then $\mathbf{x}$ is contained in every V_F for which $F \preccurlyeq G$ but is orthogonal to every other V-space, and hence $\mathbf{C}\mathbf{x} = \left(\sum_{F \preccurlyeq G} k_F \theta_F\right) \mathbf{x}$. Thus W_G is an eigenspace of $\mathbf{C}$. ∎

In principle we can calculate the eigenvalues of $\mathrm{Cov}(\mathbf{Y})$ from the covariances $\rho_F \sigma^2$ by using the Hasse diagram twice, but we shall not do so. We simply call the eigenvalues of $\mathrm{Cov}(\mathbf{Y})$ the *stratum variances*, and denote them ξ_F for F in $\mathcal{F}$. We usually assume that if $F \prec G$ then $\xi_F < \xi_G$.

10.11 Randomization

For almost all orthogonal plot structures in actual use, the Hasse diagram gives the correct method of randomization. Draw the Hasse diagram for all the plot factors (ignoring treatment factors). Randomize by working down the diagram from U to E as follows.

(i) At U, do nothing, and mark U as 'done';

(ii) at F, if there is a single line coming down into F like this,

and if G has been marked 'done', then randomize whole classes of F within each class of G; then mark F as 'done';

(iii) at F, if there are two or more lines coming down into F and all the points above F have been marked 'done', then do nothing and mark F as 'done';

(iv) continue until E is 'done'.

Although there are some orthogonal plot structures for which this method does not work, the only one that I have come across in practice is the superimposed design like the one in Section 9.3.1. Even this is a problem only when the second design has not been randomized at the same time as the first.

Example 10.13 revisited (Cross-over with blocks) We randomize periods and sexes and independently. Then we randomize people within each sex independently. The periods and people define the experimental units uniquely, so there is nothing more to do.

10.12 Orthogonal designs

10.12.1 Desirable properties

Each design has three components—the plot structure, the treatment structure and the design function T which allocates treatments to plots. In Sections 10.9–10.10 we examined desirable properties of the first two. Now we can look at the design function.

The first important property of the design function is that treatment factors which are orthogonal to each other when considered as factors on $\mathcal{T}$ should remain orthogonal to each other when considered as factors on Ω. This ensures that the meaningful orthogonal decomposition of the treatment space is not changed by the choice of replications for the treatments. The simplest way to achieve this is with equal replication. If the Hasse diagram for the treatment structure is just a chain, as in Figure 10.6(c) or Example 10.6, then the treatment structure remains orthogonal on Ω no matter what the replication is. The same is true for treatment structures like that in Example 10.7, where there is one factor splitting the treatments into types and, for each type, another factor distinguishing between the treatments of that type. Any other case of unequal replication needs careful checking for orthogonality. Some possibilities are shown in Figure 10.8.

If G is a treatment factor then we want the treatment subspace W_G to lie in one stratum and hence be orthogonal to the other strata. Thus we need G to be orthogonal to all the plot factors. In particular, T itself must be orthogonal to all the plot factors. In a complicated design, this orthogonality condition needs some checking, but Corollary 10.4 can be used to cut down the work.

In Example 10.1 we saw that strip $\vee\, T =$ cultivar. We know from Section 8.3 that the subspace for cultivar is estimated in a different stratum from the rest of the treatment space. In general, if F is a plot factor and G is a treatment factor then we need $F \vee G$ to be a treatment factor so that the subspace for $F \vee G$ can be removed from W_G.

Surprisingly, this third condition is not a real constraint, as we shall now show. If F is a plot factor and G is a treatment factor then $T \preccurlyeq G \preccurlyeq F \vee G$, so $F \vee G$ does give a grouping of the treatments, and this can simply be added to the list of treatment factors. If $F \vee G$ has no natural meaning on the treatments then some people call it a treatment *pseudofactor* rather than a treatment factor.

Suppose that we simply take all suprema of the form $F \vee G$, where F is a plot factor and G is a treatment factor, and include them in the list of treatment factors. There is no need to repeat the process, because if F_1 and F_2 are plot factors then $F_1 \vee (F_2 \vee G) = (F_1 \vee F_2) \vee G$, which is in the list after the first run, because $F_1 \vee F_2$ is also a plot factor. Moreover, if G_1 and G_2 are treatment factors then so is $G_1 \vee G_2$: now $(F_1 \vee G_1) \vee (F_2 \vee G_2) = (F_1 \vee F_2) \vee (G_1 \vee G_2)$, which is also in the new list of treatment factors. So long as every plot factor is orthogonal to every (original) treatment factor, orthogonality is not violated by the inclusion of these new treatment factors: Corollary 10.4 shows first that $F_1 \vee G_1$ is orthogonal to F_2, secondly that $F_1 \vee G_1$ is orthogonal to G_2, thirdly that $F_1 \vee G_1$ is orthogonal to $F_2 \vee G_2$.

10.12.2 General definition

Definition A design whose plot structure consists of a set $\mathcal{F}$ of factors on a set Ω of plots, whose treatment structure consists of a set $\mathcal{G}$ of factors on a set $\mathcal{T}$ of treatments and whose treatments are allocated to plots according to a design function T is an *orthogonal design* if

(i) $\mathcal{F}$ is an orthogonal plot structure;

(ii) $\mathcal{G}$ is an orthogonal treatment structure;

(iii) the function T is such that

    (a) every pair of treatment factors in $\mathcal{G}$ remain orthogonal to each other on Ω;

    (b) if $F \in \mathcal{F}$ and $G \in \mathcal{G}$ then F is orthogonal to G;

    (c) if $F \in \mathcal{F}$ and $G \in \mathcal{G}$ then $F \vee G \in \mathcal{G}$.

In principle there are twelve conditions to check to verify that a design is orthogonal. Fortunately, good statistical computing packages can do most of this for you. I recommend always doing such a computer check before giving out a plan for a complicated experiment. Unfortunately, most statistical computing packages give no facility to construct suprema, or check for their inclusion.

The following theorem shows that we can often reduce the amount of checking. Its proof is contained in the discussion at the end of Section 10.12.1.

Theorem 10.10 *If $\mathcal{F}$ is an orthogonal plot structure on Ω and the treatment factor T is orthogonal to F for every factor F in $\mathcal{F}$, then $(\mathcal{F}, \mathcal{G}, T)$ is an orthogonal design, where $\mathcal{G} = \{T \vee F : F \in \mathcal{F}\}$.*

All the designs that we have met so far are orthogonal.

10.12.3 Locating treatment subspaces

We need to know which stratum contains each treatment subspace W_G. This is straightforward if the treatment subspaces are all estimated in the same stratum, as happens in a complete-block design or the second version of the calf-feeding trial in Chapter 8. In other cases we do not want to have to go through an argument like the proof of Theorem 8.2. Fortunately, the next result shows that the Hasse diagrams can be used to locate the treatment subspaces in the appropriate strata. This theorem is a more general version of Theorem 8.2, but its proof is scarcely more difficult.

Theorem 10.11 *Let G be a treatment factor in an orthogonal design. Then there is a unique plot factor F finer than or equivalent to G which is coarsest in the sense that every other plot factor which is finer than or equivalent to G is also finer than or equivalent to F. Moreover, $W_G \subseteq W_F$.*

10.12. Orthogonal designs

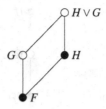

Fig. 10.21. Part of the proof that $W_G \subseteq W_F$ in Theorem 10.11

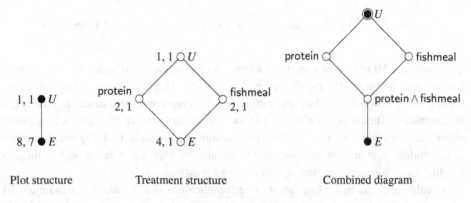

Fig. 10.22. Three Hasse diagrams for Example 5.12

Proof The equality factor E is in $\mathcal{F}$ and $E \preccurlyeq G$ so there is at least one plot factor finer than or equivalent to G. Let $F_1, F_2, \ldots, F_n$ be the plot factors finer than, or equivalent to, G. Put $F = F_1 \vee F_2 \vee \cdots \vee F_n$. Then $F \in \mathcal{F}$ and $F_i \preccurlyeq F \preccurlyeq G$ for $i = 1, 2, \ldots, n$, so F is the coarsest plot factor finer than or equivalent to G.

Since $F \preccurlyeq G$, we have $W_G \subseteq V_G \subseteq V_F$. Suppose that H is a plot factor with $F \prec H$. Then we cannot have $H \preccurlyeq G$, and so $H \vee G \not\equiv G$. See Figure 10.21.

Since the design is orthogonal, $H \vee G$ is a treatment factor and so $W_G \subseteq V_G \cap V_{H \vee G}^\perp$, which is orthogonal to V_H because the factors G and H are orthogonal to each other. Therefore $W_G \subseteq V_H^\perp$. This is true for all plot factors H with $F \prec H$. Hence

$$W_G \subseteq V_F \cap \bigcap_{H \succ F,\ H \in \mathcal{F}} V_H^\perp = W_F. \quad \blacksquare$$

Thus we locate treatment subspaces by combining the two Hasse diagrams into one (omitting all the numbers), using a composite type of dot to indicate any factor, such as U, that is in both $\mathcal{F}$ and $\mathcal{G}$. For each treatment factor, simply find the coarsest plot factor which is below it or equivalent to it.

When the plots are unstructured the combined Hasse diagram is very simple. Figure 10.22 shows the three diagrams for Example 5.12.

At first sight, the complete-block design in Example 4.11, the first version of the calf-feeding trial in Example 8.1 and the improperly replicated experiment in Example 10.4 are

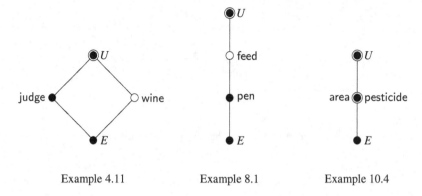

Fig. 10.23. Combined Hasse diagrams for three superficially similar experiments

quite similar. All have plot structure like the one in Figure 10.10(a) and treatment structure like the one in Figure 10.9(b). However, the combined Hasse diagrams, which are shown in Figure 10.23, immediately show the differences between these three cases. In Example 4.11 all treatment differences are estimated in the bottom stratum. In Example 8.1 all treatment differences are estimated in the middle (pen) stratum. In Example 10.4 they are also estimated in the middle stratum, but now treatments are aliased with a plot factor and so there is no residual mean square for testing significance or for estimating variance.

Combined diagrams for four rather more complicated examples are shown in Figure 10.24. It can be verified that each combined Hasse diagram leads to the skeleton analysis of variance given previously.

Example 10.1 revisited (Rye-grass) We have seen that strip $\vee T$ = cultivar. Each treatment occurs in each field so field $\vee T = U$. Therefore field $\vee$ cultivar and field $\vee$ nitrogen are also equal to U. Every level of nitrogen occurs on each strip, so strip $\vee$ nitrogen $= U$. Hence we obtain the combined Hasse diagram in Figure 10.25. This is consistent with the skeleton analysis of variance in Table 8.15, because it shows that $W_{\text{cultivar}} \subseteq W_{\text{strip}}$, $W_{\text{nitrogen}} \subseteq W_E$ and $W_T \subseteq W_E$.

10.12.4 Analysis of variance

An orthogonal plot structure determines the null analysis of variance. There is one line for each stratum, showing the name F and the number of degrees of freedom d_F. Conventionally, the line for F is written lower than the line for G if $F \prec G$. In particular, E is always written at the bottom, which explains why W_E is sometimes called the *bottom stratum*.

The null analysis-of-variance table is expanded to calculations ignoring treatments by writing in the sum of squares, $SS(F)$, for each F in $\mathcal{F}$. These sums of squares are calculated from the crude sums of squares by using the Hasse diagram for $\mathcal{F}$.

Use the treatment structure $\mathcal{G}$ and Theorem 10.6 to obtain a list of the treatment subspaces. Obtain their degrees of freedom by using the Hasse diagram for treatments. Use Theorem 10.11 to decide which stratum contains each treatment subspace. This information is used to expand the null analysis of variance to the skeleton analysis of variance.

10.12. Orthogonal designs

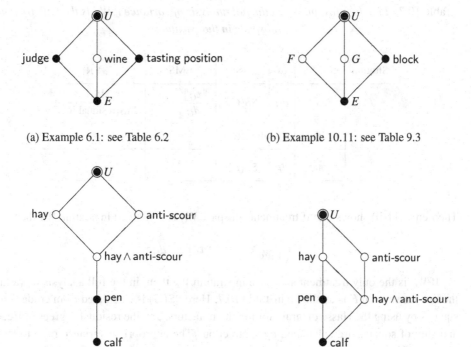

(a) Example 6.1: see Table 6.2

(b) Example 10.11: see Table 9.3

(c) Example 8.1, second version: see Table 8.5

(d) Example 8.1, third version: see Table 8.8

Fig. 10.24. Combined Hasse diagrams with reference to the relevant skeleton analyses of variance

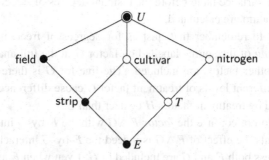

Fig. 10.25. Combined Hasse diagram in Example 10.1

The skeleton analysis of variance is an excellent summary of the design. Good statistical computing packages can calculate the skeleton analysis of variance before there are any data.

To turn the skeleton analysis of variance into the full analysis of variance, we need to calculate expected mean squares (for the theoretical version) and treatment sums of squares and residual sums of squares (for arithmetic on the data). Write τ for $\mathbb{E}(\mathbf{Y})$ and τ_G for $\mathbf{P}_{W_G}\tau$.

Table 10.7. *Line for stratum W_F in the full analysis of variance if W_G is the only treatment subspace in that stratum*

Stratum	Source	df	SS	EMS	VR
F	G	d_G	SS(G)	$\dfrac{\|\tau_G\|^2}{d_G} + \xi_F$	$\dfrac{\text{MS}(G)}{\text{MS}(\text{residual in }F)}$
	residual	by subtraction		ξ_F	–
	total	d_F	SS(F)		

Theorem 2.11(ii) shows that if treatment subspace W_G is contained in stratum W_F then

$$\mathbb{E}\left(\|\mathbf{P}_{W_G}\mathbf{Y}\|^2\right) = \|\tau_G\|^2 + d_G \xi_F.$$

If W_G is the only treatment subspace in stratum W_F then, in the full analysis of variance, the line for stratum F is as shown in Table 10.7. Here $SS(G)$ is calculated from crude sums of squares by using the Hasse diagram for treatment factors, and the residual degrees of freedom and sum of squares are calculated by subtraction. The appropriate variance ratio to test the effect of G is

$$\frac{\text{MS}(G)}{\text{MS}(\text{residual in }F)}.$$

The estimator of ξ_F is $\text{MS}(\text{residual in }F)$: this estimate is used in the formula for standard errors of contrasts in W_G.

If there is more than one treatment subspace in stratum W_F then they are all written in the full analysis-of-variance table before the residual degrees of freedom and residual sum of squares for that stratum are calculated.

It is important to remember that, just as for degrees of freedom, effects and sums of squares, the *meaning* of the source labelled by factor G in the full analysis-of-variance table depends on what other factors are included. The line for G is there to help us answer the question 'do the different levels of treatment factor G cause differences in response over and above those caused by treatment factors H coarser than G?'

In particular, do not confuse the *factor $F \wedge G$* with the F-by-G interaction. It is true that in many experiments the effect of $F \wedge G$ is indeed the F-by-G interaction. However, it is not the interaction unless both F and G are included in $\mathcal{G}$. Even when F and G are both in $\mathcal{G}$, the effect of $F \wedge G$ is not the whole of the F-by-G interaction if there is another treatment factor coarser than $F \wedge G$. An example of this will be given at the end of the next section.

10.13 Further examples

This section contains a collection of new examples of orthogonal designs. It is not exhaustive, but does demonstrate the versatility of the methods developed in this chapter.

10.13. Further examples

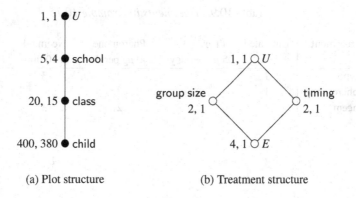

Fig. 10.26. Hasse diagrams for Example 10.14

Example 10.14 (Example 1.10 continued: Mental arithmetic) The treatments are all combinations of group size (whole class or four children) and timing (one hour per week or 12 minutes per day). The Hasse diagram for the treatment factors is in Figure 10.26(b).

Suppose that five schools take part in the experiment, each of these schools having four classes of children of the appropriate age. If there are 20 children per class, we obtain the Hasse diagram for plot factors in Figure 10.26(a).

Teaching methods can be applied only at the whole-class level, but we do not want differences between schools to bias our conclusions. Therefore, the four treatments are randomized to the four classes independently within each school. Then class $\prec T$ but T is orthogonal to school and $T \vee \text{school} = U$. Thus we obtain the skeleton analysis of variance in Table 10.8.

The one unrealistic assumption here is that all the classes have exactly 20 children. A possible way forward if the classes have different numbers of children is to use the average test mark per class as the response, since treatments are in any case estimated in the class stratum. However, these averages will not have the same variance if they are taken over different numbers of children, so this method should not be used unless the class sizes are reasonably similar.

Example 10.15 (Bean weevils) An experiment was conducted to investigate the effects of

Table 10.8. *Skeleton analysis of variance for Example 10.14*

Stratum	Source	Degrees of freedom
mean	mean	1
school	school	4
class	group size	1
	timing	1
	group size ∧ timing	1
	residual	12
	total	15
child	child	380
Total		400

Table 10.9. *Treatments in Example 10.15*

treatment	Untreated	Pheromone 5 μg per day	Pheromone 240 μg per day	Neem oil in 10 litres	Neem oil in 25 litres
type	1	2	2	3	3
pheromone	1	2	3	4	4
neem	1	2	2	3	4

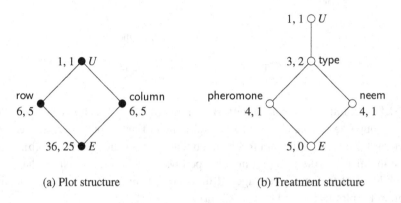

(a) Plot structure (b) Treatment structure

Fig. 10.27. Hasse diagrams for Example 10.15

certain behaviour-modifying chemicals on bean weevils. One chemical was a pheromone intended to attract the insects. This was released at either 5 μg per day or 240 μg per day. The other chemical was neem oil, which is intended to deter the insects from feeding and so make them move elsewhere. This was applied at 4 kg/ha, dissolved in water at the rate of either 10 or 25 litres per hectare. For comparison there was an untreated control.

The treatments are summarized in Table 10.9. Since some treatments are designed to *increase* the number of weevils relative to the control, while others are designed to *decrease* it, it is not appropriate to have a treatment factor which distinguishes between the control and the rest. However, it is appropriate to have a treatment factor type which distinguishes between the three types of treatment, and two further factors, pheromone and neem, which distinguish between treatments of a single type. These are shown in Table 10.9 and Figure 10.27(b).

Beans were sown in a field in the spring. It was impossible to predict the direction from which the insects would arrive, so a row–column design was used. The plot structure is in Figure 10.27(a). The experimenters wanted double the replication for the control treatment, and so the plots were arranged in a 6×6 square. The design chosen was like a Latin square, except that the control treatment occurred twice in each row and twice in each column. In fact, the design was constructed and randomized using a 6×6 Latin square in which two of the letters had been assigned to the control treatment.

Theorem 4.1 shows that all treatment factors are orthogonal to rows and columns and that $T \vee \text{row} = T \vee \text{column} = U$. Hence we obtain the skeleton analysis of variance in Table 10.10.

This experiment has a further complication. If one treatment is successful in attracting insects then the weevils may spread to neighbouring plots. To avoid bias, we should have the

10.13. Further examples

Table 10.10. *Skeleton analysis of variance for Example 10.15*

Stratum	Source	Degrees of freedom
mean	mean	1
row	row	5
column	column	5
plot	type	2
	pheromone	1
	neem	1
	residual	21
	total	25
Total		36

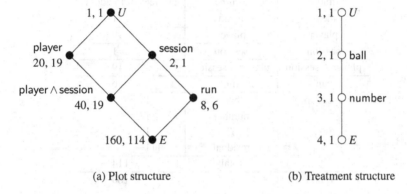

(a) Plot structure   (b) Treatment structure

Fig. 10.28. Hasse diagrams for Example 10.16

successful treatment next to each other treatment equally often. There are ways of achieving such *neighbour balance*, but they are beyond the scope of this book.

Example 10.16 (Rugby) A rugby coach wants to test whether rugby players run faster with the ball in their left hand, in their right hand, or held in both hands. He selects twenty right-handed rugby players and gives them two sessions of test runs. In each session, each player makes four runs of 50 metres, being timed from the 10-metre line to the 40-metre line. In one run he carries the ball in his left hand; in one run in his right; in one run in both. For comparison, he makes one run with no ball. In each session, the treatments are allocated to the four runs by twenty players according to a row–column design made of five Latin squares of order four.

The plot factors are U, player, session, run, player $\wedge$ session and player $\wedge$ run, which is E. These are shown on the Hasse diagram in Figure 10.28(a), which gives the null analysis of variance in Table 10.11.

The four treatments are best regarded not as a 2×2 factorial but as being grouped by the factor number according to the number of hands used; the levels of this factor are in turn grouped by the factor ball according to whether or not a ball is carried. These give the

Table 10.11. *Null analysis of variance in Example 10.16*

Stratum	df
mean	1
player	19
session	1
player ∧ session	19
run	6
player ∧ run	114
Total	160

Table 10.12. *Skeleton analysis of variance for Example 10.16*

Stratum	Source	Degrees of freedom	
mean	mean	1	
player	player	19	
session	session	1	
player ∧ session	player ∧ session	19	
run	run	6	
player ∧ run	ball	1	
	number	1	
	T	1	
	residual	111	
	total		114
Total			160

Hasse diagram in Figure 10.28(b). The single degree of freedom for ball is for the comparison between carrying the ball and not; that for number for the comparison between one and two hands; that for T for the comparison between right and left hands.

The Latin square construction in each session ensures that each treatment occurs once in each class of player ∧ session and five times in each run. Thus treatments also occur equally often in each session and in each player, and so the treatment space lies in the bottom stratum, which is W_E. Thus we obtain the skeleton analysis of variance in Table 10.12.

Example 10.17 (Carbon dating) An archaeology organization wishes to examine the consistency of three different methods of carbon dating: liquid scintillation counting, gas proportional counting and accelerator mass spectrometry. Thirty laboratories are willing to take part in the experiment. The organization has available 11 sets of equipment for liquid scintillation counting, six for gas proportional counting and 13 for accelerator mass spectrometry. These sets of equipment are allocated at random to the 30 laboratories, and technicians at each are trained in their use.

The organization has eight test items for dating, taken from different archaeological sites and composed of different substances. At each laboratory, a single trained technician will use its equipment to carbon-date each of the test items.

10.13. Further examples

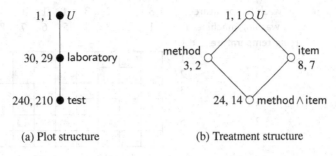

(a) Plot structure   (b) Treatment structure

Fig. 10.29. Hasse diagrams for Example 10.17

The plot factors are U, laboratory and test, which is E. The Hasse diagram for plots is given in Figure 10.29(a).

The treatment factors are U, method, item and method $\wedge$ item, which is T. The archaeologists are chiefly interested in the consistency of the methods in dating the different samples, that is, in the method-by-item interaction. The Hasse diagram for treatments is shown in Figure 10.29(b).

Treatments are not equally replicated. They have the following replications.

method	item							
	1	2	3	4	5	6	7	8
liquid scintillation counting	11	11	11	11	11	11	11	11
gas proportional counting	6	6	6	6	6	6	6	6
accelerator mass spectrometry	13	13	13	13	13	13	13	13

Theorem 10.5 shows that method is orthogonal to item on the set of plots. Within each method, every possible combination of T and laboratory occurs just once, so T is orthogonal to laboratory and $T \vee$ laboratory $=$ method. Each item is tested once by each laboratory and so item is orthogonal to laboratory and item $\vee$ laboratory $= U$. Thus we obtain the skeleton analysis of variance in Table 10.13.

Apart from the unequal replication, this example is essentially the same as the third version of Example 8.1.

Table 10.13. *Skeleton analysis of variance for Example 10.17*

Stratum	Source	Degrees of freedom
mean	mean	1
laboratory	method	2
	residual	27
	total	29
test	item	7
	method $\wedge$ item	14
	residual	189
	total	210
Total		240

wash temperature	3	4	2	3	4	1	1	2
washing machine	1	2	3	4	5	6	7	8

drying temperature	dryer							
1	1							
3	2							
2	3							
3	4							
2	5							
1	6							

Fig. 10.30. One possible randomized plan in the criss-cross version of Example 10.19

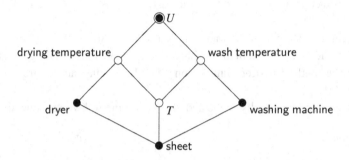

Fig. 10.31. Combined Hasse diagram in the criss-cross version of Example 10.19

Example 10.18 (Example 5.13 continued: Park Grass) After the Park Grass experiment had been running for many decades, the soil became over-acidic. In 1903 the field was split into two columns, running right across all the plots, which were rows. Thereafter, lime was applied to one whole column in every fourth year.

This type of design, where levels of one treatment factor are applied to rows and levels of another treatment factor are applied to columns, is called a *criss-cross* design. Of course, the criss-cross version of Park Grass still has no replication.

Example 10.19 (Unwrinkled washing) A manufacturer of household appliances wants to find the best combination of wash temperature and drying temperature to produce unwrinkled cotton sheets at the end of the laundry session. He wants to compare four different wash temperatures and three different drying temperatures. He uses eight similar washing machines and six similar dryers. First, 48 cotton sheets are randomly allocated to the washing machines, six per machine. The wash temperatures are randomly allocated to the washing machines so that two machines are run at each temperature. After the wash, the six sheets in each machine are randomly allocated to the dryers, one per dryer. Then the drying temperatures are randomly allocated to the dryers so that two machines are run at each temperature. After the drying, all 48 sheets are scored by experts for how wrinkled they are. One possible randomized plan is shown in Figure 10.30. This is a criss-cross design.

The plot factors are U, washing machine, dryer and sheet. The treatment factors are U, wash temperature, drying temperature and wash temperature $\wedge$ drying temperature, which

10.13. Further examples

Table 10.14. *Null analysis of variance in the criss-cross version of Example 10.19*

Stratum	df
mean	1
dryer	5
washing machine	7
sheet	35
Total	48

Table 10.15. *Skeleton analysis of variance for the criss-cross version of Example 10.19*

Stratum	Source	Degrees of freedom
mean	mean	1
dryer	drying temperature	2
	residual	3
	total	5
washing machine	wash temperature	3
	residual	4
	total	7
sheet	T	6
	residual	29
	total	35
Total		48

is equal to T. The combined Hasse diagram is in Figure 10.31. The null and skeleton analyses of variance are in Tables 10.14 and 10.15 respectively. Here the source labelled T represents the interaction between the wash temperature and the drying temperature.

Alternatively, he might choose to use half as many sheets, using twelve sheets in four washing machines and three dryers, then another twelve sheets in the remaining washing machines and dryers, as shown in Figure 10.32. Now the two-level factor block must be included, where block = (washing machine) $\vee$ dryer. The combined Hasse diagram is shown in Figure 10.33, and the null and skeleton analyses of variance in Tables 10.16 and 10.17. This type of design is sometimes called a *strip-plot design*.

```
         wash temperature      4  3  2  1  4  2  1  3
         washing machine       1  2  3  4  5  6  7  8
drying temperature   dryer
         2             1
         1             2
         3             3
         3             4
         2             5
         1             6
```

Fig. 10.32. One possible randomized plan in the strip-plot version of Example 10.19

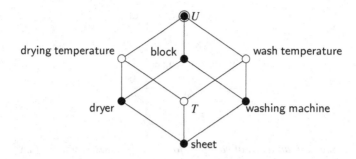

Fig. 10.33. *Combined Hasse diagram in the strip-plot version of Example 10.19*

Table 10.16. *Null analysis of variance in the strip-plot version of Example 10.19*

Stratum	df
mean	1
block	1
dryer	4
washing machine	6
sheet	12
Total	24

Table 10.17. *Skeleton analysis of variance for the strip-plot version of Example 10.19*

Stratum	Source	Degrees of freedom
mean	mean	1
block	block	1
dryer	drying temperature	2
	residual	2
	total	4
washing machine	wash temperature	3
	residual	3
	total	6
sheet	T	6
	residual	6
	total	12
Total		24

10.13. Further examples

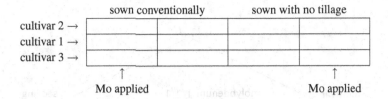

Fig. 10.34. Typical block in Example 10.20

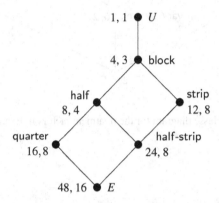

Fig. 10.35. Hasse diagram for the plot factors in Example 10.20

Example 10.20 (Molybdenum) An agronomy institute in Brazil conducted a factorial experiment on beans. The treatment factors were cultivar (there were three cultivars), seeding (either conventional or with no tillage), and molybdenum (either applied or not). The experimental area was divided into four blocks. Each block was a 3×4 rectangle. The three rows were called strips; cultivars were applied to these. The four columns, which were called quarters, were grouped into two halves. The different seeding methods were used in the two halves of each block, and the Mo (molybdenum) was applied to one quarter in each half. Figure 10.34 shows a typical block.

The plot factors are shown in Figure 10.35 and the treatment factors in Figure 10.36. These give the skeleton analysis of variance in Table 10.18.

Example 10.21 (Example 4.2 continued: Irrigated rice) Figure 4.1 shows 32 plots in a rice paddy in an 8×4 rectangular array. The columns correspond to irrigation channels, so should be used as a system of blocks. The rows correspond to distance from the main irrigation channel, but this is a continuous source of variation so we have some freedom over how many rows to put together to give blocking in this direction. If there are eight treatments then it is convenient to put the rows into pairs, so that each treatment can occur once in each column and once in each pair of rows. A design with this property is called a *semi-Latin square*.

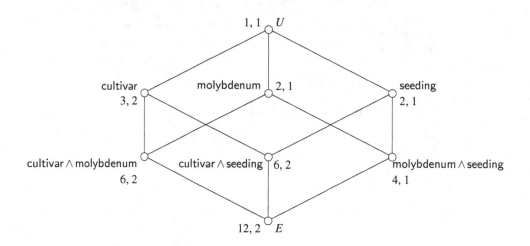

Fig. 10.36. Hasse diagram for the treatment factors in Example 10.20

Table 10.18. *Skeleton analysis of variance for Example 10.20*

Stratum	Source	Degrees of freedom
mean	mean	1
block	block	3
strip	cultivar	2
	residual	6
	total	8
half	seeding	1
	residual	3
	total	4
half-strip	cultivar ∧ seeding	2
	residual	6
	total	8
quarter	molybdenum	1
	seeding ∧ molybdenum	1
	residual	6
	total	8
plot	cultivar ∧ molybdenum	2
	cultivar ∧ seeding ∧ molybdenum	2
	residual	12
	total	16
Total		48

10.13. Further examples

A	B	C	D
E	F	G	H
D	A	B	C
H	E	F	G
C	D	A	B
G	H	E	F
B	C	D	A
F	G	H	E

(a) A semi-Latin square for eight treatments

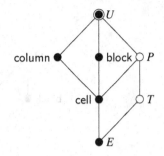

(b) Combined Hasse diagram

Fig. 10.37. Diagrams for Example 10.21

The Hasse diagram for the plot factors is similar to the one in Figure 10.19(b). Let us call a pair of rows a *block* and each of the 16 classes of block ∧ column a *cell*. Theorem 10.5 shows that the only way that we can have the treatment factor orthogonal to cells is for any two cells to either contain the same two treatments or have no treatments in common. In other words, we must have a design like the one in Figure 10.37(a), which is made from a Latin square of order four by replacing each letter by two letters.

This design should be randomized by randomizing blocks, randomizing columns and then randomizing the pair of plots within each cell.

To analyse the design, we need the treatment pseudofactor P which groups the eight treatments into the four pairs $\{A,E\}$, $\{B,F\}$, $\{C,G\}$ and $\{D,H\}$. Then $P = \text{cell} \vee T$ and $\text{cell} \prec P$. However, $\text{block} \vee T = \text{column} \vee T = U$, so we obtain the combined Hasse diagram shown in Figure 10.37(b) and the skeleton analysis of variance in Table 10.19.

The three degrees of freedom for P are estimated less precisely than the remaining treatment degrees of freedom, so this design may not be a good choice. In fact, there are better non-orthogonal designs, but they will not be covered in this book.

Table 10.19. *Skeleton analysis of variance for Example 10.21*

Stratum	Source	Degrees of freedom
mean	mean	1
column	column	3
block	block	3
cell	P	3
	residual	6
	total	9
plot	T	4
	residual	12
	total	16
Total		32

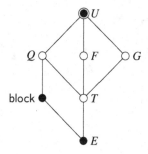

Fig. 10.38. Combined Hasse diagram for Example 10.11 revisited

Example 10.11 revisited (Main-effects only design in blocks) The original version of this design was a single replicate in four blocks of four plots each. We had to assume that the F-by-G interaction was zero. However, if we double up the design by repeating the original four blocks then we can also estimate the interaction. Note that all eight blocks should be randomized amongst themselves.

Let Q be the factor that groups the sixteen treatments into the four sets of four that were allocated to blocks in the original design. In the new design, block $\vee T = Q$. Now, $Q \vee F = Q \vee G = U$ and Q is orthogonal to both F and G. Thus if the treatment factors are U, F, G, Q and T then W_Q is part of the F-by-G interaction. This gives the combined Hasse diagram in Figure 10.38 and the skeleton analysis of variance in Table 10.20.

Chapter 12 builds on this idea of splitting up interactions.

Table 10.20. *Skeleton analysis of variance for Example 10.11 revisited*

Stratum	Source	Degrees of freedom
mean	mean	1
block	Q (part of the F-by-G interaction)	3
	residual	4
	total	7
plot	F	3
	G	3
	$F \wedge G$ (rest of the F-by-G interaction)	6
	residual	12
	total	24
Total		32

Questions for discussion

10.1 In each of the following examples, draw the Hasse diagram for the treatment factors, showing numbers of levels and degrees of freedom.

(a) Example 1.16.

(b) Example 1.17.

(c) Example 3.3.

(d) Example 3.5.

(e) Question 3.5.

(f) Example 5.3.

(g) Example 5.10.

(h) Example 8.7.

(i) Example 10.12.

10.2 Consider the data set in Question 4.3. Use the factors fumigant, dose and type to decompose the treatment sum of squares into four parts, and calculate the corresponding tables of means. Use this information with that calculated in Question 4.3 to interpret the complete analysis of variance.

10.3 Complete the analysis of the data in Question 2.5, using the full information about treatments given in Question 3.2.

10.4 Consider the experiment in Example 8.7.

(a) Draw the Hasse diagram for the factors on the plots (ignoring treatment factors).

(b) Draw the Hasse diagram for the factors on the treatments.

(c) Hence write down the skeleton analysis-of-variance table.

10.5 In an experiment into the digestibility of stubble, four feed treatments are to be applied to sheep. There are 16 sheep, in four rooms of four animals each. There are four test periods of four weeks each, separated by two-week recovery periods. Each sheep is to be fed all treatments, one in each test period. During the recovery periods all animals will receive their usual feed, so that they will return to normal conditions before being subjected to a new treatment.

Draw the Hasse diagram for the non-treatment factors involved. Describe how you would construct a suitable design and randomize it. Write down the skeleton analysis-of-variance table, showing stratum, source and degrees of freedom.

10.6 An experiment was carried out to find out if so-called 'non-herbicidal' pesticides affect photosynthesis in plants. Six pesticides were compared: diuron, carbofuran, chlorpyrifos, tributyltin chloride, phorate and fonofos. Each of these was dissolved in water at five different concentrations. In addition, plain water was used as a control treatment.

Two petri dishes were used for each treatment. Each petri dish was filled with pesticide solution or water. Five freshly cut leaves from mung beans were floated on the surface of the solution in each dish. After two hours, the chlorophyll fluorescence of each leaf was measured.

(a) Draw the Hasse diagram for the factors on the plots (ignoring treatment factors).

(b) Draw the Hasse diagram for the factors on the treatments.

(c) Write down the skeleton analysis-of-variance table.

10.7 Consider the word-processing experiment in Question 6.2. Now suppose that only one copy of each word-processor is available, so that the experiment must last three weeks. The typists will be split into three groups according to experience; each week one group will try out the word-processors.

Furthermore, we now know that the word-processors are of two kinds. Three of them are WYSIWYG and the other two are SGML.

(a) Draw the Hasse diagram for the factors on the plots (ignoring treatment factors).

(b) Draw the Hasse diagram for the factors on the treatments.

(c) Describe how to construct the design for the experiment.

(d) Describe how to randomize the design.

(e) Construct the skeleton analysis-of-variance table.

10.8 An experiment on the response of sugar beet to nitrogen and to time of lifting used the following treatments.

treatment	1	2	3	4	5	6
lifting	early	early	early	late	late	late
nitrogen	nil	early	late	nil	early	late

Introduce a factor to distinguish zero nitrogen from applied nitrogen, and another factor to account for the interaction of this with time of lifting. Draw the Hasse diagram for the factors on the treatments.

Investigate which patterns of replication are allowable if these treatment factors are to remain mutually orthogonal.

10.9 In a modification of the previous experiment, the experimenter said that there were three two-level treatment factors: early nitrogen (which may be applied or not), late nitrogen (which may be applied or not), and time of lifting (which may be early or late). Draw two possible Hasse diagrams for the factors on the treatments.

Which do you think is more appropriate?

Questions for discussion

10.10 A large-scale trial was carried out in the UK in the three years 2000–2002 to find out if genetically modified crops would alter the diversity and quantity of wildlife. Sixty farms were involved in the experiment. At each farm, one field was chosen and split into two halves. One half was randomly chosen for the genetically modified crop, the other for the normal crop. In each of the three years, ecologists recorded the numbers of several species of wildlife present on each half of the field.

(a) Draw the Hasse diagram for the factors on the plots (ignoring treatment factors).

(b) Draw the Hasse diagram for the factors on the treatments.

(c) Write down the skeleton analysis-of-variance table.

10.11 Young eucalyptus trees are susceptible to the 'damping-off' fungus. In a fungicide trial there are three leaf treatments: two commercial chemicals—Dithane M-45 and Thiran—and 'no treatment'. These are combined with the five soil fungicides in Table 10.2, to give fifteen treatments in all.

Draw the Hasse diagram for the factors on the treatments.

10.12 In 2002 the BBC science series *Horizon* carried out an experiment to test the claims of homeopathy. Histamine was diluted in water by one part in 100: this dilution is called 1C. This was diluted again by one part in 100, to give a 2C dilution. The dilution process was continued until homeopathic dilutions of histamine were available at dilutions 15C, 16C, 17C and 18C. There were five test tubes of each of these dilutions, and also 20 test tubes of pure water, which was to be used as a control. There were thus 40 test tubes in total. They were relabelled with random numbers, and the key to the relabelling locked in a safe until the end of the experiment, to ensure blindness.

Two laboratories took part in the experiment. The contents of each of the 40 test tubes was split into two parts, and one part sent to each laboratory. Each laboratory recruited five volunteers to give blood. Sufficient blood was taken from each donor for 40 samples to be taken from it, each sample to be used to test the contents of one test tube. The liquid was added to the blood sample, then the number of activated and inactivated basophil cells was counted.

(a) Draw the Hasse diagram for the factors on the plots (ignoring treatment factors).

(b) Draw the Hasse diagram for the factors on the treatments.

(c) Write down the skeleton analysis-of-variance table.

10.13 Consider a variant of Example 10.19 in which there are four washing machines and four dryers, and in which a laundry session, using all the machines, is held on each of four days. On each day, one cotton sheet is used in each combination of washing machine and dryer.

Draw the Hasse diagram for the plot structure.

Suppose that the experimenter wants to investigate two different wash temperatures, two different spin speeds for the washing machines, and four different dryer temperatures. Describe how you would construct a suitable design and randomize it. Write down the skeleton analysis-of-variance table, showing stratum, source and degrees of freedom.

10.14 Consider Example 10.12. Suppose that the quantity of detergent is not altered, so that there are only four treatments.

(a) Show how to allocate treatments to plots in such a way that every judge assesses all four treatments, all treatments are assessed by the same number of judges per day, and each judge in each day sees both levels of solubility and both levels of coarseness.

(b) Draw the combined Hasse diagram for the plot and treatment factors. Hence write down the skeleton analysis-of-variance table, showing stratum, source and degrees of freedom.

(c) Describe how you would randomize the design.

10.15 How would the plot structure in Example 10.17 be changed if

(a) at each laboratory, eight technicians are trained to use the new equipment and each of them carbon-dates a single test item;

(b) at each laboratory, four technicians are trained to use the new equipment and each of them carbon-dates two test items?

In each case, what are the implications for the design of the experiment? How would the skeleton analysis of variance change?

10.16 A research scientist on a water management project wants to conduct an experiment on drip systems of irrigation of the sesame crop. One treatment factor is the distance between the drip-lines, which he proposes to set at 90 cm, 135 cm or 180 cm. Since the plots are 5.4 m wide, this will result in six, four or three drip-lines per plot respectively. The other treatment factor is the irrigation level, which he will control by setting the cumulative pan evaporation ratio at 0.4, 0.6 or 0.8. The available experimental area permits him to have 27 plots, each 5.4 m wide by 4.0 m long, in a 3×9 rectangle.

He has thought about the following possible ways of laying out the experiment.

(a) Divide the rectangle into three 3×3 squares. Use each square as a block.

   (i) Assign the nine treatment combinations to the plots in each block in a randomized complete-block design.

   (ii) Use a split-plot design in which the levels of irrigation are applied to the columns within each square and the distances are applied to plots within columns.

   (iii) Use a strip-plot design, with the distances applied to columns within each square and the levels of irrigation applied to the rows within each square.

(b) Amalgamate the three plots in each column into a single 5.4 m $\times$ 12.0 m 'large plot', randomly allocate the nine treatments to the nine 'large plots', and then measure three random samples taken from each 'large plot'.

Construct the skeleton analysis of variance for each of these. Hence make notes on how to advise the research scientist.

Chapter 11

Incomplete-block designs

11.1 Introduction

In Chapter 4 we saw that natural blocks may have a fixed size irrespective of the number of treatments. In that chapter we considered block designs where each treatment occurred at least once per block. Here we deal with block designs where the blocks are not large enough to contain all treatments.

Assume that there are b blocks of k plots each, and t treatments each replicated r times. Thus
$$N = bk = tr. \qquad (11.1)$$
Also assume that blocks are *incomplete* in the sense that (i) $k < t$ and (ii) no treatment occurs more than once in any block.

Definition For distinct treatments i and j, the *concurrence* λ_{ij} of i and j is the number of blocks which contain both i and j.

Example 11.1 (Concurrence) The four blocks of a design with $t = 6$, $r = 2$, $b = 4$ and $k = 3$ are as follows.
$$\{1,2,3\}, \{1,4,5\}, \{2,4,6\}, \{3,5,6\}$$
Here $\lambda_{12} = 1$ and $\lambda_{16} = 0$.

This chapter gives a very brief introduction to the topic of incomplete-block designs, covering the classes of design which are most useful in practice. Proofs are deliberately omitted.

11.2 Balance

Definition An incomplete-block design is *balanced* if there is an integer λ such that $\lambda_{ij} = \lambda$ for all distinct treatments i and j.

The name 'balanced incomplete-block design' is often abbreviated to BIBD.

Theorem 11.1 *In a balanced incomplete-block design,*
$$\lambda(t-1) = r(k-1)$$
and therefore $t-1$ divides $r(k-1)$.

Proof Treatment 1 occurs in r blocks, each of which contains $k-1$ other treatments. So the total number of concurrences with treatment 1 is $r(k-1)$. However, this number is the sum of the λ_{1j} for $j \neq 1$, each of which is equal to λ, so the sum is $(t-1)\lambda$. ■

Here are some methods of constructing balanced incomplete-block designs.

Unreduced designs There is one block for each k-subset of $\mathcal{T}$. Thus

$$b = {}^tC_k = \frac{t!}{k!(t-k)!},$$

which is large unless $k=2$, $k=t-2$ or $k=t-1$. Also,

$$r = \frac{bk}{t} = \frac{(t-1)!}{(k-1)!(t-k)!} = {}^{t-1}C_{k-1}$$

and

$$\lambda = \frac{r(k-1)}{t-1} = \frac{(t-2)!}{(k-2)!(t-k)!} = {}^{t-2}C_{k-2}.$$

For example, the unreduced design for five treatments in blocks of size two has the following ten blocks.

$$\{1,2\},\ \{1,3\},\ \{1,4\},\ \{1,5\},\ \{2,3\},\ \{2,4\},\ \{2,5\},\ \{3,4\},\ \{3,5\},\ \{4,5\}$$

Cyclic designs from difference sets This method gives designs with $b=t$. Identify the treatments with the integers modulo t. Choose an *initial* block $B = \{i_1, i_2, \ldots, i_k\}$. The other blocks of the *cyclic* design are $B+1, B+2, \ldots, B+(t-1)$, where

$$B+1 = \{i_1+1, i_2+1, \ldots, i_k+1\},$$

and so on, all arithmetic being modulo t.

For example, when $t=7$ the initial block $\{1,2,4\}$ gives the design in Figure 11.1.

Not every choice of initial block will lead to a balanced design. To check whether an initial block B is suitable, construct its *difference table*.

$-$	i_1	i_2	$\ldots$	i_k
i_1	0	$i_2 - i_1$	$\ldots$	$i_k - i_1$
i_2	$i_1 - i_2$	0	$\ldots$	$i_k - i_2$
$\vdots$	$\vdots$	$\vdots$	$\ddots$	$\vdots$
i_k	$i_1 - i_k$	$i_2 - i_k$	$\ldots$	0

All subtraction in this table is done modulo t. It can be shown that λ_{ij} is equal to the number of occurrences of $i-j$ in the body of the table. Of course, all the diagonal entries are equal to 0. If every nonzero integer modulo t occurs equally often in the non-diagonal entries of the table then B is said to be a *difference set* and the cyclic design is balanced.

The difference table for the initial block in Figure 11.1 is as follows.

$-$	1	2	4
1	0	1	3
2	6	0	2
4	4	5	0

11.3. Lattice designs

$$\{1,2,4\}, \{2,3,5\}, \{3,4,6\}, \{4,5,0\}, \{5,6,1\}, \{6,0,2\}, \{0,1,3\}$$

Fig. 11.1. Cyclic design generated from the initial block $\{1,2,4\}$ modulo 7.

Every nonzero integer modulo 7 occurs once in the body of the table, so the cyclic design should be balanced with $\lambda = 1$. This can be verified directly from Figure 11.1.

Complements Given a balanced incomplete-block design with parameters t, r, b, k and λ, construct another one for the same numbers of treatments and blocks by replacing each block by its complement, that is, the new block contains all the treatments not in the original block. The new block size is $t - k$, and the new replication is $b - r$. Let i and j be any two distinct treatments. The original design has λ blocks containing both i and j, $r - \lambda$ blocks containing i but not j, and $r - \lambda$ further blocks containing j but not i. Therefore the number of blocks in the original design which omit both i and j is $b - \lambda - 2(r - \lambda)$, which is $b - 2r + \lambda$. This number is the concurrence of i and j in the complementary design.

Designs from Latin squares If q is a power of a prime number and either $t = q^2$ and $k = q$ or $t = q^2 + q + 1$ and $k = q + 1$, there are constructions of balanced incomplete-block designs which use mutually orthogonal Latin squares of size q. These are given at the end of Section 11.3.

Balanced designs are obviously desirable, and in some sense fair. They are optimal, in a sense to be defined in Section 11.8. However, there may not exist a balanced incomplete-block design for given values of t, r, b and k. Equation (11.1) and Theorem 11.1 give two necessary conditions. A more surprising necessary condition is given by the following theorem.

Theorem 11.2 (Fisher's Inequality) *In a balanced incomplete-block design, $b \geq t$.*

Practical constraints such as costs often force us to have fewer blocks than treatments, so balanced incomplete-block designs are not as useful as they seem at first sight.

11.3 Lattice designs

Definition An incomplete-block design is *resolved* if the blocks are grouped into larger blocks and the large blocks form a complete-block design.

When natural conditions force us to use small blocks, and hence an incomplete-block design, use a resolved design if possible so that large blocks can be used for management. Of course, this is not possible unless k divides t.

In Example 1.7, the strips form a resolved incomplete-block design because the fields are large blocks.

Theorem 11.3 (Bose's Inequality) *If a balanced incomplete-block design is resolved then $b \geq t + r - 1$.*

Most resolved designs used in practice are not balanced, because too many blocks would be needed for balance.

A *lattice design* is a special sort of resolved incomplete-block design in which $t = k^2$. The construction method is as follows.

(i) Write the treatments in a $k \times k$ square array.

(ii) For the first large block, the rows of the square are the blocks.

(iii) For the second large block, the columns of the square are the blocks.

(iv) For the third large block (if any), write down a $k \times k$ Latin square and use its letters as the blocks.

(v) For the fourth large block (if any), write down a $k \times k$ Latin square orthogonal to the first one and use its letters as the blocks.

(vi) And so on, using $r - 2$ mutually orthogonal Latin squares.

Example 11.2 (Lattice design) Suppose that $t = 9$ and $k = 3$. We can take the treatment array

1	2	3
4	5	6
7	8	9

Then the blocks in the first large block are

$$\{1,2,3\}, \{4,5,6\}, \{7,8,9\}$$

and the blocks in the second large block are

$$\{1,4,7\}, \{2,5,8\}, \{3,6,9\}.$$

For a third large block we can use the Latin square

A	B	C
B	C	A
C	A	B

If this is superimposed on the treatment array then letter A occurs in the same positions as treatments 1, 6 and 8, letter B occurs in the same positions as treatments 2, 4 and 9, and letter C occurs in the same positions as treatments 3, 5 and 7. This gives the blocks

$$\{1,6,8\}, \{2,4,9\}, \{3,5,7\}.$$

Finally, for a fourth large block we can use the following Latin square, orthogonal to the previous one:

α	β	γ
γ	α	β
β	γ	α

11.4. Randomization

The blocks in the fourth large block are

$$\{1,5,9\}, \{2,6,7\}, \{3,4,8\}.$$

Thus we obtain a design with two, three or four large blocks.

If two treatments in a lattice design are in the same block in one large block then they are in different blocks in every other large block. Therefore, for each treatment i, there are $r(k-1)$ treatments j with $\lambda_{ij} = 1$. The number of remaining treatments j is $t - 1 - r(k-1) = k^2 - 1 - r(k-1) = (k+1-r)(k-1)$: these all have concurrence λ_{ij} equal to zero.

If there exist $k-1$ mutually orthogonal $k \times k$ Latin squares then we obtain a lattice design with $k+1$ large blocks. Now $r(k-1) = (k+1)(k-1) = k^2 - 1 = t - 1$, and so the design is balanced with $\lambda = 1$. If k is a prime number or a power of a prime number, the methods in Section 9.2 give $k-1$ mutually orthogonal $k \times k$ Latin squares.

In this case, $k+1$ extra treatments can be inserted to give a balanced incomplete-block design with $k^2 + k + 1$ treatments, block size $k+1$ and $\lambda = 1$; the new design is not resolved. Label the new treatments $j_1, \ldots, j_{k+1}$. Insert treatment j_s into every block in the s-th large block. Then add one further block containing $\{j_1, j_2, \ldots, j_{k+1}\}$.

Example 11.3 (Balanced design) For a balanced incomplete-block design for 13 treatments in 13 blocks of size four, start with the version of Example 11.2 which has four large blocks. Insert treatment 10 into every block in the first large block; insert treatment 11 into every block in the second large block; insert treatment 12 into every block in the third large block; insert treatment 13 into every block in the fourth large block. Finally, add one new block containing all the new treatments. This gives the following blocks.

$$\{1,2,3,10\}, \{4,5,6,10\}, \{7,8,9,10\},$$
$$\{1,4,7,11\}, \{2,5,8,11\}, \{3,6,9,11\},$$
$$\{1,6,8,12\}, \{2,4,9,12\}, \{3,5,7,12\},$$
$$\{1,5,9,13\}, \{2,6,7,13\}, \{3,4,8,13\},$$
$$\{10,11,12,13\}$$

11.4 Randomization

To randomize a resolved incomplete-block design, proceed as follows.

(i) Randomize large blocks; that is, randomize the numbering of the large blocks.

(ii) Within each large block independently, randomize blocks; that is, randomly order the blocks within each large block (this will affect which treatments occur in which real block).

(iii) Within each block independently, randomize plots; that is, randomly order the treatments which have been assigned to that block.

Do *not* randomize the treatment labels. As we shall see in Section 11.7, if the design is not balanced then different treatment contrasts have different variances. A good designer chooses

Housewife 1				
Washload	1	2	3	4
Detergent	1	2	3	4

Housewife 2				
Washload	1	2	3	4
Detergent	1	2	3	5

Housewife 3				
Washload	1	2	3	4
Detergent	1	2	4	5

Housewife 4				
Washload	1	2	3	4
Detergent	1	3	4	5

Housewife 5				
Washload	1	2	3	4
Detergent	2	3	4	5

Housewife 6				
Washload	1	2	3	4
Detergent	1	2	3	4

Housewife 7				
Washload	1	2	3	4
Detergent	1	2	3	5

Housewife 8				
Washload	1	2	3	4
Detergent	1	2	4	5

Housewife 9				
Washload	1	2	3	4
Detergent	1	3	4	5

Housewife 10				
Washload	1	2	3	4
Detergent	2	3	4	5

Fig. 11.2. Systematic design in Example 11.4

the treatment labels in such a way that the most important contrasts have lower variance. Randomizing treatment labels would undo the effect of careful matching of real treatments to treatment labels.

A common mistake is to randomize treatment labels independently within each large block. This can have the effect of making the concurrences very unequal, and so making the design much less efficient than it need be, even on average, not just for the important contrasts.

Randomization of non-resolved incomplete-block designs is similar, except that the first step is omitted.

(i) Randomize blocks.

(ii) Within each block independently, randomize plots.

Example 11.4 (Example 1.11 continued: Detergents) Here ten housewives each do four washloads in an experiment to compare new detergents. Thus $b = 10$ and $k = 4$. If there are five new detergents then $t = 5$. The obvious balanced incomplete-block design for five treatments in blocks of size four is the unreduced design, which has five blocks. To obtain a balanced design in ten blocks, simply take two copies of the unreduced design. This gives the systematic design in Figure 11.2.

One of the many possible randomized plans for this experiment is shown in Figure 11.5. Both blocks and plots have been randomized by the method given at the end of Section 2.2. The stream of random numbers (to three decimal places) used to randomize the blocks is in Figure 11.3, while the stream of random digits used to randomize the plots within blocks is in Figure 11.4.

11.4. Randomization

```
0.653  0.686  0.911  0.402  0.130  0.486  0.678  0.995  0.707  0.808
  4      6      9      2      1      3      5     10      7      8
```

Fig. 11.3. Stream of random numbers, used to randomize the blocks in Example 11.4

```
0 5 9 3 | 6 2 3 5 | 0 3 9 5 |
1 4 5 3 | 4 1 2 3 | 1 3 5 4 |

1 1 9 2 9 6 | 9 8 7 6 | 7 2 2 5 6 |
1 X 5 2 X 3 | 4 3 2 1 | 5 1 X 2 4 |

5 9 2 6 | 7 3 8 3 0 | 2 0 3 9 |
3 5 2 4 | 4 3 5 X 2 | 2 1 3 5 |

8 9 1 9 2
4 5 1 X 2
```

Fig. 11.4. Stream of random digits, used to randomize the plots within blocks in Example 11.4

Housewife 1				
Washload	1	2	3	4
Detergent	1	4	5	3

Housewife 2				
Washload	1	2	3	4
Detergent	4	1	2	3

Housewife 3				
Washload	1	2	3	4
Detergent	1	3	5	4

Housewife 4				
Washload	1	2	3	4
Detergent	1	5	2	3

Housewife 5				
Washload	1	2	3	4
Detergent	4	3	2	1

Housewife 6				
Washload	1	2	3	4
Detergent	5	1	2	4

Housewife 7				
Washload	1	2	3	4
Detergent	3	5	2	4

Housewife 8				
Washload	1	2	3	4
Detergent	4	3	5	2

Housewife 9				
Washload	1	2	3	4
Detergent	2	1	3	5

Housewife 10				
Washload	1	2	3	4
Detergent	4	5	1	2

Fig. 11.5. Randomized plan in Example 11.4

11.5 Analysis of balanced incomplete-block designs

Assume that blocks have fixed effects, so that

$$\mathbb{E}(Y_\omega) = \tau_i + \zeta_j \qquad (11.2)$$

if plot ω is in block j and receives treatment i; moreover, $\text{Cov}(\mathbf{Y}) = \xi \mathbf{I}$. As we showed in Section 4.5, we can replace every treatment parameter τ_i by $\tau_i + c$ for some fixed constant c, so long as we replace every block parameter ζ_j by $\zeta_j - c$. The algorithm that follows estimates $\tau_i + c$ where $c = -\sum_{i=1}^{t} \tau_i / t$.

Suppose that $\mathbf{y}$ is the vector of observations from an experiment carried out in a balanced incomplete-block design. To analyse the data by hand, proceed as follows.

(i) In $\mathbf{y}$, subtract the block mean from every observation in each block, to give a new vector $\mathbf{y}^{(B)}$.

(ii) Calculate treatment means from $\mathbf{y}^{(B)}$.

(iii) Multiply each treatment mean by

$$\frac{t-1}{t} \frac{k}{k-1}$$

to obtain the treatment estimates $\hat{\tau}_i$.

(iv) In $\mathbf{y}$, subtract $\hat{\tau}_i$ from every observation on treatment i, to give a new vector $\mathbf{y}^{(T)}$.

(v) Obtain $\hat{\zeta}_j$, the estimate of the parameter for block j, as the mean of the entries in block j in $\mathbf{y}^{(T)}$.

(vi) In $\mathbf{y}^{(T)}$, subtract $\hat{\zeta}_j$ from every entry in block j, to obtain the vector of residuals.

(vii) Estimate the plots stratum variance ξ by

$$\frac{\text{sum of squares of residuals}}{N - t - b + 1}.$$

Theorem 11.4 *If $\theta_1, \ldots, \theta_t$ are real numbers such that $\sum_{i=1}^{t} \theta_i = 0$ and the blocks have fixed effects then the above procedure gives $\sum \theta_i \hat{\tau}_i$ as the best linear unbiased estimator of $\sum \theta_i \tau_i$. Moreover, the variance of this estimator is equal to*

$$\left(\frac{1}{r} \sum_{i=1}^{t} \theta_i^2 \right) \left(\frac{t-1}{t} \frac{k}{k-1} \right) \xi,$$

where ξ is the plots stratum variance.

Example 11.5 (Lithium carbonate) In Table 1.1, it is clear that the recorded data for drugs A and D are of the form $m/300$ where $m = 60, 70, \ldots, 220$. Similarly, those for B have the form $m/250$ for $m = 40, 50, \ldots, 200$ and those for C the form $m/450$ for $m = 50, 60, \ldots, 130$. Thus the recorded figure of 0.667 for A could be anything from $195/300 = 0.650$ to

11.5. Analysis of balanced incomplete-block designs

Table 11.1. *Calculations in Example 11.5*

Person	Day	Drug	y	$\mathbf{y}^{(B)}$	$\hat{\boldsymbol{\tau}}$	$\mathbf{y}^{(T)}$	$\hat{\boldsymbol{\zeta}}$	$\mathbf{x}_1$	$\mathbf{x}_2$
1	1	A	0.67	0.095	0.154	0.516	0.425	0.091	0.074
1	2	B	0.48	−0.095	0.146	0.334	0.425	−0.091	−0.074
2	1	D	0.70	0.285	0.098	0.602	0.565	0.037	0.020
2	2	C	0.13	−0.285	−0.398	0.528	0.565	−0.037	−0.020
3	1	C	0.16	−0.285	−0.398	0.558	0.567	−0.009	−0.026
3	2	A	0.73	0.285	0.154	0.576	0.567	0.009	0.026
4	1	B	0.68	0.260	0.146	0.534	0.546	−0.012	−0.029
4	2	C	0.16	−0.260	−0.398	0.558	0.546	0.012	0.029
5	1	D	0.73	0.030	0.098	0.632	0.574	0.058	0.041
5	2	A	0.67	−0.030	0.154	0.516	0.574	−0.058	−0.041
6	1	D	0.60	−0.040	0.098	0.502	0.518	−0.016	−0.033
6	2	B	0.68	0.040	0.146	0.534	0.518	0.016	−0.033
7	1	B	0.80	0.035	0.146	0.654	0.615	0.039	0.022
7	2	A	0.73	−0.035	0.154	0.576	0.615	−0.039	−0.022
8	1	B	0.80	0.050	0.146	0.654	0.628	0.026	0.009
8	2	D	0.70	−0.050	0.098	0.602	0.628	−0.026	−0.009
9	1	C	0.11	−0.230	−0.398	0.508	0.490	0.018	0.001
9	2	D	0.57	0.230	0.098	0.472	0.490	−0.018	−0.001
10	1	A	0.70	0.065	0.154	0.546	0.509	0.037	0.020
10	2	D	0.57	−0.065	0.098	0.471	0.509	−0.037	−0.020
11	1	A	0.67	0.235	0.154	0.516	0.557	−0.041	−0.058
11	2	C	0.20	−0.235	−0.398	0.598	0.557	0.041	0.058
12	1	C	0.13	−0.295	−0.398	0.528	0.551	−0.023	−0.040
12	2	B	0.72	0.295	0.146	0.574	0.551	0.023	0.040

$205/300 = 0.683$. Therefore the third decimal place is completely spurious, and we shall truncate the data to two decimal places.

The vector **y** in Table 11.1 gives the quantity of lithium in the blood after two hours, as a proportion of the quantity of lithium carbonate in the dose. If we ignore the days, the design of this experiment is a balanced incomplete-block design with the people as blocks. The remaining columns of Table 11.1 show the calculations with the data, using one more decimal place than the data, which is a good rule of thumb. First the block means are calculated and subtracted to give $\mathbf{y}^{(B)}$. Each treatment mean is a treatment total divided by 6. Since $t = 4$ and $k = 2$, this has to be multiplied by $3/2$. Thus the treatment estimates in $\hat{\boldsymbol{\tau}}$ are the treatment totals from $\mathbf{y}^{(B)}$ divided by 4. Then $\mathbf{y}^{(T)} = \mathbf{y} - \hat{\boldsymbol{\tau}}$. The block means of this give the block estimates $\hat{\boldsymbol{\zeta}}$, from which the residual vector $\mathbf{x}_1$ is given by $\mathbf{x}_1 = \mathbf{y}^{(T)} - \hat{\boldsymbol{\zeta}}$. We can now verify that we have the least-squares estimates of $\boldsymbol{\tau}$ and $\boldsymbol{\zeta}$, because $\mathbf{x}_1$ is orthogonal to both blocks and treatments (to the accuracy to which we are working).

Now we can incorporate the information about days. Since days are orthogonal to both blocks (people) and treatments (drugs), we estimate the day parameters from the appropriate means in $\mathbf{x}_1$, giving 0.017 for Day 1 and -0.017 for Day 2. Subtracting these day means from $\mathbf{x}_1$ gives the final vector of residuals $\mathbf{x}_2$. The residual mean square is the sum of the squares in $\mathbf{x}_2$, divided by 8, which is 0.004023. Hence the standard error of each treatment

difference is
$$\sqrt{\frac{2}{6} \times \frac{3}{2} \times 0.004023} \approx 0.045.$$

Of course, in practice everyone leaves these calculations to their statistical software, but this example does demonstrate that the multiplicative factor in step (iii) is necessary.

Although we shall not prove the whole of Theorem 11.4, we shall show that it gives unbiased estimators of τ_i and ζ_j under the assumption that the parameters in Equation (11.2) have been chosen so that $\sum_{i=1}^{t} \tau_i = 0$. Suppose that $k = 3$ and that block 1 contains treatments 1, 2 and 3. Then the expected values of the responses in block 1 are $\tau_1 + \zeta_1$, $\tau_2 + \zeta_1$ and $\tau_3 + \zeta_1$, whose mean is $(\tau_1 + \tau_2 + \tau_3)/3 + \zeta_1$. After subtracting this mean, the expected values are $\tau_1 - (\tau_1 + \tau_2 + \tau_3)/3$, $\tau_2 - (\tau_1 + \tau_2 + \tau_3)/3$ and $\tau_3 - (\tau_1 + \tau_2 + \tau_3)/3$.

In general, the expected value of the mean on the block $B(\omega)$ containing plot ω is
$$\left(\sum_{\ell \in B(\omega)} \tau_\ell + k\zeta_{B(\omega)} \right) \bigg/ k = \sum_{\ell \in B(\omega)} \tau_\ell / k + \zeta_{B(\omega)}.$$

After subtracting the block mean, the expected value on a plot ω with treatment i is
$$\tau_i - \sum_{\ell \in B(\omega)} \tau_\ell / k. \tag{11.3}$$

Since treatment i occurs r times altogether, and λ times with each other treatment, the mean of the values of expression (11.3) over all plots which receive treatment i is

$$\begin{aligned}
\tau_i - \frac{1}{rk}\left(r\tau_i + \lambda \sum_{\ell \neq i} \tau_\ell \right) &= \tau_i - \frac{1}{rk}\left((r-\lambda)\tau_i + \lambda \sum_{\ell=1}^{t} \tau_\ell \right) \\
&= \frac{(rk - r + \lambda)\tau_i}{rk}, \qquad \text{because } \sum_{\ell=1}^{t} \tau_\ell = 0, \\
&= \frac{(r(k-1)(t-1) + r(k-1))\tau_i}{rk(t-1)}, \qquad \text{by Theorem 11.1,} \\
&= \frac{(k-1)t}{k(t-1)} \tau_i.
\end{aligned}$$

Thus the estimators $\hat{\tau}_i$ are unbiased.

Because $\mathbb{E}(\hat{\tau}_i) = \tau_i$ for all i, the act of subtracting $\hat{\tau}_i$ from $\mathbb{E}(Y_\omega)$ for all plots ω which receive treatment i leaves an expected value of ζ_j for every plot in block j. Hence the mean of these values is also equal to ζ_j, and so the estimators $\hat{\zeta}_j$ are unbiased.

Note that the foregoing argument does not work for incomplete-block designs which are not balanced.

What about the random-effects model? Here the assumptions are that $\mathbb{E}(Y_\omega) = \tau_i$ if $T(\omega) = i$ and that
$$\text{Cov}(\mathbf{Y}) = \xi_0 \mathbf{P}_{V_U} + \xi_B(\mathbf{P}_{V_B} - \mathbf{P}_{V_U}) + \xi(\mathbf{I} - \mathbf{P}_{V_B}),$$
where $\mathbf{P}_{V_U}$ is the matrix of orthogonal projection onto V_U, which is $N^{-1}\mathbf{J}$, and $\mathbf{P}_{V_B}$ is the matrix of orthogonal projection onto V_B, which is equal to $k^{-1}\mathbf{J}_B$. Often we assume that ξ_B is so large relative to ξ that we will base all estimation on the data orthogonal to blocks, which is $\mathbf{y} - \mathbf{P}_{V_B}\mathbf{y}$; that is, we will begin with the first step of the preceding algorithm.

11.6. Efficiency

Theorem 11.5 *If* $\theta_1, \ldots, \theta_t$ *are real numbers such that* $\sum_{i=1}^{t} \theta_i = 0$ *and blocks have random effects then the preceding procedure gives* $\sum \theta_i \hat{\tau}_i$ *as the best linear unbiased estimator of* $\sum \theta_i \tau_i$ *which uses only the data* $\mathbf{y} - \mathbf{P}_{V_B} \mathbf{y}$ *orthogonal to blocks. Moreover, the variance of this estimator is equal to*

$$\left(\frac{1}{r} \sum_{i=1}^{t} \theta_i^2 \right) \left(\frac{t-1}{t} \frac{k}{k-1} \right) \xi,$$

where ξ *is the plots stratum variance.*

11.6 Efficiency

Definition The *efficiency* for a treatment estimator $\sum \theta_i \hat{\tau}_i$ in an incomplete-block design Δ relative to a complete-block design Γ with the same values of t and r is

$$\frac{\operatorname{Var}\left(\sum \theta_i \hat{\tau}_i\right) \text{ in } \Gamma}{\operatorname{Var}\left(\sum \theta_i \hat{\tau}_i\right) \text{ in } \Delta}.$$

Thus efficiency is large when variance is small.

Let σ^2 be the plots stratum variance in Γ. Then the variance of the estimator $\sum \theta_i \hat{\tau}_i$ in Γ is equal to $\left(\sum \theta_i^2 / r\right) \sigma^2$. Suppose that the variance of the estimator $\sum \theta_i \hat{\tau}_i$ in Δ is $c\xi$, where ξ is the plots stratum variance in Δ. Then the efficiency for this estimator is

$$\frac{\left(\sum \theta_i^2 / r\right) \sigma^2}{c \xi} = \frac{\sum \theta_i^2}{rc} \times \frac{\sigma^2}{\xi}.$$

The right-hand factor, σ^2/ξ, depends only on the variability of the experimental material in the blocks of the two sizes. It is usually unknown, but a reasonable guess at its size may often be made in advance of the experiment. If the blocking is good then ξ should be less than σ^2, so this factor should be greater than 1. The left-hand factor, $\sum \theta_i^2 / (rc)$, depends only on the linear combination $\sum \theta_i \tau_i$ and properties of the design Δ; it is called the *efficiency factor* for the treatment contrast corresponding to the linear combination $\sum \theta_i \tau_i$ in the design Δ. It can be shown that efficiency factors are always between 0 and 1.

Theorem 11.4 shows that if Δ is a balanced incomplete-block design then

$$c = \frac{\sum \theta_i^2}{r} \times \frac{t-1}{t} \times \frac{k}{k-1}$$

and so the efficiency factor for every treatment contrast is equal to

$$\frac{t}{t-1} \frac{k-1}{k}.$$

Since $tk - t < tk - k$, this is less than 1.

Example 11.6 (Comparing block designs) Suppose that we want to do an experiment to compare seven treatments, and that we have enough treatment material for three replicates. We might have a choice between seven blocks of size three, with plots stratum variance ξ, and three blocks of size seven, with plots stratum variance σ^2. In the first case we can use the balanced incomplete-block design in Figure 11.1. Now

$$\frac{t-1}{t} \times \frac{k}{k-1} = \frac{6}{7} \times \frac{3}{2} = \frac{9}{7},$$

and so the variance of each estimator of a difference between two treatments is equal to

$$\frac{2}{r} \times \frac{9}{7} \times \xi = \frac{2}{3} \times \frac{9}{7} \times \xi = \frac{6}{7}\xi.$$

In the second case we can use a complete-block design, and the variance of each estimator of a difference between two treatments is equal to $2\sigma^2/r$, which is $2\sigma^2/3$. Thus the incomplete-block design is better if and only if $6\xi/7 < 2\sigma^2/3$; that is, if and only if $\xi < 7\sigma^2/9$.

Sometimes the notion of efficiency and the relative sizes of the plots stratum variances enable us to choose between block designs whose block sizes differ. At other times the block size is fixed and the concept of efficiency enables us to choose between two different designs for the same block size.

The efficiency factor for the simple contrast $\tau_i - \tau_j$ is

$$\frac{2\xi}{r[\text{Var}(\hat{\tau}_i - \hat{\tau}_j) \text{ in } \Delta]}.$$

If this is equal to e then the standard error of this difference is

$$\sqrt{\text{MS}(\text{residual})\frac{2}{re}}.$$

The simpler formula in Section 2.9 uses $e = 1$. Similarly, in power calculations like those in Section 2.13, the variance constant v becomes $2/(re)$ instead of $2/r$.

11.7 Analysis of lattice designs

Lattice designs also have a method of analysis by hand that is relatively straightforward. For simplicity, we consider only the fixed-effects model: the adaptation to the random-effects model is the same as in Section 11.5.

As we saw in Section 2.6, for an equireplicate design with replication r, the treatment contrast corresponding to the linear combination $\sum \theta_i \tau_i$ is $\sum (\theta_i/r) \mathbf{u}_i$, where $\mathbf{u}_i$ takes the value 1 on ω if $T(\omega) = i$ and the value 0 otherwise. In resolved designs in general, the treatment contrast $\mathbf{v}$ is said to be *confounded with blocks* in a given large block if v_ω takes a constant value throughout each block in that large block. On the other hand, the treatment contrast $\mathbf{v}$ is said to be *orthogonal to blocks* in a given large block if the sum of the values in $\mathbf{v}$ is zero in every block in that large block. Lattice designs have the property that each treatment contrast that is confounded with blocks in any large block is orthogonal to blocks in every other large block.

Example 11.2 revisited (Lattice design) In this resolved design, the contrast for estimating $\tau_1 + \tau_2 + \tau_3 - \tau_4 - \tau_5 - \tau_6$ is confounded with blocks in the first large block. It is orthogonal to blocks in each of the remaining large blocks. Similarly, the contrast for estimating $\tau_1 + \tau_4 + \tau_7 - \tau_2 - \tau_5 - \tau_8$ is confounded with blocks in the second large block and is orthogonal to blocks in all the other large blocks.

There are thus three types of treatment contrast in a lattice design.

11.7. Analysis of lattice designs

(i) If a contrast is confounded with blocks in one large block then it cannot be estimated from that large block, because there is no way of distinguishing the effect of the treatment contrast from the effect of the corresponding block contrast. However, it is then orthogonal to blocks in every other large block, so it can be estimated from all the other large blocks in the normal way. There are $r-1$ other large blocks, and so

$$\mathrm{Var}\left(\sum_i \theta_i \hat{\tau}_i\right) = \frac{\sum \theta_i^2}{r-1}\xi.$$

Thus the efficiency for this contrast is

$$\frac{\left(\sum \theta_i^2/r\right)\sigma^2}{\left(\sum \theta_i^2/(r-1)\right)\xi} = \frac{r-1}{r}\frac{\sigma^2}{\xi}$$

and the efficiency factor is $(r-1)/r$.

Contrasts like this form a subspace W_1 of W_T of dimension $r(k-1)$.

(ii) Put $W_2 = W_T \cap W_1^\perp$, whose dimension is $t-1-r(k-1) = (k+1-r)(k-1)$. If a contrast is in W_2 then it is orthogonal to blocks in every large block so it can be estimated from every large block in the normal way. Then

$$\mathrm{Var}\left(\sum_i \theta_i \hat{\tau}_i\right) = \frac{\sum \theta_i^2}{r}\xi$$

and the efficiency factor is equal to 1.

(iii) Because $W_T = W_1 \oplus W_2$, every other treatment contrast can be expressed as a sum of contrasts of the first two types. Moreover, these can be chosen to be orthogonal to each other, so their estimators are uncorrelated and the variance of their sum is the sum of their variances.

(iv) The preceding steps give unbiased estimators for the treatment parameters τ_i (up to an additive constant c). From these, obtain estimators for the block parameters ζ_j and the plots stratum variance ξ just as in Section 11.5.

Example 11.2 revisited (Lattice design) Table 11.2 shows the version of this design with two large blocks. The treatment contrast **v** corresponds to the linear combination $(\tau_1+\tau_2+\tau_3)-(\tau_4+\tau_5+\tau_6)$ while the treatment contrast **w** corresponds to $(2\tau_1-\tau_2-\tau_3)-(2\tau_4-\tau_5-\tau_6)$.

The contrast **v** is confounded with blocks in the first large block. If we tried to estimate $(\tau_1+\tau_2+\tau_3)-(\tau_4+\tau_5+\tau_6)$ from the first large block we would use $Y_1+Y_2+Y_3-Y_4-Y_5-Y_6$, whose expectation is equal to $(\tau_1+\tau_2+\tau_3)-(\tau_4+\tau_5+\tau_6)+3\zeta_1-3\zeta_2$. There is no way to disentangle the τ parameters from the ζ parameters. However, this contrast is orthogonal to blocks in the second large block, so we can use $Y_{10}-Y_{11}+Y_{13}-Y_{14}+Y_{16}-Y_{17}$ to estimate $(\tau_1+\tau_2+\tau_3)-(\tau_4+\tau_5+\tau_6)$. The variance of this estimator is equal to

$$\frac{1^2+(-1)^2+1^2+(-1)^2+1^2+(-1)^2}{1}\xi = 6\xi.$$

Table 11.2. *Treatment contrasts in Example 11.2*

Large block	Block j	Plot ω	Treatment	$\mathbb{E}(Y_\omega)$	Treatment contrasts			
					v	w	x	z
1	1	1	1	$\tau_1 + \zeta_1$	1	2	1	1
1	1	2	2	$\tau_2 + \zeta_1$	1	-1	-1	0
1	1	3	3	$\tau_3 + \zeta_1$	1	-1	0	-1
1	2	4	4	$\tau_4 + \zeta_2$	-1	-2	1	0
1	2	5	5	$\tau_5 + \zeta_2$	-1	1	-1	-1
1	2	6	6	$\tau_6 + \zeta_2$	-1	1	0	1
1	3	7	7	$\tau_7 + \zeta_3$	0	0	1	-1
1	3	8	8	$\tau_8 + \zeta_3$	0	0	-1	1
1	3	9	9	$\tau_9 + \zeta_3$	0	0	0	0
2	4	10	1	$\tau_1 + \zeta_4$	1	2	1	1
2	4	11	4	$\tau_4 + \zeta_4$	-1	-2	1	0
2	4	12	7	$\tau_7 + \zeta_4$	0	0	1	-1
2	5	13	2	$\tau_2 + \zeta_5$	1	-1	-1	0
2	5	14	5	$\tau_5 + \zeta_5$	-1	1	-1	-1
2	5	15	8	$\tau_8 + \zeta_5$	0	0	-1	1
2	6	16	3	$\tau_3 + \zeta_6$	1	-1	0	-1
2	6	17	6	$\tau_6 + \zeta_6$	-1	1	0	1
2	6	18	9	$\tau_9 + \zeta_6$	0	0	0	0

On the other hand, the treatment contrast **w** is orthogonal to blocks in both large blocks, so we use $(2Y_1 - Y_2 - Y_3 - 2Y_4 + Y_5 + Y_6 + 2Y_{10} - 2Y_{11} - Y_{13} + Y_{14} - Y_{16} + Y_{17})/2$ to estimate $(2\tau_1 - \tau_2 - \tau_3) - (2\tau_4 - \tau_5 - \tau_6)$. The variance of this estimator is equal to

$$\frac{2^2 + (-1)^2 + (-1)^2 + (-2)^2 + 1^2 + 1^2 + 2^2 + (-2)^2 + (-1)^2 + 1^2 + (-1)^2 + 1^2}{4} \xi = 6\xi.$$

The sum of these two estimators is used to estimate $3(\tau_1 - \tau_4)$. The first estimator does not use any data from the first large block, so it is obviously independent of the part of the second estimator that uses the first large block. On the second large block, the two estimators are orthogonal to each other, so Theorem 2.6 shows that they are uncorrelated. Hence the variance of their sum is equal to $6\xi + 6\xi$, and so $\mathrm{Var}(\hat\tau_1 - \hat\tau_4) = 12\xi/9 = 4\xi/3$. The efficiency for $\hat\tau_1 - \hat\tau_4$ is

$$\frac{(2/2)\sigma^2}{(4/3)\xi} = \frac{3}{4}\frac{\sigma^2}{\xi},$$

while the efficiency factor is $3/4$.

To estimate the difference $\tau_1 - \tau_5$ we use the treatment contrasts **v**, **x** and **z** shown in Table 11.2. The contrast **x** is confounded with blocks in the second large block, so we use it in the first large block only to estimate $\tau_1 - \tau_2 + \tau_4 - \tau_5 + \tau_7 - \tau_8$ with variance 6ξ. The contrast **z** is orthogonal to blocks in both large blocks, so we use it in both large blocks to estimate $\tau_1 - \tau_3 - \tau_5 + \tau_6 - \tau_7 + \tau_8$ with variance 3ξ. The sum of these three estimators gives an estimator for $3(\tau_1 - \tau_5)$ with variance $6\xi + 6\xi + 3\xi$, because the three contrasts are orthogonal to each other in each large block. Hence $\mathrm{Var}(\hat\tau_1 - \hat\tau_5) = 15\xi/9 = 5\xi/3$. The

11.8. Optimality

efficiency for $\hat{\tau}_1 - \hat{\tau}_5$ is
$$\frac{(2/2)\sigma^2}{(5/3)\xi} = \frac{3}{5}\frac{\sigma^2}{\xi},$$
while the efficiency factor is $3/5$.

Theorem 11.6 *In a lattice design with replication r and block size k, the estimator $\hat{\tau}_i - \hat{\tau}_j$ has variance*
$$\begin{cases} \dfrac{2(k+1)}{kr}\xi & \text{if } \lambda_{ij} = 1 \\[2mm] \dfrac{2(kr-k+r)}{kr(r-1)}\xi & \text{if } \lambda_{ij} = 0, \end{cases}$$
where ξ is the plots stratum variance. In particular, in a balanced lattice design $(r = k+1)$, every such estimator has variance
$$\frac{2(k+1)}{kr}\xi = \frac{2r}{(r-1)r}\xi = \frac{2}{r-1}\xi.$$

Note that
$$\frac{k+1}{kr} = \frac{(k+1)(r-1)}{kr(r-1)} = \frac{kr-k+r-1}{kr(r-1)} < \frac{kr-k+r}{kr(r-1)},$$
so the variance of the estimator of the difference $\tau_i - \tau_j$ is bigger if treatments i and j do not occur in a block together.

Warning For general incomplete-block designs, the variance of the estimator of the difference $\tau_i - \tau_j$ does not depend on the concurrence λ_{ij} in this simple way.

11.8 Optimality

Ideally we want to choose a design in which the variance of every estimator of the form $\hat{\tau}_i - \hat{\tau}_j$ is as small as possible.

Definition The *overall efficiency factor E* for an incomplete-block design for t treatments replicated r times is defined by
$$\frac{2}{t(t-1)}\sum_{i=1}^{t-1}\sum_{j=i+1}^{t}\operatorname{Var}(\hat{\tau}_i - \hat{\tau}_j) = \frac{2\xi}{rE},$$
where ξ is the plots stratum variance.

Definition An incomplete-block design is *optimal* if it has the largest value of E over all incomplete-block designs with the same values of t, r and k.

The next two results justify our concentration on balanced designs and on lattice designs.

Theorem 11.7 *For an incomplete-block design, the following conditions are equivalent.*

(i) *The design is balanced.*

(ii) All estimators of the form $\hat{\tau}_i - \hat{\tau}_j$ have the same variance.

(iii) All estimators of the form $\hat{\tau}_i - \hat{\tau}_j$ have the same efficiency factor.

(iv) The overall efficiency factor satisfies

$$E = \frac{t}{t-1}\frac{k-1}{k}$$

where t is the number of treatments and k is the block size.

Moreover, if the design is not balanced then

$$E < \frac{t}{t-1}\frac{k-1}{k}.$$

Hence balanced incomplete-block designs are optimal.

Theorem 11.8 *Lattice designs are optimal. Moreover, for these parameters, E cannot be increased by allowing block designs that are not resolved.*

11.9 Supplemented balance

Balanced designs and lattice designs are suitable whether the treatments are structured or not. Incomplete-block designs especially suitable for factorial treatments will be given in Chapter 12. The other important treatment structure has one or more control treatments.

Definition An incomplete-block design for a set of treatments which includes one or more control treatments has *supplemented balance* if every control treatment occurs once in each block and the remaining parts of the blocks form a balanced incomplete-block design for the other treatments.

Example 11.4 revisited (Detergents) Suppose that one standard detergent is included in the experiment to compare five new detergents. If each housewife does one washload with the standard detergent then she can do three with new detergents. The unreduced design for five treatments in blocks of size three has ten blocks. Inserting the standard detergent X into each block gives the following design, which has supplemented balance.

$$\{1,2,3,X\}, \quad \{1,2,4,X\}, \quad \{1,2,5,X\}, \quad \{1,3,4,X\}, \quad \{1,3,5,X\},$$
$$\{1,4,5,X\}, \quad \{2,3,4,X\}, \quad \{2,3,5,X\}, \quad \{2,4,5,X\}, \quad \{3,4,5,X\}$$

More generally, any constant number of copies of the control treatment may be inserted into each block. Then different designs with supplemented balance and the same block size may be combined.

Example 11.7 (Example 4.1 continued: Insect repellent) Here the experimental units are the arms of twelve volunteers, so they form twelve blocks of size two. Suppose that there are four insect repellents to compare. A balanced design needs the same number of blocks for each of the six pairs, as follows.

$$\{A,B\}, \quad \{A,B\}, \quad \{A,C\}, \quad \{A,C\}, \quad \{A,D\}, \quad \{A,D\},$$
$$\{B,C\}, \quad \{B,C\}, \quad \{B,D\}, \quad \{B,D\}, \quad \{C,D\}, \quad \{C,D\}$$

11.10. Row–column designs with incomplete columns

However, if treatment A is untreated control and the other three treatments are of unproven efficacity, then we may be more interested in the comparisons of each of the new treatments with A than in any comparisons among the new treatments. For a design with supplemented balance we can use s_1 copies of each pair containing A, and s_2 copies of each other block, for suitable values of s_1 and s_2. Then the replication of A is $3s_1$ while the replication of each other treatment is $s_1 + 2s_2$. Equation (3.1) shows that we should have $3s_1 \approx \sqrt{3}(s_1 + 2s_2)$, which implies that $s_2 \approx ((\sqrt{3}-1)/2)s_1 \approx 0.366s_1$. We also have $3s_1 + 3s_2 = 12$, so we should put $s_1 = 3$ and $s_2 = 1$, which gives the following design.

$$\{A,B\}, \quad \{A,C\}, \quad \{A,D\}, \quad \{A,B\}, \quad \{A,C\}, \quad \{A,D\},$$
$$\{A,B\}, \quad \{A,C\}, \quad \{A,D\}, \quad \{B,C\}, \quad \{B,D\}, \quad \{C,D\}$$

We have overlooked the fact that Equation (3.1) was derived under the assumption that the variance of the estimator of $\tau_i - \tau_j$ is proportional to $1/r_i + 1/r_j$, which is not true for incomplete-block designs in general. The calculations for the general case are beyond the scope of this book. However, for the numbers given here, the exact calculations do indeed show that the design given above minimizes the sum of the variances of the estimators of the differences between the control and the new treatments.

11.10 Row–column designs with incomplete columns

Sometimes the numbers of rows and columns in a row–column design do not satisfy the conditions given in Section 6.1. Suppose that there are k rows and b columns, where $k < t$. Then the columns should form an incomplete-block design. If we still want every treatment to occur equally often in each row then b must be a multiple of t.

A famous result called Hall's Marriage Theorem states that, if $b = t$ and the columns form any equireplicate incomplete-block design, then we can rearrange the treatments within each column in such a way that each treatment occurs once in each row. The proof of Hall's Marriage Theorem gives the following simple algorithm for finding such a rearrangement.

Technique 11.1 (Turning an incomplete-block design into a row–column design) Given an equireplicate incomplete-block design with the same number of blocks as treatments, arrange the blocks as the columns of a row–column design as follows.

(i) Choose the treatments for the first row like this. For each block in turn do the following.

   (a) If that block contains any treatments that have not yet been chosen, then choose one of them; go to the next block.

   (b) Otherwise, choose any treatment in that block. This has already been chosen for a previous block, so return to that block, and choose another treatment from that block. If possible, this should be a treatment that is not yet chosen; otherwise choose any treatment and return to the earlier block where this treatment was chosen. Repeat until a new treatment is chosen.

(ii) From each block, remove the treatment chosen for the first row.

(iii) Repeat the above two steps for each succeeding row.

1	2	3	4	5	6	0
2	3	4	5	6	0	1
4	5	6	0	1	2	3

Fig. 11.6. A Youden square for seven treatments in three rows and seven columns

| | \multicolumn{12}{c}{Volunteer} |
Arm	1	2	3	4	5	6	7	8	9	10	11	12
right	A	B	A	C	A	D	B	C	B	D	C	D
left	B	A	C	A	D	A	C	B	D	B	D	C

Fig. 11.7. Design for the experiment in Example 11.7 if it is thought that there may be a difference between right arms and left arms

If the incomplete-block design is cyclic with $b = t$, then the original cyclic construction gives the row–column design directly. If the treatment in row j of the first column is i_j then the treatment in row j and column l is just $i_j + l - 1$ modulo t.

Definition A row–column design with the same number of columns as treatments is called a *Youden square* if the rows are complete blocks and the columns form a balanced incomplete-block design.

Example 11.8 (A Youden square) Let $t = b = 7$ and $r = k = 3$. The cyclic design shown in Figure 11.1 is readily converted into the Youden square in Figure 11.6.

An extension of Hall's Marriage Theorem shows that, if t divides b and the columns form any equireplicate incomplete-block design, then we can rearrange the treatments within each column in such a way that each treatment occurs b/t times in each row.

Example 11.7 revisited (Insect repellent) If the experimenter decides that right arms should be distinguished from left arms, then we need a row–column design in which the columns are the volunteers and there are two rows, one for left arms and one for right arms. The design for four treatments and twelve volunteers on page 234 can be arranged as the row–column design in Figure 11.7.

A further extension of Hall's Theorem shows that, so long as the replication of every treatment is a multiple of the block size k, then every incomplete-block design can be arranged in a rectangle whose columns are the blocks of the original design and whose rows form an orthogonal block design in the sense defined in Section 4.2.

Example 11.9 (Microarrays) In some experiments in genomics, the blocks are glass slides called *microarrays*. Each microarray contains a rectangular grid of spots. A robot puts material from several genes onto the slides, in such a way that cDNA from each gene occurs on the same spot on every slide. Two treatments (such as tissue from cancerous and non-cancerous cells) are applied to every slide. One is coloured with a red dye, the other is coloured with a

11.10. Row–column designs with incomplete columns

red	1	2	1	3	2	3
green	2	1	3	1	3	2

(a) Smallest dye-swap design for three treatments

red	0	1	0	2	0	3	0	4
green	1	0	2	0	3	0	4	0

(b) Double reference design for five treatments of which the one labelled 0 is a reference treatment

red	1	2	3	4	5	6	7	8	9
green	2	3	4	5	6	7	8	9	1

(c) Loop design for nine treatments

Fig. 11.8. Three common classes of design for microarray experiments

green dye; then a 50:50 mixture of the two is washed over each slide. Some of this material binds with each gene, and the rest is washed off. The intensity of the fluorescence of each colour on each spot measures, on an exponential scale, how much material from that treatment has bound to the cDNA in the gene on that spot. The logarithms of these intensities are analysed.

The geneticists expect the treatments to behave differently with different genes: in fact, the purpose of the experiment is often to identify those few genes where the treatments differ. Effectively they are looking for a gene-by-treatment interaction rather than for any main effect. Thus it makes sense to analyse the data for each spot (over all slides) separately.

Now the slides can be considered as blocks of size two. There may also be a difference between the colours, so we have a row–column design in which the slides are columns and the colours are rows.

Microarray experiments are not always equireplicate. The second extension of Hall's Theorem shows that if the blocks (columns) have size two and every treatment has even replication then the treatments in each column can be arranged so that each treatment has half of its occurrences in each row.

Figure 11.8 shows three common designs. In the *multiple dye-swap* design, each pair of treatments occurs equally often as a column, coloured both ways round. In the *double reference* design, there is a control treatment, also known as the reference treatment. For each other treatment, there are two slides containing it and the reference treatment, one for each of the two ways of allocating colours. The so-called *loop* design is just a cyclic design, for any number of treatments, with initial block $\{1,2\}$.

This type of row–column design is particularly useful as a cross-over design when it is not practicable to have as many periods as treatments. One application is in Phase I clinical trials, when several doses of a new drug may be tested on healthy volunteers. For example, two copies of the Youden square in Figure 11.6 could be used for a cross-over trial comparing seven doses, using 14 volunteers and three periods.

If a placebo is included, and given special status, then it may be possible to convert an incomplete-block design with supplemented balance into a suitable row–column design.

Suppose that a balanced incomplete-block design for t treatments replicated r times in

0	4	7	1	0	3	8	2	5	9	6	0
1	0	8	0	2	6	0	9	3	5	7	4
2	5	0	4	8	9	6	0	7	1	0	3
3	6	9	7	5	0	1	4	0	0	2	8

Fig. 11.9. Cross-over design in Example 11.10

b blocks of size k has $r = k+1$. Balanced lattice designs satisfy this. Insert the placebo into every block, to give b blocks of size $k+1$. Now the replication of each original treatment is equal to the new block size $k+1$, while the replication of the placebo is equal to b. However, $b = b(k+1) - bk = b(k+1) - tr = b(k+1) - t(k+1)$, which is also divisible by $k+1$. Thus the second extension to Hall's Theorem shows that the treatments in each block can be arranged as the columns of a $(k+1) \times b$ rectangle so that each original treatment occurs once in each row, leaving the placebo to occupy the remaining $b - t$ positions in each row.

Example 11.10 (Phase I cross-over trial with a placebo) The balanced lattice design in Example 11.2 has $t = 9$, $k = 3$, $r = 4$ and $b = 12$. Inserting the placebo into each block gives twelve blocks of size four, which can be rearranged as the 4×12 rectangle in Figure 11.9. The rows form an orthogonal block design and the columns form an incomplete-block design with supplemented balance.

Since rows are orthogonal to treatments, the analysis of data from such a row–column design is no more complicated than analysis of data from the incomplete-block design for the columns. Row means may be fitted and removed from the data either before or after the procedures described in Sections 11.5 and 11.7. Of course, the number of degrees of freedom for residual must be reduced by $k - 1$. Such an analysis has already been shown in Example 11.5.

Questions for discussion

11.1 Show that $\{1,3,4,5,9\}$ is a difference set modulo 11.

11.2 For each of the following sets of values of t, b and k, *either* construct a balanced incomplete-block design for t treatments in b blocks of size k *or* prove that no such balanced incomplete-block design exists.

(a) $t = 5$ $b = 10$ $k = 3$

(b) $t = 7$ $b = 7$ $k = 4$

(c) $t = 7$ $b = 14$ $k = 3$

(d) $t = 10$ $b = 15$ $k = 3$

(e) $t = 11$ $b = 11$ $k = 5$

(f) $t = 21$ $b = 14$ $k = 6$

(g) $t = 21$ $b = 21$ $k = 5$

(h) $t = 25$ $b = 30$ $k = 5$

(i) $t = 25$ $b = 40$ $k = 5$

(j) $t = 31$ $b = 31$ $k = 6$

11.3 Verify that if an equireplicate incomplete-block design satisfies Equation (11.1) and Theorem 11.1 then so does its complement.

11.4 Construct a lattice design for 25 treatments in 20 blocks of size five.

11.5 Suppose that there are 48 experimental units, arranged in twelve blocks of size four. Construct a suitable design for

(a) 16 treatments;

(b) four new treatments and one control treatment;

(c) four new treatments and two control treatments.

11.6 Suppose that there are 66 experimental units, arranged in eleven blocks of size six. Construct a suitable design for

(a) eleven treatments;

(b) eleven new treatments and one control treatment.

11.7 Find the efficiency factor for treatment contrasts for the design in Example 11.3.

Suppose that the plots stratum variance for this design is $5/8$ of the plots stratum variance for an alternative design in four complete blocks of size 13. Which design is better?

11.8 A horticulture research institute wants to compare nine methods of treating a certain variety of houseplant while it is being grown in a greenhouse in preparation for the Christmas market. One possibility is to ask twelve small growers to test three treatments each in separate chambers in their greenhouses. A second possibility is to ask three large commercial growers to test nine methods each, also in separate greenhouse chambers.

(a) Construct a suitable design for the first possibility.

(b) Randomize this design.

(c) If the plots stratum variance is the same in both cases, which design is more efficient?

(d) Compare the designs in terms of likely cost, difficulty and representativeness of the results.

11.9 Plasma screens for television sets are among the products manufactured by a large electronics company. The engineers have devised six new types of plasma screen, and want to know how potential customers will rate these. As it is not practicable to experiment on the general public, the screens will be rated by employees from another part of the company. These employees are aware that sharpness and brightness matter, but do not have the detailed technical background of the engineers directly involved.

Each employee involved will go into a small room containing a few television sets with the new screens. After watching each television for ten minutes, they will give each screen a score between zero and ten.

(a) Alice suggests that each employee taking part in the experiment should watch television on three of the new screens. Construct a suitable design in which each new type of screen is rated by ten people.

(b) Bob thinks that it would be better to put four televisions into the test room, so that each employee involved can rate four screens. Construct a suitable design in which each new type of screen is rated by ten people.

(c) Alice points out that if people look at four types of screen rather than three then they may become a little confused, with the result that the variance of their ratings increases. Which design is better if the variance increases by 10% when they rate four screens instead of three? Which design is better if it increases by 20%?

11.10 The seven treatments in Question 3.5 are to be compared using 21 patients in a crossover trial with five periods.

(a) Construct a suitable incomplete-block design with the patients considered as blocks.

(b) Arrange these blocks as a suitable cross-over design.

11.11 Assuming that Hall's Marriage Theorem does indeed show that the algorithm given in Technique 11.1 always succeeds, prove the two extensions of Hall's Theorem stated on page 236.

Chapter 12
Factorial designs in incomplete blocks

12.1 Confounding

Definition A treatment subspace contained in stratum W_F is *confounded* with F.

Definition Factors F and G are *strictly orthogonal* to each other if they are orthogonal to each other in the sense of Section 10.6 and $F \vee G = U$.

If F is strictly orthogonal to G then $V_F \cap V_0^\perp$ is orthogonal to $V_G \cap V_0^\perp$ so all contrasts in F are orthogonal to all contrasts in G. Moreover, all combinations of levels of F and levels of G occur. Further, if G is a treatment factor and F is a plot factor then the treatment subspace W_G is not confounded with F, nor with any plot factor H for which $F \prec H$.

Example 12.1 (Example 1.7 continued: Rye-grass) The treatments are all combinations of cultivar with nitrogen. These two factors are strictly orthogonal to each other. The small blocks are strips, and cultivars are confounded with strips.

Example 12.2 (Example 10.11 continued: Main-effects-only design in blocks) The treatments are all combinations of levels of factors F and G. Blocks are incomplete, and we use a Latin square to ensure that F and G are both orthogonal to blocks and that F is strictly orthogonal to G.

In this chapter we develop a more general way of constructing incomplete-block designs for factorial experiments, which includes both of the above. We want to ensure that the single-replicate designs are orthogonal, in the sense of Section 10.12. We also want to identify which treatment subspace, if any, is confounded with blocks. In Figure 10.38 we used a factor Q with no real physical meaning in order to identify this confounding. This idea is generalized in this chapter.

Example 12.3 (Watering chicory) Chicory is grown in boxes in sheds. Each box is watered by a continuous flow of water from a pipe. The experimenter can vary two factors which affect this watering. One is the rate of flow of the water; this can be adjusted to one of three rates by opening the tap on the input pipe. The second factor is the depth of standing water which is permitted in the box; this can be adjusted to have one of three values by putting side bars of appropriate heights around the box.

241

Two sheds will be used for the experiment. Each shed contains three rooms, each of which contains three boxes of chicory. The boxes are the plots. At the end of the experiment the chicory is harvested and the saleable chicory per box is weighed.

We want each shed to be a single replicate of the treatments. How should treatments be applied?

(i) We could use a split-plot design, applying flowrate to whole rooms and applying depth to individual boxes. Then the main effect of flowrate is confounded with rooms, and so it is estimated less precisely than the main effect of depth or the flowrate-by-depth interaction. If rates of flow cannot be varied within a room then we must use such a design; otherwise, it is not a good choice unless we are less interested in the main effect of flowrate than in its interaction with depth.

(ii) We could interchange the roles of flowrate and depth, and construct a split-plot design in which depth is confounded with rooms. This has analogous disadvantages to the previous design.

(iii) We could confound flowrate with rooms in the first shed and confound depth with rooms in the second shed. Thus each main effect will be estimated more precisely from just *half* of the plots; in other words, each main effect has *efficiency factor* equal to $1/2$, in the sense of Section 11.6. The interaction is estimated better than the two main effects. Unless we are chiefly interested in the interaction, this is not a good design.

(iv) As in Section 9.1.3, we could use a Latin square to construct a main-effects-only design in each shed. Now part of the flowrate-by-depth interaction is confounded with rooms.

(v) As there are four degrees of freedom for the flowrate-by-depth interaction, and only two are confounded with rooms in any one shed, it is better if we can arrange for different parts of the interaction to be confounded in the two sheds. Thus we need to know how to decompose interactions.

12.2 Decomposing interactions

Let $F_1, F_2, \ldots, F_n$ be treatment factors with p levels each, where p is a prime number. Suppose that treatments are all combinations of the levels of $F_1, F_2, \ldots, F_n$. Code the levels of each factor by $\mathbb{Z}_p$, the integers modulo p. Then we can regard each treatment as being an n-tuple of integers modulo p.

Definition If the treatments are all combinations of levels of factors $F_1, \ldots, F_n$, each of whose levels are coded by $\mathbb{Z}_p$, then a *character* of the treatments is a function from $\mathcal{T}$ to $\mathbb{Z}_p$ which is a linear combination of the F_i with coefficients in $\mathbb{Z}_p$.

In particular, each treatment factor F_i is itself a character.

Example 12.3 revisited (Watering chicory) Here we have $p = 3$ and $n = 2$. Let A be the factor flowrate, with the three rates coded as 0, 1, 2 in any order. Similarly, let B be the factor depth, with the three depths coded as 0, 1, 2 in any order. Then treatments are ordered pairs of integers modulo 3: the first entry is the level of A and the second is the level of B. These

12.2. Decomposing interactions

Table 12.1. *Characters in Example 12.3*

Characters	Treatments								
A	0	0	0	1	1	1	2	2	2
B	0	1	2	0	1	2	0	1	2
$A+B$	0	1	2	1	2	0	2	0	1
$A+2B$	0	2	1	1	0	2	2	1	0
$2A+B$	0	1	2	2	0	1	1	2	0
$2A+2B$	0	2	1	2	1	0	1	0	2
$2A$	0	0	0	2	2	2	1	1	1
$2B$	0	2	1	0	2	1	0	2	1
I	0	0	0	0	0	0	0	0	0

are shown above the line in Table 12.1. The remaining characters are below the line. Each character takes a value on each treatment. For example

$$(A+2B)\big((2,1)\big) = 1 \times 2 + 2 \times 1 = 2 + 2 = 1.$$

(Remember that the values are all in $\mathbb{Z}_3$, so all arithmetic must be done modulo 3.) It is conventional to write I for the character whose coefficients are all zero.

Each character can be regarded as a factor. The table has some noteworthy features.

(i) There are nine treatments and nine characters.

(ii) The factor A is aliased with the factor $2A$; that is, $A \equiv 2A$. Similarly, $B \equiv 2B$.

(iii) $I \equiv U$.

(iv) The factor $A+B$ has three levels, and $A+B \prec U$, so there are two degrees of freedom for contrasts between levels of $A+B$.

(v) The factor $A+B$ is strictly orthogonal to both A and B, so the contrasts between levels of $A+B$ are orthogonal to those for the main effects of both A and B. But $A \wedge B \prec A+B$ and so the two degrees of freedom for $A+B$ are part of the A-by-B interaction.

(vi) $2A+2B \equiv A+B$.

(vii) The factor $A+2B$ also accounts for two degrees of freedom from the A-by-B interaction, by similar reasoning to that used for $A+B$.

(viii) The factors $A+B$ and $A+2B$ are strictly orthogonal to each other, so the 4-dimensional subspace of treatment contrasts belonging to the A-by-B interaction is the orthogonal sum of two 2-dimensional subspaces, one corresponding to the factor $A+B$ and the other corresponding to the factor $A+2B$.

(ix) $2A+B = 2(A+2B)$ and $2A+B \equiv A+2B$.

If we use the characters $A+B$ and $A+2B$ to decompose the interaction, we obtain the Hasse diagram in Figure 12.1.

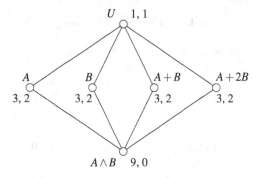

Fig. 12.1. Hasse diagram for treatment structure in Example 12.3

Although we shall not prove it here, the above pattern extends to any number of factors and any prime p, as the following theorem states. In fact, we shall not prove any of the theorems in this chapter, but the reader may easily verify them on any given example.

Theorem 12.1 *Let $F_1, F_2, \ldots, F_n$ each have p levels, coded by the integers modulo p, where p is prime. Then the following hold.*

(i) There are p^n treatments and p^n characters.

(ii) If $G = \sum_i g_i F_i$ and $H = \sum_i h_i F_i$ and $G \ne I$ and $H \ne I$ then either

  (a) there is some nonzero k in $\mathbb{Z}_p$ such that $g_i = k h_i$ for $i = 1, \ldots, n$, in which case $G \equiv H$; or

  (b) as factors, G and H are strictly orthogonal.

Hence the $(p^n - 1)$ characters different from I split into $(p^n - 1)/(p - 1)$ sets of $(p - 1)$ characters, in such a way that characters in the same set are equivalent to each other while characters in different sets are strictly orthogonal to each other.

(iii) If $G = \sum g_i F_i$ then G belongs to the interaction of those F_i for which $g_i \ne 0$.

Example 12.4 (Two factors with five levels) If A and B both have five levels then the 24 characters apart from I fall into six sets of four as follows.

$$
\begin{array}{llll}
A \equiv 2A \equiv 3A \equiv 4A & \text{main effect of } A \\
B \equiv 2B \equiv 3B \equiv 4B & \text{main effect of } B \\
A + B \equiv 2A + 2B \equiv 3A + 3B \equiv 4A + 4B & \text{part of the } A\text{-by-}B \text{ interaction} \\
A + 2B \equiv 2A + 4B \equiv 3A + B \equiv 4A + 3B & \text{part of the } A\text{-by-}B \text{ interaction} \\
A + 3B \equiv 2A + B \equiv 3A + 4B \equiv 4A + 2B & \text{part of the } A\text{-by-}B \text{ interaction} \\
A + 4B \equiv 2A + 3B \equiv 3A + 2B \equiv 4A + B & \text{part of the } A\text{-by-}B \text{ interaction}
\end{array}
$$

Thus the interaction is the orthogonal sum of four parts, each with four degrees of freedom. These parts are defined by the characters

$$A + B, \quad A + 2B, \quad A + 3B \quad \text{and} \quad A + 4B.$$

12.3. Constructing designs with specified confounding

Table 12.2. *Characters in Example 12.5*

A	$\equiv$	$2A$	main effect of A
B	$\equiv$	$2B$	main effect of B
C	$\equiv$	$2C$	main effect of C
$A+B$	$\equiv$	$2A+2B$	part of the A-by-B interaction
$A+2B$	$\equiv$	$2A+B$	part of the A-by-B interaction
$A+C$	$\equiv$	$2A+2C$	part of the A-by-C interaction
$A+2C$	$\equiv$	$2A+C$	part of the A-by-C interaction
$B+C$	$\equiv$	$2B+2C$	part of the B-by-C interaction
$B+2C$	$\equiv$	$2B+C$	part of the B-by-C interaction
$A+B+C$	$\equiv$	$2A+2B+2C$	part of the A-by-B-by-C interaction
$A+B+2C$	$\equiv$	$2A+2B+C$	part of the A-by-B-by-C interaction
$A+2B+C$	$\equiv$	$2A+B+2C$	part of the A-by-B-by-C interaction
$A+2B+2C$	$\equiv$	$2A+B+C$	part of the A-by-B-by-C interaction

This representation is not unique. For example, the second part is equally well defined by the characters $2A+4B$ (which can be written as $2A-B$), $3A+B$ or $4A+3B$. It is conventional to use the character whose first nonzero coefficient is equal to 1.

Warning Many people write the characters multiplicatively. For example, they write AB^2 in place of $A+2B$. Then the notation AB may mean $A+B$ (the character) or $A \wedge B$ (the factor) or A-by-B (the interaction).

Example 12.5 (Three factors with three levels) If there are three treatment factors A, B and C, each with three levels, then the 26 characters other than I fall into 13 sets of two as shown in Table 12.2. We have already seen how to decompose each of the two-factor interactions. There are eight degrees of freedom for the A-by-B-by-C interaction. This is the eight-dimensional space of contrasts in $A \wedge B \wedge C$ whose coordinates sum to zero on each class of each of the factors $A \wedge B$, $A \wedge C$ and $B \wedge C$. By using characters we can decompose this interaction into four orthogonal subspaces, each of dimension two. The subspaces correspond to the characters

$$A+B+C, \quad A+B+2C, \quad A+2B+C \quad \text{and} \quad A+2B+2C.$$

12.3 Constructing designs with specified confounding

If G is a character and we want to construct a block design in which G is confounded with blocks, the obvious method is to evaluate G on every treatment, then put into the first block all those treatments u with $G(u) = 0$, into the second block all those treatments u with $G(u) = 1$, and so on.

Example 12.3 revisited (Watering chicory) The values of $A+B$ and $A+2B$ have already been calculated in Table 12.1. Confounding $A+B$ with rooms in the first shed and $A+2B$ with rooms in the second shed gives the design in Figure 12.2.

	First Shed			Second Shed		
	Room 1	Room 2	Room 3	Room 4	Room 5	Room 6
	$A+B=0$	$A+B=1$	$A+B=2$	$A+2B=0$	$A+2B=1$	$A+2B=2$
A	0 1 2	0 1 2	0 1 2	0 1 2	0 1 2	0 1 2
B	0 2 1	1 0 2	2 1 0	0 1 2	2 0 1	1 2 0

Fig. 12.2. Design confounding $A+B$ in one shed and confounding $A+2B$ in the other in Example 12.3

However, there is a simpler method, that does not demand evaluation of the characters.

Each treatment u can be written as an n-tuple of integers modulo p, as $u = (u_1, u_2, \ldots, u_n)$, where $F_i(u) = u_i$. Then treatments can be added by the rule

$$u + v = (u_1 + v_1, u_2 + v_2, \ldots, u_n + v_n).$$

Equivalently, $F_i(u+v) = F_i(u) + F_i(v)$ for each i. Then the set $\mathcal{T}$ of treatments forms an *abelian group*. This means that:

(i) for all u and v in $\mathcal{T}$, the sum $u+v$ is also in $\mathcal{T}$;

(ii) for all u and v in $\mathcal{T}$, we have $u+v = v+u$;

(iii) $(u+v)+w = u+(v+w)$ for all u, v, w in $\mathcal{T}$;

(iv) there is a treatment $(0,0,\ldots,0)$ which satisfies

$$u + (0,0,\ldots,0) = u \quad \text{for all } u;$$

(v) for each u in $\mathcal{T}$ there is a treatment $(-u_1, -u_2, \ldots, -u_n)$, which we write as $-u$, such that

$$u + (-u) = (0,0,\ldots,0).$$

Property (ii) says that the operation $+$ is *commutative*; property (iii) says that it is *associative*. For property (v), note that in $\mathbb{Z}_p$ the element $-u_i$ is the same as $(p-1)u_i$ and as $p - u_i$: for example, if $p = 5$ and $u_1 = 2$ then $-u_1 = 3$.

Definition In a single-replicate design, the block containing the treatment $(0,0,\ldots,0)$ is the *principal* block.

Theorem 12.2 *If the blocks of a single-replicate design are defined by the levels of one or more characters then the principal block is a subgroup of $\mathcal{T}$; this means that if u and v are in the principal block then so is $u+v$. Moreover, every block is a coset of the principal block, in other words a subset of the form*

$$\{v + u : u \in \textit{principal block}\}$$

for some fixed v. Thus, to construct the block containing v, simply add v to every element of the principal block.

12.3. Constructing designs with specified confounding

Example 12.5 revisited (Three factors with three levels) To confound $A + B + 2C$ (which is part of the three-factor interaction) with blocks, the principal block consists of those treatments u for which
$$A(u) + B(u) + 2C(u) = 0.$$
This equation can be rewritten as
$$C(u) = A(u) + B(u), \tag{12.1}$$
because $2 = -1$ modulo 3. Thus every pair of levels of A and B uniquely determines a level of C so that Equation (12.1) is satisfied. The principal block therefore contains the following nine treatments:

$(0,0,0),\ (0,1,1),\ (0,2,2),\ (1,0,1),\ (1,1,2),\ (1,2,0),$
$(2,0,2),\ (2,1,0),\ (2,2,1).$

The treatment $(1,1,1)$ does not appear in the principal block. To obtain the block containing it, we add $(1,1,1)$ to every treatment in the principal block. This gives a second block, with the following nine treatments:

$(1,1,1),\ (1,2,2),\ (1,0,0),\ (2,1,2),\ (2,2,0),\ (2,0,1),$
$(0,1,0),\ (0,2,1),\ (0,0,2).$

The block containing $(2,2,2)$ is constructed similarly.

Since each block contains nine of the 27 treatments, it would be possible to construct the third block as simply 'all those treatments which are not in the first two blocks'. This method gives no check on mistakes. It is better to construct every block as a coset of the principal block, and then check that every treatment appears in one and only one block.

Example 12.6 (Sugar beet) An experiment on sugar beet investigated three three-level treatment factors, whose real levels were coded by the integers modulo 3 as follows.

Factor	0	1	2
Sowing date (D)	18 April	9 May	25 May
Spacing between rows (S)	10 inches	15 inches	20 inches
Nitrogen fertilizer (N)	nil	0.3 cwt/acre	0.6 cwt/acre

There were three blocks of nine plots each. These were constructed to confound $D + S + 2N$ with blocks. Thus the allocation of treatments to blocks can be deduced from Example 12.5. Table 12.3 shows the layout after randomization.

Example 12.7 (Four factors with two levels) Suppose that A, B, C and D each have two levels. To confound $A + B + C + D$ (which is the whole of the four-factor interaction) with blocks, the principal block contains those treatments u for which
$$A(u) + B(u) + C(u) + D(u) = 0.$$
Solving such equations is relatively easy when $p = 2$, because the only possible values are 0 and 1. Thus the principal block consists of all those treatments for which an *even* number of the factors A, B, C, D have level 1:

$(0,0,0,0),\ (0,0,1,1),\ (0,1,0,1),\ (0,1,1,0),\ (1,0,0,1),\ (1,0,1,0),$
$(1,1,0,0),\ (1,1,1,1).$

Table 12.3. *Layout of the sugar-beet experiment in Example 12.6*

Block 1	$D+S+2N=1$		
Plot	Date	Spacing	Nitrogen
1	25 May	15in	0.6
2	9 May	15in	0.3
3	9 May	10in	0
4	25 May	20in	0
5	9 May	20in	0.6
6	18 April	15in	0
7	25 May	10in	0.3
8	18 April	20in	0.3
9	18 April	10in	0.6

Block 2	$D+S+2N=0$		
Plot	Date	Spacing	Nitrogen
10	18 April	15in	0.3
11	18 April	10in	0
12	25 May	10in	0.6
13	9 May	10in	0.3
14	9 May	15in	0.6
15	25 May	20in	0.3
16	25 May	15in	0
17	9 May	20in	0
18	18 April	20in	0.6

Block 3	$D+S+2N=2$		
Plot	Date	Spacing	Nitrogen
19	18 April	15in	0.6
20	25 May	15in	0.3
21	9 May	10in	0.6
22	9 May	20in	0.3
23	25 May	10in	0
24	18 April	10in	0.3
25	9 May	15in	0
26	25 May	20in	0.6
27	18 April	20in	0

When the context is clear, we sometimes miss out the commas and the parentheses in the notation for factorial treatments. In this abbreviated notation, the eight treatments in the principal block in Example 12.7 are as follows:

$$0000, \quad 0011, \quad 0101, \quad 0110, \quad 1001, \quad 1010, \quad 1100, \quad 1111.$$

The treatments in the other block are those in which an odd number of the factors A, B, C, D have level 1:

$$1000, \quad 0100, \quad 0010, \quad 0001, \quad 1110, \quad 1101, \quad 1011, \quad 0111.$$

Example 12.8 (Pill manufacture) Medicinal tablets are composed largely of excipients (inactive ingredients) which serve to hold the active ingredients and release them into the body. In the first stage of making the tablets, the inactive ingredients are mixed. A certain volume of ingredients is put into a blender. They are blended for a length of time. Then the mixed ingredients are milled to a fine powder. Finally they are blended again. In a simple experiment each of these factors has two levels: the volume is small or large; each blending time is short or long; the milling speed is slow or fast. Only eight batches can be made up per day. The design in Example 12.7 can be used to confound the four-factor interaction with days.

12.4 Confounding more than one character

Let G and H be two characters. We can define $G+H$ by putting

$$(G+H)(u) = G(u) + H(u)$$

for all treatments u, continuing to do all arithmetic modulo p. Then we find that:

(i) if G and H are characters then so is $G+H$;

(ii) if G and H are characters then $G+H = H+G$;

(iii) if G, H and J are characters then $(G+H)+J = G+(H+J)$;

(iv) if G is any character then $G+I = G$;

(v) if G is any character then there is another character $-G$ such that $G+(-G) = I$ (in fact, $-G = (p-1)G$).

This means that the set of characters also forms an abelian group. We shall use this to work out more general confounding schemes.

Example 12.7 revisited (Four factors with two levels) Suppose that we have four blocks of four plots (instead of two blocks of eight plots). Each character partitions the treatments into two sets of eight. Since linearly independent characters are strictly orthogonal to each other, any pair of linearly independent characters partitions the treatments into four sets of four. So we can create four blocks of size four by confounding two linearly independent characters with blocks.

We could choose to confound $A+B+C+D$ and $A+B+C$ with blocks. Then

$$A+B+C+D \quad \text{is constant on each block and}$$
$$A+B+C \quad \text{is constant on each block};$$

therefore D is also constant on each block and so the main effect of D is confounded with blocks. This is probably not advisable.

Theorem 12.3 *If characters G and H are confounded with blocks then so is $G+H$. In fact, the set of all confounded characters, together with I, forms a subgroup of the group of all characters.*

Block	Treatments				Level of	
					$A+B+C$	$B+C+D$
Principal	0000	1011	1101	0110	0	0
Block 2	1000	0011	0101	1110	1	0
Block 3	0100	1111	1001	0010	1	1
Block 4	0001	1010	1100	0111	0	1

Fig. 12.3. Design for four two-level treatment factors in four blocks of four plots

Example 12.7 revisited (Four factors with two levels) For four blocks of size four, it might be better to confound the characters $A+B+C$, $B+C+D$ and $A+D$. If we confound $A+B+C$ and $B+C+D$ then the principal block is defined by

$$A+B+C = B+C+D = 0,$$

so

$$A = -(B+C) = B+C = D.$$

The design is shown in Figure 12.3, using the abbreviated notation for the treatments.

The procedures of this section and the preceding one are summarized in Technique 12.1.

Technique 12.1 (Single-replicate factorial designs in incomplete blocks) Given n treatment factors, each with p levels, where p is prime, construct a single-replicate block design with p^s blocks of p^{n-s} plots as follows.

(i) Find a subgroup $\mathcal{G}$ of the set of all characters such that

   (a) if any main effect has to be applied to whole blocks then the corresponding character is in $\mathcal{G}$;

   (b) if a character G belongs to an effect that must be estimated precisely then $G \notin \mathcal{G}$;

   (c) $|\mathcal{G}| = p^s$.

(ii) Form the principal block as those p^{n-s} treatments u satisfying

$$G(u) = 0 \quad \text{for all } G \text{ in } \mathcal{G}.$$

(iii) The remaining blocks are cosets of the principal block.

Note that Step (i) requires trial and error, and may be impossible. Step (ii) is straightforward, and Step (iii) is entirely automatic.

Example 12.9 (Field beans) In an experiment on field beans, there are five two-level factors, whose levels are coded by the integers modulo 2 as follows.

Factor	0	1
Row spacing (S)	18 inches	24 inches
Dung (D)	nil	10 ton/acre
Nitrochalk (N)	nil	0.4 cwt/acre
Superphosphate (P)	nil	0.6 cwt/acre
Potash (K)	nil	1 cwt/acre

12.5. Pseudofactors for mixed numbers of levels

Block	Treatments factors in order S, D, N, P, K							
Principal	00000	00101	01010	01111	10011	10110	11001	11100
Block 2	10000	10101	11010	11111	00011	00110	01001	01100
Block 3	01000	01101	00010	00111	11011	11110	10001	10100
Block 4	11000	11101	10010	10111	01011	01110	00001	00100

Fig. 12.4. Design for the experiment on field beans in Example 12.9

The experiment is a single replicate, in four blocks of eight plots each. Thus three characters of the form G, H and $G+H$ must be confounded. Because we are working modulo 2, if any two of these characters have an odd number of letters then the third has an even number of letters. Thus at least one of them has an even number of letters, and we should choose this number to be four, to avoid confounding any two-factor interaction. If we confound the five-letter character $S+D+N+P+K$ and *any* four-letter character then we also confound a main effect, which is probably not desirable. Any two four-letter characters have three letters in common, so if we confound two of them then we also confound a two-factor interaction. Thus the best we can do is to confound one four-letter character and two three-letter characters. For example, we can confound $D+N+P+K$, $S+D+P$ and $S+N+K$. This gives the design in Figure 12.4.

12.5 Pseudofactors for mixed numbers of levels

Treatment factors with different numbers of levels can be accommodated together so long as their numbers of levels are all powers of the same prime, say p. A factor with p^m levels is represented by m pseudofactors, each with p levels.

Example 12.10 (A mixed factorial) Factors A, B and C have two levels each, and factor D has the four levels 1, 2, 3, 4. One possible correspondence between D and its pseudofactors D_1 and D_2 is shown below.

D	1	2	3	4
D_1	0	0	1	1
D_2	0	1	0	1

Suppose that we want a single-replicate design in four blocks of eight plots each which confounds no main effect or two-factor interaction with blocks. We need to confound two different characters and their sum, and all three confounded characters must involve at least three factors. One possible choice is to confound

$A+B+D_1$, which is part of the A-by-B-by-D interaction, and
$A+C+D_2$, which is part of the A-by-C-by-D interaction, and
$B+C+D_1+D_2$, which is part of the B-by-C-by-D interaction.

The principal block is constructed by putting $D_1 = A+B$ and $D_2 = A+C$. The other three blocks are constructed as cosets of the principal block. Finally, the ordered pairs of levels of D_1 and D_2 are translated back into levels of D. The first two blocks, showing levels of both pseudofactors and genuine factors, are in Figure 12.5.

	Principal block	Block 2
A	0 0 0 0 1 1 1 1	0 0 0 0 1 1 1 1
B	0 0 1 1 0 0 1 1	0 0 1 1 0 0 1 1
C	0 1 0 1 0 1 0 1	0 1 0 1 0 1 0 1
D_1	0 0 1 1 1 1 0 0	1 1 0 0 0 0 1 1
D_2	0 1 0 1 1 0 1 0	0 1 0 1 1 0 1 0
D	1 2 3 4 4 3 2 1	3 4 1 2 2 1 4 3

Fig. 12.5. *Two blocks of the design in Example 12.10*

We might use such a design if we can assume that the *A*-by-*B*-by-*D* interaction, the *A*-by-*C*-by-*D* interaction, the *B*-by-*C*-by-*D* interaction and the *A*-by-*B*-by-*C*-by-*D* interaction are all zero. The skeleton analysis-of-variance table is in Table 12.4.

Notice that the assumption that the *A*-by-*B*-by-*D* interaction is zero is equivalent to the assumption that there are no differences between the different levels of the factor $A \wedge B \wedge D$ *apart from* those differences which are already accounted for by treatment factors coarser than $A \wedge B \wedge D$ (that is, A, B, D, $A \wedge B$, $A \wedge D$ and $B \wedge D$). But $A \wedge B \wedge D$ is itself coarser than $A \wedge B \wedge C \wedge D$, so it would be nonsense to assume that the latter is nonzero while the former is zero. The general form of this restriction is in Principle 12.1.

Principle 12.1 (Ordering Principle) If F and G are treatment factors with $F \prec G$ and the effect of F is assumed to be nonzero then the effect of G should also be assumed to be nonzero.

Table 12.4. *Skeleton analysis of variance in Example 12.10*

Stratum	Source	Degrees of freedom
mean	mean	1
blocks		3
plots	A	1
	B	1
	C	1
	D	3
	$A \wedge B$	1
	$A \wedge C$	1
	$B \wedge C$	1
	$A \wedge D$	3
	$B \wedge D$	3
	$C \wedge D$	3
	$A \wedge B \wedge C$	1
	residual	9
	total	28
Total		32

12.6. Analysis of single-replicate designs

Table 12.5. *Skeleton analysis of variance for Example 12.6 if we can assume that the three-factor interaction is zero*

Stratum	Source	Degrees of freedom
mean	mean	1
blocks	blocks	2
plots	D	2
	S	2
	N	2
	$D \wedge S$	4
	$D \wedge N$	4
	$S \wedge N$	4
	residual	6
	total	24
Total		27

12.6 Analysis of single-replicate designs

There are three ways to analyse data from a single-replicate factorial experiment in blocks. The first method was implied in Section 9.1.3. We assume that some treatment effects are zero (in accord with Principle 12.1) and then analyse in the usual way.

Example 12.6 revisited (Sugar beet) If we can assume that the three-factor interaction is zero then we obtain the skeleton analysis of variance in Table 12.5. All main effects and two-factor interactions can be estimated. Their presence can be tested for, using the residual mean square in the plots stratum. That mean square can also be used to find estimates of the variances of those estimators.

On the other hand, if we cannot assume that any treatment effects are zero then we can still estimate effects and obtain the full analysis-of-variance table, but there is no residual line in any stratum.

Example 12.6 revisited (Sugar beet) If we cannot assume that the three-factor interaction is zero then we obtain the skeleton analysis of variance in Table 12.6. We can still estimate the main effects and two-factor interactions, but there are no residual degrees of freedom in the plots stratum so we cannot carry out hypothesis tests or estimate variances. We cannot realistically estimate the three-factor interaction, because we expect the blocks stratum variance to be high.

In this situation, where we cannot assume in advance that any effects are zero, a third method is helpful. It is graphical. We assume the usual normal model, whether blocks are fixed or random, with plots stratum variance ξ. If all effects in the plots stratum are zero, then the sums of squares for the p-valued characters in the plots stratum are independent random variables, each of which is ξ times a χ^2 random variable on $p-1$ degrees of freedom, by Theorem 2.11(vi). There are m such characters, where

$$m = \frac{p^n - p^s}{p-1} = \frac{p^s(p^{n-s}-1)}{p-1} = p^s(p^{n-s-1} + p^{n-s-2} + \cdots + p + 1).$$

Table 12.6. *Skeleton analysis of variance for Example 12.6 if we cannot assume that the three-factor interaction is zero*

Stratum	Source	Degrees of freedom
mean	mean	1
blocks	$D+S+2N$	2
	(part of D-by-S-by-N interaction)	
plots	D	2
	S	2
	N	2
	$D \wedge S$	4
	$D \wedge N$	4
	$S \wedge N$	4
	rest of D-by-S-by-N interaction	6
	total	24
Total		27

Let $S_1, \ldots, S_m$ be these sums of squares, in increasing order of magnitude. For $i = 1, \ldots, m$, let Q_i be the $(i - \frac{1}{2})/m$-th quantile of the χ^2 distribution on $p - 1$ degrees of freedom; that is, if X is a χ^2 random variable with $p - 1$ degrees of freedom then

$$\Pr[X \leq Q_i] = \frac{i - \frac{1}{2}}{m} = \frac{2i - 1}{2m}.$$

If these sums of squares are not inflated by any nonzero treatment effects, then the graph of S_i against Q_i should be a straight line through the origin with slope ξ. This graph is called a *quantile plot*. Typically, such a graph shows most points on a line through the origin, with a few points at the top end lying above the line. These should be removed, m reduced accordingly, and the graph redrawn. When sufficient points have been removed for the graph to be (approximately) a straight line through the origin, the conclusion is that the effects corresponding to the removed points are nonzero, and should be investigated further. Of course, Principle 12.1 has to be observed: do not remove a two-letter character without removing the single letters in it; and do not remove part of an interaction without removing all of it.

Example 12.6 revisited (Sugar beet) Table 12.7 shows the yields of beet, in cwt/acre, for the experiment in Table 12.3. There is a column for each character not confounded with blocks, showing the total yield on each level of the character, and hence the sum of squares for the character. For example, the sum of squares for the character $D + S$ is $(410.6^2 + 404.1^2 + 410.0^2)/9 - 1224.7^2/27$, which is 2.87 to two decimal places. These twelve sums of squares, in increasing order, are the initial values of $S_1, \ldots, S_{12}$.

Since the χ^2 distribution with two degrees of freedom is just the exponential distribution with mean 2, the quantiles are readily calculated as $Q_i = -2\ln((25 - 2i)/24)$. Figure 12.6 shows the graph. It is clear that the main effects of D and N are nonzero, and possibly also the main effect of S.

After removing these three largest effects, the quantile plot becomes the one shown in Figure 12.7. Now the uppermost points are no longer above the line.

12.6. Analysis of single-replicate designs

Table 12.7. *Yields of beet, in cwt/acre, and calculations of sums of squares for the characters, for the experiment in Table 12.3*

Plot	Yield	D	S	N	D+S	D+2S	D+N	D+2N	S+N	S+2N	D+S+N	D+2S+N	D+2S+2N
1	52.2	2	1	2	0	1	1	0	0	2	2	0	2
2	52.7	1	1	1	2	0	2	0	2	0	0	1	2
3	47.8	1	0	0	1	1	1	1	0	0	1	1	1
4	35.2	2	2	0	1	0	2	2	2	2	1	0	0
5	45.4	1	2	2	0	2	0	2	1	0	2	1	0
6	44.6	0	1	0	1	2	0	0	1	1	1	2	2
7	46.0	2	0	1	2	2	0	1	1	2	0	0	1
8	51.4	0	2	1	2	1	1	2	0	1	0	2	0
9	50.5	0	0	2	0	0	2	1	2	1	2	2	1
10	49.7	0	1	1	1	2	1	2	2	0	2	0	1
11	47.8	0	0	0	0	0	0	0	0	0	0	0	0
12	44.1	2	0	2	2	2	1	0	2	1	1	1	0
13	52.5	1	0	1	1	1	2	0	1	2	2	2	0
14	49.3	1	1	2	2	0	0	2	0	2	1	2	1
15	46.2	2	2	1	1	0	0	1	0	1	2	1	2
16	47.1	2	1	0	0	1	2	2	1	1	0	1	1
17	47.2	1	2	0	0	2	1	1	2	2	0	2	2
18	56.0	0	2	2	2	1	2	1	1	0	1	0	2
19	50.9	0	1	2	1	2	2	1	0	2	0	1	0
20	38.2	2	1	1	0	1	0	1	2	0	1	2	0
21	43.0	1	0	2	1	1	0	2	2	1	0	0	2
22	36.5	1	2	1	0	2	2	0	0	1	1	0	1
23	38.0	2	0	0	2	2	2	2	0	0	2	2	2
24	45.7	0	0	1	0	0	1	2	1	2	1	1	2
25	37.1	1	1	0	2	0	1	1	1	1	2	0	0
26	34.2	2	2	2	1	0	1	0	1	0	0	2	1
27	35.4	0	2	0	2	1	0	0	2	2	2	1	1
Total on 0		432.0	415.4	380.2	410.6	398.7	395.9	400.0	420.1	409.8	420.3	403.5	402.6
Total on 1		411.5	421.8	418.9	404.1	423.6	409.4	419.9	408.6	400.5	397.4	415.3	396.5
Total on 2		381.2	387.5	425.6	410.0	402.4	419.4	404.8	396.0	414.4	407.0	405.9	425.6
SS		145.15	73.92	133.47	2.87	40.12	30.91	23.96	32.29	11.14	29.29	8.64	52.33

Overall total = 1224.7

In fact these data are only half of a two-replicate design to be described in Section 12.7. The full analysis of all the data does confirm that only the three main effects are needed to model the yields.

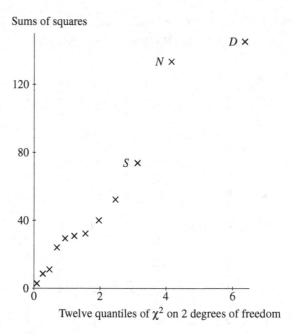

Fig. 12.6. Plot of sums of squares from Table 12.7

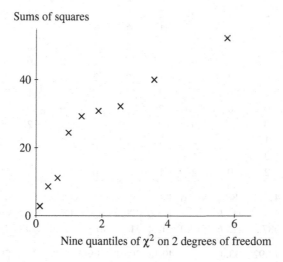

Fig. 12.7. Plot of sums of squares from Table 12.7, omitting the three largest

If $p > 2$ then every interaction consists of more than one character. The sums of squares for each interaction should be separated into those for the individual characters to draw the quantile plot. However, for the other two methods of analysis the sums of squares for the individual characters should be pooled into sums of squares for the various interactions. For example, in an analysis of variance based on either Table 12.5 or Table 12.6, the sum of

12.7 Several replicates

squares for the *D*-by-*N* interaction should be reported as 54.87 on four degrees of freedom.

12.7 Several replicates

Confounding can also be used in factorial block designs which have more than one replicate. Typically we use Technique 12.1 in each replicate, but confound different characters in each replicate, unless there is any factor that cannot be applied to smaller units than blocks.

The argument in Section 11.7 shows that if there are r replicates and a character is confounded with blocks in q of those replicates then the efficiency factor for that character is $(r-q)/r$. If we can arrange for every character to be confounded in the same number of replicates then the design is balanced.

Often we choose to confound higher-order interactions, believing them to be less important. However, sometimes we already know about main effects and want to know about the interactions: then we might choose to confound main effects more.

Example 12.3 revisited (Watering chicory) The design in Figure 12.2 has two replicates. It confounds part of the interaction in one replicate and the other part of the interaction in the other replicate. Thus both main effects have full efficiency while the interaction has efficiency factor $1/2$.

Example 12.6 revisited (Sugar beet) In fact, this experiment also had two replicates. One is shown in Table 12.3, confounding $D+S+2N$. There was also a second replicate, confounding $D+S+N$.

Example 12.11 (Balanced interactions) Suppose that there are three two-level treatment factors A, B and C, and that the experimental material consists of eight blocks of four plots each. We can divide these into four replicates, and then confound each of $A+B+C$, $A+B$, $A+C$ and $B+C$ in one replicate. Then there is full efficiency on all main effects, and all of the interactions have efficiency factor $3/4$ because they can all be estimated from three of the four replicates.

The analysis of data from these designs is conducted just as in Section 11.7.

Even if the treatments are not factorial, we can use the methods of this chapter to construct incomplete-block designs so long as the number of treatments and the block size are both powers of the same prime. We simply impose an artificial factorial treatment structure on the treatments. To obtain an efficient design, a good strategy is to confound all characters in the same number of replicates as nearly as possible.

Example 12.12 (Example 1.11 continued: Detergents) Here the ten housewives are the blocks, each of size four. If there are eight detergents then we can arbitrarily assign them to the combinations of levels of three two-level factors A, B and C. We need five replicates, and there are seven characters apart from I, so we should choose to confound five different characters, one in each replicate. For example, we can label the eight detergents as follows.

detergent	1	2	3	4	5	6	7	8
A	0	0	0	0	1	1	1	1
B	0	0	1	1	0	0	1	1
C	0	1	0	1	0	1	0	1

If we then confound the characters A, B, C, $A+B$ and $A+C$ in different replicates we obtain the following blocks.

$$\{1,2,3,4\}, \quad \{5,6,7,8\}, \quad \{1,2,5,6\}, \quad \{3,4,7,8\}, \quad \{1,3,5,7\},$$
$$\{2,4,6,8\}, \quad \{1,2,7,8\}, \quad \{3,4,5,6\}, \quad \{1,3,6,8\}, \quad \{2,4,5,7\}$$

Questions for discussion

12.1 Construct a resolved design for three three-level treatments factors in nine blocks of size nine in such a way that no main effect or two-factor interaction is confounded in any replicate and that no treatment contrast is totally confounded with blocks.

12.2 Construct a single-replicate design for six two-level treatment factors in four blocks of 16 plots each, assuming that all interactions involving four or more factors are zero. Write down the skeleton analysis-of-variance table, showing stratum, source and degrees of freedom.

12.3 Construct the remaining two blocks in Example 12.10, giving levels of both pseudo-factors and genuine factors.

12.4 There are three treatment factors, A, B and C, each with three levels. There are 54 plots, grouped into 18 blocks of size three. Suppose that factor A must be confounded with blocks and that a resolved design is required.

   (a) Construct one replicate (nine blocks) in such a way that no main effect apart from A is confounded with blocks.

   (b) List the characters which are confounded with blocks in this replicate (ignoring characters which are multiples of those already listed).

   (c) Explain how to construct the second replicate in such a way that no treatment effect other than A is confounded with blocks in both replicates, and no other main effect is confounded with blocks in the second replicate.

12.5 The experiment on cider apples in Example 8.3 used twelve blocks. Suppose that each block contains sixteen plots, and that we want to investigate combinations of four varieties of apple, four methods of pruning and four sprays. Moreover, suppose that levels of each of these factors can be applied to individual plots.

Explain how to construct a design in which all the main effects and two-factor interactions have full efficiency and in which all parts of the three-factor interaction have efficiency factor at least $2/3$.

12.6 Construct a balanced incomplete-block design for eight treatments in fourteen blocks of size four, where the treatments are all combinations of three treatment factors with two levels each.

Chapter 13

Fractional factorial designs

13.1 Fractional replicates

A factorial design is a fractional replicate if not all possible combinations of the treatment factors occur. A fractional replicate can be useful if there are a large number of treatment factors to investigate and we can assume that some interactions are zero. In Chapter 9 we constructed some fractional replicate designs from Latin squares. Here we use characters to give us more types of fractional replicate.

Definition Let $\mathcal{T}'$ be any subset of the set of treatments. Two characters G and H are *aliased* on $\mathcal{T}'$ if $G \equiv H$ on $\mathcal{T}'$. If the design has only the treatments in $\mathcal{T}'$ and G is aliased with I on $\mathcal{T}'$ (in other words, if G has a single level throughout $\mathcal{T}'$) then G is a *defining contrast*.

If G and H are both defining contrasts then G and H both have constant levels throughout the subset $\mathcal{T}'$, so $G+H$ also has a constant level throughout $\mathcal{T}'$. Thus the set of defining contrasts (including I) forms a subgroup of the group of all characters.

This suggests that we can use as a fractional replicate one block alone from a confounded block design.

Example 13.1 (Example 12.5 continued: Three factors with three levels) On any block from Example 12.5 we have
$$A + B + 2C = \text{constant}.$$
So, on that block,
$$A = \text{constant} - (B+2C) = \text{constant} - B + C = \text{constant} + 2B + C,$$
and thus
$$A \equiv 2B + C = C - B \equiv B + 2C.$$
Hence we cannot distinguish between the effect of A and the effect of $B+2C$. A large effect shown by the data might be due to either alone or a combination of both. A small effect might appear if each has a large effect but these cancel each other out. Hence we can estimate one only if the other is known to be zero. But $B+2C$ is a rather artificial part of the B-by-C interaction, so we can estimate A only if the B-by-C interaction is known to be zero. Similarly, we can estimate B and C only if the A-by-C and A-by-B interactions are known to be zero. So this fraction consisting of nine treatments is a main-effects-only design.

Table 13.1. *Alias sets for the one-third-replicate in Example 13.1*

$$
\begin{aligned}
\mathcal{G} &= \{I, A+B+2C, 2A+2B+C\} \\
\mathcal{G}+A &= \{A, 2A+B+2C, 2B+C\} \\
\mathcal{G}+2A &= \{2A, B+2C, A+2B+C\} \\
\mathcal{G}+B &= \{B, A+2B+2C, 2A+C\} \\
\mathcal{G}+2B &= \{2B, A+2C, 2A+B+C\} \\
\mathcal{G}+A+B &= \{A+B, 2A+2B+2C, C\} \\
\mathcal{G}+A+2B &= \{A+2B, 2A+2C, B+C\} \\
\mathcal{G}+2A+B &= \{2A+B, 2B+2C, A+C\} \\
\mathcal{G}+2A+2B &= \{2A+2B, 2C, A+B+C\}
\end{aligned}
$$

Theorem 13.1 *Let $\mathcal{G}$ be a subgroup of the characters, and let $\mathcal{T}'$ be the set of treatments in any one block of the single-replicate block design which confounds $\mathcal{G}$. Then*

(i) *two characters G and H are aliased on $\mathcal{T}'$ if and only if they are in the same coset of $\mathcal{G}$, that is, if $G - H \in \mathcal{G}$;*

(ii) *if the fraction consists just of the treatments in $\mathcal{T}'$ then a character is a defining contrast if and only if it is in $\mathcal{G}$;*

(iii) *if two characters are neither defining contrasts nor aliased with each other then they are strictly orthogonal to each other on $\mathcal{T}'$.*

We shall not prove this theorem here.

13.2 Choice of defining contrasts

Many choices of $\mathcal{G}$ are not suitable as defining-contrasts subgroups. If G is a defining contrast then its effect cannot be estimated. If G is aliased with H then the effect of one can be estimated only if the other can be assumed to be zero. Thus, in any coset of $\mathcal{G}$, *either* we must be able to assume that at most one character has a nonzero effect *or* we must accept that we cannot estimate the effect of anything in that coset. In this context, the cosets are known as *alias sets*.

Example 13.1 revisited (Three factors with three levels) The alias sets for this fraction are shown in Table 13.1. Each contains at most one character belonging to a main effect, so we can use this fraction for a main-effects-only design. Table 13.2 gives the skeleton analysis of variance.

Example 13.2 (A two-level fraction) If A, B, C, D and E are factors with two levels, then we can construct a half-replicate in 16 plots by using $A + B + C + D + E$ as a defining contrast. The alias sets are shown in Table 13.3.

Now $A \equiv B + C + D + E$, and similarly for the other main effects, so we can estimate all main effects if we assume that all the four-factor interactions are zero. Also $A + B \equiv C + D + E$, so

13.2. Choice of defining contrasts

Table 13.2. *Skeleton analysis of variance for the one-third-replicate in Example 13.1*

Stratum	Source	Degrees of freedom
mean	mean	1
plots	A	2
	B	2
	C	2
	residual	2
	total	8
Total		9

Table 13.3. *Alias sets for the half-replicate in Example 13.2*

$$\begin{aligned}
\mathcal{G} &= \{I, A+B+C+D+E\} \\
\mathcal{G}+A &= \{A, B+C+D+E\} \\
\mathcal{G}+B &= \{B, A+C+D+E\} \\
\mathcal{G}+C &= \{C, A+B+D+E\} \\
\mathcal{G}+D &= \{D, A+B+C+E\} \\
\mathcal{G}+E &= \{E, A+B+C+D\} \\
\mathcal{G}+(A+B) &= \{A+B, C+D+E\} \\
\mathcal{G}+(A+C) &= \{A+C, B+D+E\} \\
\mathcal{G}+(A+D) &= \{A+D, B+C+E\} \\
\mathcal{G}+(A+E) &= \{A+E, B+C+D\} \\
\mathcal{G}+(B+C) &= \{B+C, A+D+E\} \\
\mathcal{G}+(B+D) &= \{B+D, A+C+E\} \\
\mathcal{G}+(B+E) &= \{B+E, A+C+D\} \\
\mathcal{G}+(C+D) &= \{C+D, A+B+E\} \\
\mathcal{G}+(C+E) &= \{C+E, A+B+D\} \\
\mathcal{G}+(D+E) &= \{D+E, A+B+C\}
\end{aligned}$$

either we can assume that the *C*-by-*D*-by-*E* interaction is zero, and estimate the *A*-by-*B* interaction

or we can assume that the *A*-by-*B* interaction is zero, and estimate the *C*-by-*D*-by-*E* interaction (of course, this implies that we do not assume that the *C*-by-*D* interaction or the *C*-by-*E* interaction or the *D*-by-*E* interaction is zero)

or we assume that neither is nonzero so we can estimate neither.

Similar arguments apply to all the two-factor interactions.

The design consists *either* of the 16 treatments *u* satisfying

$$(A+B+C+D+E)(u) = 0$$

or of the 16 treatments *u* satisfying

$$(A+B+C+D+E)(u) = 1.$$

00000, 11000, 10100, 10010, 10001, 01100, 01010, 01001,
00110, 00101, 00011, 11110, 11101, 11011, 10111, 01111

Fig. 13.1. First half-replicate in Example 13.2

The first set is shown in Figure 13.1: it consists of those treatments which have an even number of coordinates equal to 1. The second set consists of those treatments with an odd number of coordinates equal to 1.

Technique 13.1 (Fractional factorial designs) If there are n treatment factors, each with p levels, where p is prime, construct a fractional replicate design with p^{n-s} treatments as follows.

(i) Find a subgroup $\mathcal{G}$ of the set of all characters such that

(a) every defining contrast (character in $\mathcal{G}$) *either* has zero effect *or* is not of interest;

(b) if G and H are aliased with each other (in the same coset of $\mathcal{G}$) and are not defining contrasts then *either* at most one of G, H has nonzero effect *or* neither of G, H is of interest;

(c) $|\mathcal{G}| = p^s$.

(ii) Form the set $\mathcal{T}''$ consisting of those p^{n-s} treatments u satisfying

$$G(u) = 0 \quad \text{for all } G \text{ in } \mathcal{G}.$$

(iii) The desired fraction $\mathcal{T}'$ may be taken as $\mathcal{T}''$ or as any other coset of $\mathcal{T}''$ in $\mathcal{T}$.

Step (i) can be very hard, or even impossible. Steps (ii) and (iii) are routine.

Only a limited amount of information may be obtained from a fractional replicate, so the combination of too few plots with too many nonzero treatment effects may lead to an insoluble design problem.

13.3 Weight

The following concept is often useful in the search for suitable groups of defining contrasts.

Definition The *weight* $w(G)$ of a character G is equal to m if G belongs to an m-factor interaction.

If there are no pseudofactors then $w(G)$ is just the number of nonzero coefficients in G. Weight must be defined more carefully if there are pseudofactors. In Example 12.10 the character $B + C + D_1 + D_2$ has weight three, because only three letters are involved.

Example 13.3 (Searching for a two-level fraction) Suppose that A, B, C, D and E are factors with two levels, and that all interactions among them are zero except the A-by-B and

13.3. Weight

C-by-D interaction. How small a fraction can we use to estimate all main effects and those two interactions?

Because all the treatment factors have two levels, all main effects and interactions have one degree of freedom. We want to estimate five main effects and two interactions, so we need at least seven degrees of freedom, and hence at least eight plots. A fraction with only eight treatments out of the possible 32 would be a quarter-replicate, so would require a defining-contrasts subgroup $\mathcal{G}$ of order four. Can we find such a subgroup $\mathcal{G}$ in which none of the seven characters

$$A, \quad B, \quad C, \quad D, \quad E, \quad A+B, \quad C+D$$

is aliased with another one?

If $\mathcal{G}$ contains A then A has a single level in the fraction (A is a defining contrast) so the main effect of A cannot be estimated. Thus $\mathcal{G}$ should contain no character of weight one. If $\mathcal{G}$ contains $A+B$ then

$$A = \text{constant} - B$$

so $A \equiv B$; that is, A and B are aliased and so their main effects cannot be disentangled. (Another way to see this is to note that

$$A = (A+B) + B,$$

so that A and B are in the same coset of $\mathcal{G}$.) Thus $\mathcal{G}$ should contain no character of weight two. Thus $\mathcal{G}$ must contain only characters of weight three or more, apart from the single character I of weight zero.

Suppose that G and H are different characters in $\mathcal{G} \setminus \{I\}$. The weight of $G+H$ has the same parity as $w(G) + w(H)$ (because we are working modulo 2), and is at most five. If $w(G) = 5$ then $w(G+H) = 5 - w(H)$, so one of H and $G+H$ has weight less than or equal to two. If $w(G) = 4$ and $w(H) = 4$ then $w(G+H) = 2$. Thus the only possibility for the three non-I characters in $\mathcal{G}$ is

$$F_1 + F_2 + F_3$$
$$F_1 + F_4 + F_5$$
$$F_2 + F_3 + F_4 + F_5,$$

where $(F_1, \ldots, F_5)$ is some reordering of (A, B, C, D, E).

If $\mathcal{G}$ contains $A+B+E$ then $A+B$ is aliased with E, because

$$A+B = (A+B+E) + E.$$

This is not allowed, because the A-by-B interaction is nonzero. Thus A and B are not both in $\{F_1, F_2, F_3\}$ or both in $\{F_1, F_4, F_5\}$. In particular, F_1 cannot be equal to A or to B. A similar argument shows that F_1 cannot be equal to C or to D, because the C-by-D interaction is nonzero. Hence $F_1 = E$, and so

$$F_2 + F_3 + F_4 + F_5 = A + B + C + D.$$

But now $\mathcal{G}$ contains $A+B+C+D$, and

$$A+B = (A+B+C+D) + (C+D),$$

Table 13.4. *Skeleton analysis of variance for the half-replicate in Example 13.3*

Stratum	Source	Degrees of freedom
mean	mean	1
plots	A	1
	B	1
	C	1
	D	1
	E	1
	$A \wedge B$	1
	$C \wedge D$	1
	residual	8
	total	15
Total		16

and so $A+B$ is aliased with $C+D$, which is not allowed if these two nonzero interactions are to be estimated. Hence no quarter-replicate will suffice.

We now try a fraction with 16 treatments, that is, a half-replicate. If we take $A+B+C+D+E$ as the single defining contrast (apart from I) then we obtain the alias sets in Table 13.3. The eight nonzero effects are in different alias sets and so are strictly orthogonal to each other. Therefore we can use the half-replicate in Figure 13.1, or its complementary half-replicate.

For either fraction, the skeleton analysis of variance is shown in Table 13.4.

That example was given at some length, because it well demonstrates the kind of trial-and-error that is needed when specific effects are given as assumed nonzero. Several computer programs exist for carrying out such a search.

If G is a defining contrast and $w(G) = m$ then there is a main effect aliased with (part of) an $(m-1)$-factor interaction, there is (part of) a two-factor interaction aliased with (part of) an $(m-2)$-factor interaction, and so on. As we argued in Example 13.3, we never use defining contrasts of weight one or two.

If $w(G) = 3$ then a main effect is aliased with (part of) a two-factor interaction. As in Example 13.1, we must assume that the two-factor interaction is zero (or very small) if we want to estimate the main effect.

If $w(G) = 4$ then a main effect is aliased with (part of) a three-factor interaction, and some two-factor interactions are aliased with each other. If the three-factor interaction is assumed to be zero then we can estimate the main effect. There is no need to assume that all two-factor interactions are zero, but we cannot disentangle the effects of any aliased pair of these.

For a $(1/p)$ fraction, we usually use a defining contrast of weight n.

Suppose that $p = 2$ and we want a quarter-replicate with defining contrasts G, H and $G+H$. Let the number of factors where G, H and $G+H$ have each combination of coefficients be a, b, c, d as follows, with $a+b+c+d = n$, where n is the number of factors.

$$
\begin{array}{cccc}
G & 0 & 0 & 1 & 1 \\
H & 0 & 1 & 0 & 1 \\
G+H & 0 & 1 & 1 & 0 \\
 & a & b & c & d
\end{array}
$$

13.4. Resolution

Then $w(G) = c+d$, $w(H) = b+d$ and $w(G+H) = b+c$, so

$$w(G) + w(H) + w(G+H) = 2(b+c+d) \leq 2n.$$

If $n = 5$ the only possible weights are four, three and three, as we have already seen in the discussion of Example 13.3. If $n = 6$ we can make the three weights all equal to four: see Example 13.6. For general n, we normally take the three weights to be approximately equal to $2n/3$, to avoid small weights.

We can do a similar analysis for a one-ninth fraction when $p = 3$. Now let the number of factors with the various combinations of coefficients be as follows.

$$
\begin{array}{c|ccccccccc}
G & 0 & 0 & 0 & 1 & 2 & 1 & 2 & 1 & 2 \\
H & 0 & 1 & 2 & 0 & 0 & 2 & 1 & 1 & 2 \\
G+H & 0 & 1 & 2 & 1 & 2 & 0 & 0 & 2 & 1 \\
G+2H & 0 & 2 & 1 & 1 & 2 & 2 & 1 & 0 & 0 \\
 & \underbrace{}_{a} & \underbrace{}_{b} & \underbrace{}_{c} & \underbrace{}_{d} & \underbrace{}_{e}
\end{array}
$$

Then $w(G) = c+d+e$, $w(H) = b+d+e$, $w(G+H) = b+c+e$ and $w(G+2H) = b+c+d$, so

$$w(G) + w(H) + w(G+H) + w(G+2H) = 3(b+c+d+e) \leq 3n.$$

When $n = 4$ the only possible weights are all three. The corresponding fraction is the same as one made from a Graeco-Latin square as in Section 9.3.4. When $n = 5$ the only possibilities are 3, 3, 3, 3 and 4, 4, 4, 3.

13.4 Resolution

In an experiment with several factors, sometimes the experimenter has no prior idea about which interactions are nonzero. Then it is common to assume that interactions involving more than a certain number of factors are all zero.

Definition A fraction has *resolution* M if every defining contrast (apart from I) has weight at least M.

Thus in a fraction of resolution $2m+1$ any character of weight m or less is aliased only with characters of weight at least $m+1$. Therefore all main effects and interactions between at most m factors can be estimated if all interactions between $m+1$ or more factors are assumed zero. In a fraction of resolution $2m$, any character of weight $m-1$ or less is aliased only with characters of weight $m+1$ or more, but some characters of weight m are aliased with each other. Therefore all main effects and interactions between at most $m-1$ factors can be estimated if all interactions between $m+1$ or more factors are assumed zero. Some authors take these properties as the definition of resolution, especially for fractions which are not subgroups of $\mathcal{T}$.

With either definition, a fraction of resolution M also has resolution $M-1$. Some authors limit the word *resolution* to the *maximum* value that we allow here for the resolution. Fractions with a given resolution are often tabulated.

000000, 011234, 022413, 033142, 044321,
101111, 112340, 123024, 134203, 140432,
202222, 213401, 224130, 230314, 241043,
303333, 314012, 320241, 331420, 342104,
404444, 410123, 421302, 432031, 443210

Fig. 13.2. Fractional design in Example 13.4: factors in order A, B, C, D, E, F

Fractions with resolution three can be constructed simply. To estimate n main effects we need $n(p-1)$ degrees of freedom, so if there are p^d plots then $n \leq (p^d - 1)/(p-1)$. First form a complete replicate of the combinations of levels of factors $F_1, \ldots, F_d$. By Theorem 12.1, the characters on $F_1, \ldots, F_d$ other than I split into $(p^d - 1)/(p-1)$ sets, of which d correspond to main effects of $F_1, \ldots, F_d$. Alias each of $F_{d+1}, \ldots, F_n$ with a character in a different set of those remaining.

Example 13.4 (Fraction with resolution three) Suppose that factors A, B, C, D, E and F each have five levels. To form a resolution-three fraction in twenty-five plots we first form all combinations of levels of A and B. Then put $C = A + B$, $D = A + 2B$, $E = A + 3B$ and $F = A + 4B$. This gives the design in Figure 13.2.

Fractional factorial designs are normally useful only if we can assume that some interactions are zero, or have much smaller effects than the main effects. However, when factors are quantitative, fractional factorial designs also provide a collection of treatments that are well spread out in n-dimensional space.

Example 13.5 (Glass insulators) A factory in Nanjing had always manufactured porcelain insulators. In 1973 they were instructed to change production to glass insulators. They had no experience of doing this. There were six quantitative factors which they could vary. They needed to find a combination of the levels of these which would ensure that five output responses all achieved at least their target value.

A statistician was assigned to the factory for three weeks. From discussions there, he rapidly learnt that there were likely to be many interactions between the six factors. There was not time to perform enough runs to estimate all of these. Therefore he advised the managers to choose five equally-spaced levels of each factor, encompassing a practical range, and to use a fractional design like the one in Figure 13.2. They did so. One of the twenty-five combinations produced glass insulators that met all of the five targets. That was all that they needed to know. They were able to begin production with that particular combination of levels of the six factors.

13.5 Analysis of fractional replicates

The analysis of data from fractional factorial designs follows the same lines as the analysis of single-replicate factorial designs. If there are any alias sets in which all characters can be assumed to have zero effect then these can be used to estimate the plots stratum variance, which can in turn be used to test for the presence of effects. If there is an alias set in which

13.5. Analysis of fractional replicates

Table 13.5. *Treatments and responses in Example 13.6*

F	T	L	V	C	M	Response
0	1	1	1	1	0	220
0	0	0	0	0	0	174
1	1	0	0	1	1	172
1	0	1	1	0	1	353
1	1	0	1	0	0	176
0	0	0	1	1	1	192
0	1	1	0	0	1	280
1	0	1	0	1	0	246
1	0	0	1	1	0	197
0	1	0	1	0	1	192
0	0	1	0	1	1	261
1	1	1	0	0	0	340
0	1	0	0	1	0	200
1	0	0	0	0	1	233
0	0	1	1	0	0	280
1	1	1	1	1	1	234

only one character is assumed to have a nonzero effect then that effect can be estimated. If there are two or more potentially nonzero effects in an alias set then there is ambiguity: a large apparent effect may mean that one is large while the other is zero (but we cannot tell which is the large one!) or it may mean that both have small effects in the same direction. On the other hand, an apparently zero effect may mean that both are zero, or it may mean that they have effects of the same size but the opposite sign.

Example 13.6 (Chromatograph) The levels of six factors were altered in an experiment on a chromatograph. The correspondence between the actual levels and the integers modulo 2 was as follows.

Factor	0	1
Flow rate (F)	0.5 ml/min	1.0 ml/min
Temperature (T)	25°C	45°C
Linear gradient time (L)	5 min	15 min
Injection volume (V)	0.5 ml	10 ml
Initial organic phase concentration (C)	0%	10%
%TFA in mobile phase (M)	0.01%	0.2%

It was assumed that all interactions between three or more factors were zero, but that two-factor interactions might not be. Therefore a fraction with resolution four was required. Defining contrasts $T+L+V+C$, $F+T+C+M$ and $F+L+V+M$ were chosen.

The experiment consisted of the sixteen runs u for which

$$(T+L+V+C)(u) = 0 = (F+T+C+M)(u).$$

The chromatographic response function was measured on each run. The treatments and responses are shown in Table 13.5.

Table 13.6. *Difference between total response on level 1 and total response on level 0*

F	152	$F+T$	92	$F+T+L$	124
T	-122	$F+L$	-112	$F+T+V$	-68
L	678	$F+V$	0		
V	-62	$F+C$	200		
C	-306	$F+M$	18		
M	84	$T+L$	10		
		$T+V$	278		

Table 13.7. *Analysis of variance in Example 13.6*

Stratum	Source	SS	df	MS	VR
mean	mean	878906.25	1	878906.25	–
runs	F	1444.00	1	1444.00	2.31
	T	930.25	1	930.25	1.49
	L	28730.25	1	28730.25	45.97
	V	240.25	1	240.25	0.38
	C	5852.25	1	5852.25	9.36
	M	441.00	1	441.00	0.71
	$F+T \equiv C+M$	529.00	1	529.00	0.85
	$F+L \equiv V+M$	784.00	1	784.00	1.25
	$F+V \equiv L+M$	0.00	1	0.00	0.00
	$F+C \equiv T+M$	2500.00	1	2500.00	4.00
	$F+M \equiv T+C \equiv L+V$	20.25	1	20.25	0.03
	$T+L \equiv V+C$	6.25	1	6.25	0.01
	$T+V \equiv L+C$	4830.25	1	4830.25	7.73
	residual	1250.00	2	625.00	–
	total	47557.75	15		
Total		926464.00	16		

Table 13.6 shows, for one character H in each alias set, the difference $\text{sum}_{H=1} - \text{sum}_{H=0}$. The sum of squares for this character is the square of this difference, divided by 16. Hence we obtain the analysis of variance in Table 13.7. The characters $F+T+L$ and $F+T+V$ both belong to alias sets with no characters of weight one or two, so these are both used for residual. Each main effect contributes one line to the analysis-of-variance table. Each of the remaining lines corresponds to two or three of the two-factor interactions.

The problem with such an analysis-of-variance table is that the F_2^1 distribution is very heavy-tailed, so it is hard to reject any null hypothesis of zero effect. The 95% point is 18.51 and the 90% point is 8.53. Thus the main effect of L is nonzero, and probably also the main effect of C and the effect caused by the aliased characters $L+C$ and $T+V$. Since neither the main effect of T nor the main effect of V seems to be nonzero, it is fairly safe to attribute this third nonzero effect to the interaction between C and L.

13.5. Analysis of fractional replicates

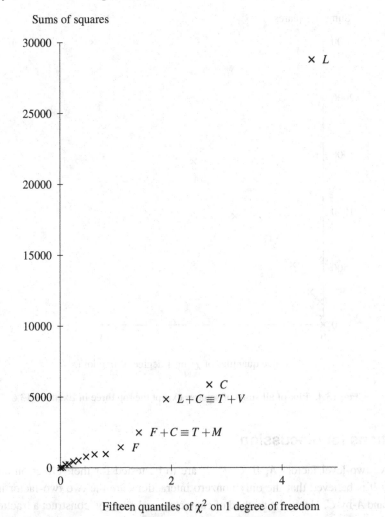

Fig. 13.3. Plot of all sums of squares in Example 13.6

The quantile plot gives a similar conclusion. The sums of squares for all fifteen characters are shown in Figure 13.3: the points do not lie on a straight line through the origin. However, removing the top three points gives the graph in Figure 13.4, which looks much more like such a line.

The conclusion from this experiment may well be that the experimenter should investigate factors L and C further. In this case, he might do well to include F in these further studies. The fourth largest effect is for $T+M$, which is aliased with $C+F$, and the fifth is for the main effect of F. These five effects are consistent with the model $V_{C \wedge L} + V_{C \wedge F}$.

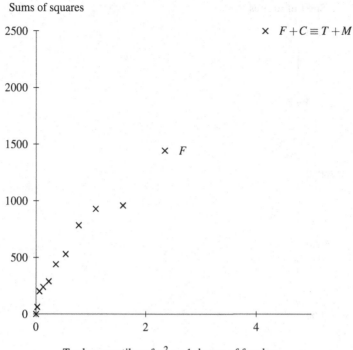

Twelve quantiles of χ^2 on 1 degree of freedom

Fig. 13.4. Plot of all sums of squares except the top three in Example 13.6

Questions for discussion

13.1 Five two-level factors A, B, C, D, E are to be tested for their effect on an industrial process. It is believed that the only nonzero interactions are the two two-factor interactions A-by-B and A-by-C. Using the method of characters and aliasing, construct a fractional design which can be used to estimate all main effects and both nonzero two-factor interactions, and which is as small as possible.

13.2 Construct a quarter-replicate main-effects-only design for two two-level treatment factors and two four-level treatment factors.

13.3 Construct a resolution-three fraction for seven two-level factors in eight plots.

13.4 Generalize the argument in Section 13.3 to find the largest possible average weight of a defining contrast for a fractional replicate design for n treatment factors, each with p levels, where p is prime, in p^{n-s} plots.

Chapter 14

Backward look

14.1 Randomization

Although randomization is generally regarded as essential, there is less agreement on how to carry it out and on how to verify whether any proposed randomization procedure achieves its objectives. Here I discuss some possibilities.

14.1.1 Random sampling

Many statistics textbooks give the impression that every collection of observations is a random sample. This is rarely the case for designed experiments. Experimental resources are too valuable for us to choose a random subset and ignore the rest. We use the experimental units which are to hand in the laboratory or the field, or we use all suitable volunteers in a clinical trial. In a small trial it is more important that the experimental units be reasonably similar than that they be random, and in a large trial it may be more important that they be representative. If the experimental units are to be working farms, then stratified random sampling can give a representative sample, but in practice we can still use only those farms whose farmer is willing to participate in the experiment.

Example 14.1 (Small trial on volunteers) In Question 2.1, the psychologist needs to choose as his experimental units eleven of the students listed in Table 2.7. For such a small trial, he should choose people of the same sex and approximately the same age. A random sample of eleven people would almost certainly be more variable, giving him less power to detect any treatment difference.

The chief situation in which random sampling plays a role in experimentation is when observational units are much smaller than experimental units. For example, in a field trial, if the response is something like the percentage of diseased plants, it is quite common for the observational units to be a random sample of very small areas within each plot. Example 1.3 is similar.

14.1.2 Random permutations of the plots

In Section 1.5, we assumed that the response Y_ω on plot ω satisfies

$$Y_\omega = Z_\omega + \tau_{T(\omega)}, \tag{14.1}$$

where Z_ω is a random variable depending only on the plot ω. In Chapter 2, I recommended randomizing a systematic design by randomly choosing a permutation g of all the plots and applying it to the design. For more complicated plot structures, choice of g must be restricted to those permutations which preserve the plot structure: for example, if there are blocks and $B(\alpha) = B(\beta)$ then we must have $B(g(\alpha)) = B(g(\beta))$. Call such permutations *allowable*. For orthogonal plot structures, the method of randomization given in Section 10.11 does restrict choice of g in this way; moreover, it gives all allowable permutations for almost all orthogonal plot structures in use in practice.

If we randomize in this way, it is appropriate to replace Z_ω in Equation (14.1) by the mixture $\tilde{Z}_\omega$ of $Z_{g(\omega)}$ over all permutations g that might be chosen. Therefore, if α and β are plots such that there is some allowable permutation g with $g(\alpha) = \beta$ then $\tilde{Z}_\alpha$ and $\tilde{Z}_\beta$ have the same distribution. All the analyses in this book, so far, assume that the plot random variables have identical distributions. Thus we may further assume that they all have expectation zero, by adding a suitable constant to all the τ parameters in Equation (14.1).

However, if α and β are in blocks of different sizes, or in blocks with potentially different fixed block effects, then there is no such permutation g, so we cannot necessarily assume that $\tilde{Z}_\alpha$ and $\tilde{Z}_\beta$ have the same expectation, let alone the same variance. Such an assumption now becomes an act of faith, not a consequence of the randomization procedure.

Example 14.2 (Blocks of unequal size) In a block design, the assumption that covariance depends only on whether or not plots are in the same block leads to the covariance matrix **C** in Equation (4.2). However, if the blocks have different sizes then the blocks subspace W_B is no longer an eigenspace of **C**.

If we are in the desirable situation where every pair of plots α and β has some allowable permutation taking one to the other, then we can ask about the joint distribution of pairs of plot random variables. If (α, β) and (γ, δ) are two pairs of distinct plots and there is an allowable permutation g such that $g(\alpha) = \gamma$ and $g(\beta) = \delta$ then $(\tilde{Z}_\alpha, \tilde{Z}_\beta)$ has the same joint distribution as $(\tilde{Z}_\gamma, \tilde{Z}_\delta)$: in particular, $\text{cov}(\tilde{Z}_\alpha, \tilde{Z}_\beta) = \text{cov}(\tilde{Z}_\gamma, \tilde{Z}_\delta)$.

Example 14.3 (Randomizing a row–column design) Suppose that we have a row–column design and α, β, γ and δ are four plots such that $\alpha \neq \beta$, $\gamma \neq \delta$ and

α is in row i_1 and column j_1,
β is in row i_2 and column j_2,
γ is in row i_3 and column j_3,
δ is in row i_4 and column j_4.

If $i_1 = i_2$ then $g(\alpha)$ and $g(\beta)$ are always in the same row if g is allowable; hence if $i_3 \neq i_4$ then we cannot have $g(\alpha) = \gamma$ and $g(\beta) = \delta$. If $i_1 = i_2$ and $i_3 = i_4$ then $j_1 \neq j_2$ and $j_3 \neq j_4$, because $\alpha \neq \beta$ and $\gamma \neq \delta$. Hence there is some permutation of the columns which takes column j_1 to

14.1. Randomization

column j_3 and column j_2 to column j_4. There is also a permutation of the rows which takes row i_1 to row i_3. The combination of these two permutations takes α to γ and β to δ.

We argue similarly if $j_1 = j_2$.

If $i_1 \neq i_2$, $i_3 \neq i_4$, $j_1 \neq j_2$ and $j_3 \neq j_4$ then there is some permutation of the rows taking row i_1 to row i_3 and row i_2 to row i_4, and there is some permutation of the columns taking column j_1 to column j_3 and column j_2 to column j_4. The combination of these two permutations takes α to γ and β to δ.

Hence we obtain the model assumed in Section 6.6.

There is a general version of the argument in Example 14.3 which covers many orthogonal plot structures, but it is too abstract to present here. The individual plot structures given in this book can be done on a case-by-case basis.

14.1.3 Random choice of plan

If we use the method in Section 14.1.2, we could argue that we are effectively using permutations to obtain a large number of plans and then choosing one of those plans at random. Why not go further, and simply choose at random from all possible plans that satisfy certain restrictions?

For example, if we have six treatments in a 6×6 row–column design and the systematic plan is a cyclic Latin square of order six then simply randomizing rows and randomizing columns will never produce the Latin square in Figure 6.2. Some people argue that it is better to choose randomly from a larger set of plans than from a smaller, so that we should choose randomly from among all 6×6 Latin squares.

In fact, there are some complicated situations where the only satisfactory method of randomization is random choice from a carefully specified set of plans. However, in general I do not recommend this method. First, it is much harder to check that desirable conditions are satisfied overall. For example, using the set of all plans that satisfy certain restrictions does not always allow us to assume that all the plot random variables have the same distribution. Secondly, it is sometimes a good precaution to choose some special systematic plan and then randomize it by the method of Section 14.1.2. As suggested in Section 9.3.1, a Latin square which has an orthogonal mate can be a safer choice than one which does not; this property is not destroyed by randomizing rows and columns.

14.1.4 Randomizing treatment labels

There are some situations where randomization of the treatment labels is a necessary part of the procedure, but I usually do not advise it. It is not necessary for any of the structures in this book.

Experimenters who are most familiar with complete-block designs may think of randomizing in terms of randomizing treatment labels independently within each block. Even split-plot designs can be naively considered in this way. If there is some hidden subtlety in the design then this method of randomizing can undo all the statistician's work.

Crop	oil-seed rape		3rd wheat		1st wheat		2nd wheat	
Date	2	1	1	2	2	1	1	2
Timing	1	2	2	1	1	2	2	1

Crop	3rd wheat		1st wheat		2nd wheat		oil-seed rape	
Date	2	1	2	1	1	2	2	1
Timing	1	2	2	1	2	1	1	2

Crop	oil-seed rape		2nd wheat		1st wheat		3rd wheat	
Date	2	1	2	1	2	1	2	1
Timing	2	1	2	1	1	2	2	1

Crop	oil-seed rape		2nd wheat		3rd wheat		1st wheat	
Date	1	2	2	1	1	2	1	2
Timing	1	2	2	1	1	2	1	2

Fig. 14.1. Actual plan of the rotation experiment in Example 14.4

Example 14.4 (Factorial rotation experiment) A rotation experiment had the following four treatment factors:

R: crop in the rotation, with four levels (oil seed rape and first, second and third wheats);
N: quantity of nitrogen fertilizer (five levels);
T: timing of the application of nitrogen (two levels);
D: date of drilling (two levels).

The experimental area was divided into four blocks, each of which was divided into four whole plots. Levels of R were applied to whole plots. Each whole plot was divided into two subplots. Levels of T and D were both applied to subplots, in such a way that $T + D$ (modulo 2) was confounded with blocks, using the methods of Chapter 12. Subplots were each split into five sub-subplots, to which levels of nitrogen were applied.

Unfortunately, the experimenter ignored the randomized plan which the statistician had provided. He correctly randomized the crops to the whole plots in each block. He correctly randomized the levels of nitrogen to the sub-subplots within each subplot. He randomized the levels of Date and Timing to the subplots within each whole plot in such a way that each treatment had overall replication two, but without any regard for the confounding of the Date-by-Timing interaction. This gave the actual plan in Figure 14.1: levels of nitrogen are omitted, as they all occurred once on each subplot.

The statistician discovered this mistake when he came to analyse the first set of data and found out that his prepared program would not run correctly. The confounding of the Date-by-Timing and Crop-by-Date-by-Timing interactions were both more complicated than he had allowed for. Moreover, the relative precision of different contrasts was not what he had intended.

Similarly, if a design in incomplete blocks is needed and there is no balanced design, the design is usually chosen to have higher efficiency factors for more important contrasts: see

14.1. Randomization

Chapter 11. Randomizing the treatment labels, even if done consistently throughout the whole design, has the effect of randomizing the efficiency factors to the treatment contrasts.

Example 14.5 (Acacia) An experiment on acacia trees compared 64 varieties in blocks of size eight in a lattice design with four replications. In the first large block, the blocks were $\{1,\ldots,8\}$, $\{2,\ldots,16\}$, and so on, following the construction in Section 11.3. Unfortunately, the list of varieties that the foresters were using had the varieties grouped into species, with the first so many belonging to the first species, and so on. Thus differences between species were largely confounded with blocks in the first large block, and so these differences were estimated less precisely than they could have been.

It could be argued that this would not have happened if the foresters had first randomized the treatment labels. A better solution, if high precision was required for the comparisons between species, would have been to deliberately construct the design in such a way that species were as orthogonal to blocks as possible. This could have been done by retaining the natural labelling and constructing the design by using the columns of the square array in Section 11.3 for one large block and the letters of three mutually orthogonal Latin squares for the other three.

14.1.5 Randomizing instances of each treatment

The methods discussed so far all assume that all the instances of a single treatment are indistinguishable, or virtually so. This is true if the treatment is an instruction, such as 'paint this component twice', or each instance is drawn from a very uniform collection, such as wheat seed or commercial medical tablets. However, sometimes the individual instances of a treatment are distinguishable before they are applied. For example, this happens when seedlings are grown in the nursery before being planted out in the experimental location. In such cases, the instances of the treatment should be randomized among themselves before being applied to their allocated experimental units. This is partly to avoid bias and partly to avoid introducing any extra pattern in the variability.

Example 14.6 (Chalk grazing) Groups of different breeds of sheep are to be put into large enclosures on public chalk downs, to investigate their effect on the diversity of plant life. The treatments are the breeds. If each group is a flock lent by one farmer then there are probably knowable differences between flocks of the same breed before the experiment starts, so all the flocks of each breed must be randomized among themselves. On the other hand, if each breed comes from a single flock, which is to be split into smaller groups for allocating to enclosures, then this splitting up should be done in such a way that groups are similar in properties like weight, age, vigour and sex. There is no need to form the groups at random.

14.1.6 Random allocation to position

Sometimes each experimental unit needs to be assigned to a position in time or space, or to some kind of group for management. Does this need to be done randomly? In general, I think not. If the positions have no more structure than the experimental units then nothing is gained by random allocation. If management groups are necessary then either they are predetermined (such as patients of particular general practitioners) or they should be matched to any existing blocking of the experimental units.

Example 14.1 revisited (Small trial on volunteers) Once the eleven volunteers have been chosen, there is no need to put them in a random order. However, they should be assigned to their positions in order before the systematic plan is randomized. Otherwise, the psychologist may be tempted to conduct all the observations on one type of pill before all of those on the other, with the result that any apparent difference between the types of pill may be caused by other conditions that change with time.

In the larger version of this trial in Question 4.2, the students should be grouped by age and sex into complete blocks. Only one observation can be taken per day, so days should be grouped into consecutive batches, one batch per block. There is no need to randomize the blocks of students to the blocks of days.

Example 14.7 (Glass jars) In Question 4.5 the treatments are 36 insect species. The experimental units are 180 glass jars containing leaf mixture. After the insects have been put into them, the jars are arranged in a 5×36 rectangle, whose rows receive different amounts of sunlight.

As in Question 2.3, it is not advisable to simply put the first species into jars 1–5, the second species into jars 6–10, and so on. That would confound species differences with any time effects, such as change in temperature or in leaf composition, or a learning curve of the experimenter. On the other hand, it is not very practicable to completely randomize the 180 jars to the 180 positions before the treatments are allocated. In fact, it is not necessary. It is a good idea to use the rows of the rectangle as blocks. Then the first 36 jars can be used for the first row, in order. A random permutation of the numbers 1–36 is used to allocate the species to these jars. I do not think that it is necessary to put the jars themselves in a random order before the species are randomized. A similar procedure can be used for each of the other rows.

On the other hand, there is some argument for random allocation of observational units to their positions if each experimental unit consists of several observational units. If this is not done then time or space may introduce an extra source of variability, which should be incorporated by using an appropriate plot factor.

The main need for random allocation of experimental units is for situations not covered by Section 14.1.2. The simplest orthogonal plot structure for which the randomization procedure in Section 10.11 does not justify the assumed covariance in Equation (10.9) is the one defined by three systems of n blocks of size n having the relationships of the rows, columns and letters of a Latin square. If we are free to choose the allocation of experimental units to at least one of the systems of blocks, then doing so randomly can restore Equation (10.9).

Example 14.8 (Silicon wafers) In the manufacture of integrated circuits, silicon wafers are processed in batches in each of several stages. Experimentation on such a process should therefore use similar stages and batches. Three treatment factors F, G and H each have two levels, each of which must be applied to a group of four wafers at the same time. Levels of F are applied at Stage 1, levels of G at Stage 2, and levels of H at Stage 3.

If the groups of four wafers are regarded as experimental units then 32 wafers are needed if all combinations of levels of F, G and H are to be tested; moreover, all treatment comparisons

14.1. Randomization

1	2	3	4
A	B	C	D
5	6	7	8
D	A	B	C
9	10	11	12
C	D	A	B
13	14	15	16
B	C	D	A

Fig. 14.2. Latin square used to allocate wafers to groups in each of three stages in Example 14.8

Wafer	Stage 1	Stage 2	Stage 3
1	f_2	g_1	h_2
2	f_2	g_2	h_2
3	f_2	g_2	h_1
4	f_2	g_1	h_1
5	f_1	g_1	h_1
6	f_1	g_2	h_2
7	f_1	g_2	h_2
8	f_1	g_1	h_1
9	f_2	g_1	h_1
10	f_2	g_2	h_1
11	f_2	g_2	h_2
12	f_2	g_1	h_2
13	f_1	g_1	h_2
14	f_1	g_2	h_1
15	f_1	g_2	h_1
16	f_1	g_1	h_2

Fig. 14.3. One possible outcome of the randomization in Example 14.8

are assessed against the group-to-group variability, so there are no degrees of freedom to estimate this variability unless even more wafers are used: see Section 8.1. However, there is no physical need to keep the wafers in the same groups of four. Instead, each batch of 16 wafers is split into groups as shown in Figure 14.2. The four rows are used as groups in Stage 1, with the two levels of factor F being randomized to the rows. The four columns are used as groups in Stage 2, with the two levels of G being randomized to the columns. In Stage 3, the four letters are used as groups, and the levels of H are randomized to the letters. Figure 14.3 shows one possible outcome of the randomization.

Initially, all 16 wafers should be completely randomized. This justifies the assumption of 16 identical random variables with the same pairwise correlations. Then the Latin square in Figure 14.2 can be superimposed on the wafers. At the first stage of processing, the rows are treated as groups, in a random order. This introduces an extra correlation between units in the same row. Something similar happens for columns at the second stage, and for letters at the

Table 14.1. *Skeleton analysis of variance for Example 14.8*

Stratum	Source	Degrees of freedom
mean	mean	1
Stage 1 groups	F	1
W_R	residual	2
	total	3
Stage 2 groups	G	1
W_C	residual	2
	total	3
Stage 3 groups	H	1
W_L	residual	2
	total	3
wafers	$F \wedge G$	1
	$F \wedge H$	1
	$G \wedge H$	1
	$F \wedge G \wedge H$	1
	residual	2
	total	6
Total		16

third stage. Thus we obtain the covariance matrix

$$\mathbf{C} = \sigma^2(\mathbf{I} + \rho_R(\mathbf{J}_R - \mathbf{I}) + \rho_C(\mathbf{J}_C - \mathbf{I}) + \rho_L(\mathbf{J}_L - \mathbf{I}) + \rho_U(\mathbf{J} - \mathbf{J}_R - \mathbf{J}_C - \mathbf{J}_L + 2\mathbf{I})),$$

where the subscripts R, C and L denotes rows, columns and letters respectively. Then the proof of Theorem 10.9 shows that the strata are W_U, W_R, W_C, W_L and W_E, in the notation of Theorem 10.6. This gives the skeleton analysis of variance in Table 14.1.

14.1.7 Restricted randomization

Suppose that you carefully follow all the above guidance on randomization, but then the experimenter does not like the randomized layout that you produce. What should you do? Some people advocate simply throwing that one away and randomizing again. They argue that the proportion of rejected layouts will be very small, and hence that such rejection will have little effect on bias or assumptions.

In practice, the proportion to be rejected can be not at all negligible. Once an experimenter has realised that he can keep asking you for fresh randomizations, he can scrutinize every layout you give him, and can easily find reasons to object to many of them. This is partly because his idea of 'haphazard' is different from your idea of 'random', and partly because he is aware of practical considerations that he has not told you.

Sometimes, complying with an apparently innocuous objection may undermine crucial assumptions. If an agronomist is worried that some variety appears too often on edge plots then he is regarding edge plots as different from interior plots. Either he should force them to be alike by using guard plots (which all have some standard variety) or border plots (which have experimental varieties but are not measured) around the experiment, or his system of blocking should recognize the distinction between edge plots and interior ones. Likewise, if the designer of a cross-over trial for tasting five makes of orange juice says that no taster should taste juice B immediately after juice A, she is telling you that the simple model (14.1), involving only one parameter per treatment, cannot be correct.

Simply telling the experimenter that he must use your layout may be counter-productive. He may go away and randomize the design himself, with consequences like those reported in Examples 14.4–14.5. It is better to discuss with the experimenter why he does not like the layout. The outcome might be like that in Example 6.1: you both realize that there is another important blocking factor, and the experiment must be redesigned. Similarly, if he says that levels of one of the treatment factors cannot be changed very often, a design of split-plot type is called for. Alternatively, the discussion may lead him to agree that his initial objection is outweighed by other design considerations.

A third possibility is to use a specialized procedure called *restricted randomization*. Here the experimenter has to specify in advance what features of a layout make it undesirable, and the statistician has to use a randomization procedure which is restricted enough to avoid the undesirable layouts but generous to justify the usual model assumptions. This is a very specialized topic, beyond the scope of this book, and not worth investigating unless there will be a large number of experiments of this sort.

14.2 Factors such as time, sex, age and breed

Are these factors part of the treatment structure or part of the plot structure? The answer depends on the type of experiment. The different roles that time can play as a factor were discussed on page 147. Attributes of living beings, such as age, sex and breed, are similar to each other. The remainder of this section concentrates on breed as a factor, but the arguments apply just as well to sex, age and country of origin.

For a treatment factor, we must be able to choose which experimental units to apply the levels to. For example, if the experimental units are paddocks and we are trying to see what affects the total weight of beef that can be raised on a paddock in a year, treatment factors could be variety of grass, breed of cattle and stocking rate. We randomize all combinations of these to paddocks. There might be an interaction between breed and variety.

On the other hand, if the experimental unit is a cow and we are trying to see what type of feed increases milk yield, then we can choose the type of feed but we cannot choose the breed, which is an inherent property of the cow. So breed is not a treatment factor.

Our first choice now is to use breed as a way of grouping the cows into blocks. This is fine if we can assume one of the two models for block designs given in Chapter 4. In the fixed-effects model, cows of one breed consistently produce more milk than those of another breed, irrespective of the feed. In the random-effects model, the mean milk yield is the same for both breeds, but yields are correlated within breeds.

What should we do if we think that there might be an interaction between breed and feed? This breaks our basic assumption, from Equation (1.1), that

$$Y_\omega = Z_\omega + \tau_{T(\omega)},$$

where Z_ω is the effect of cow ω and $\tau_{T(\omega)}$ is the effect of the feed $T(\omega)$ given to that cow. In that case, other basic assumptions may also be broken. For example, $\text{Var}(Y_\omega)$ may also

Table 14.2. *Average weight of fat in milk yield (grams per day)*

Feed	Breed	
	Jersey	Holstein
without monensin	1294	1254
with monensin	1216	1339

depend on the breed of cow ω. It may be better to analyse such an experiment as two different experiments, one for each breed, with no intention of generalizing the results to other breeds.

Example 14.9 (Jerseys and Holsteins) Jerseys and Holsteins are very different breeds of dairy cattle. Holsteins produce much larger quantities of milk, but it is much less creamy than milk from Jerseys. An experiment was done to find the effect of adding monensin to their diets. Jersey farmers predicted that it would interfere with the special cream-making ability of their cows.

Table 14.2 shows the overall results, in grams of fat in the daily milk yield. It is clear that there is an interaction between feed and breed.

We *could* analyse the data under the assumption that there are four expectation parameters, one for each combination of breed and feed, that there is a different variance for each breed and that all the cows are independent. However, there is no randomization justification for this, because we cannot randomly allocate breeds to cows, so the assumptions are very strong.

Let $B(\omega)$ denote the breed of cow ω. If we randomize treatments within each breed of cow independently, then we can assume that

$$\mathbb{E}(Y_\omega) = \begin{cases} \tau_{J,T(\omega)} & \text{if } B(\omega) = \text{Jersey} \\ \tau_{H,T(\omega)} & \text{if } B(\omega) = \text{Holstein} \end{cases}$$

and

$$\text{Cov}(Y_\alpha, Y_\beta) = \begin{cases} \sigma_J^2 & \text{if } \alpha = \beta \text{ and } B(\alpha) = \text{Jersey} \\ \rho_J \sigma_J^2 & \text{if } \alpha \neq \beta \text{ and } B(\alpha) = B(\beta) = \text{Jersey} \\ \sigma_H^2 & \text{if } \alpha = \beta \text{ and } B(\alpha) = \text{Holstein} \\ \rho_H \sigma_H^2 & \text{if } \alpha \neq \beta \text{ and } B(\alpha) = B(\beta) = \text{Holstein} \\ \gamma & \text{if } B(\alpha) = \text{Jersey and } B(\beta) = \text{Holstein, or vice versa.} \end{cases}$$

Now the separate analyses for the different breeds are just like the one in Section 2.14. Suppose that there are n_J Jersey cows, of whom r_J receive each feed treatment, so that $n_J = 2r_J$. Put

$$\xi_{J,0} = \sigma_J^2(1 - \rho_J + n_J \rho_J),$$
$$\xi_{J,1} = \sigma_J^2(1 - \rho_J).$$

The analysis of variance is in Table 14.3. Here $\tau_{J,0}$ is the vector indexed by Jersey cows with every entry equal to $(\tau_{J,1} + \tau_{J,2})/2$, while $\tau_{J,T}$ is the vector indexed by Jersey cows whose entry for cow ω is $\tau_{J,T(\omega)} - (\tau_{J,1} + \tau_{J,2})/2$. From this, we estimate the difference between the feeds on Jerseys with variance $2\xi_{J,1}/r_J$.

14.2. Factors such as time, sex, age and breed

Table 14.3. *Analysis of variance for Jersey cows*

Stratum	Source	df	EMS
mean	mean	1	$\|\|\tau_{J,0}\|\|^2 + \xi_{J,0}$
cows	feeds	1	$\|\|\tau_{J,T}\|\|^2 + \xi_{J,1}$
	residual	$n_J - 2$	$\xi_{J,1}$
Total		n_J	

Table 14.4. *Analysis of variance for Holstein cows*

Stratum	Source	df	EMS
mean	mean	1	$\|\|\tau_{H,0}\|\|^2 + \xi_{H,0}$
cows	feeds	1	$\|\|\tau_{H,T}\|\|^2 + \xi_{H,1}$
	residual	$n_H - 2$	$\xi_{H,1}$
Total		n_H	

Table 14.4 gives the analysis of variance for Holstein cows, using similar notation. Thus we can estimate the difference between the feeds on Holsteins, and the variance of the estimator is $2\xi_{H,1}/r_H$.

To estimate the difference between Jerseys and Holsteins, we would use $\mathbf{x} \cdot \mathbf{Y}$, where $x_\omega = 1/n_J$ if ω is a Jersey cow and $x_\omega = -1/n_H$ if ω is a Holstein cow. Now,

$$\text{Var}(\mathbf{x} \cdot \mathbf{Y}) = \frac{\sigma_J^2}{n_J^2}(n_J + n_J(n_J-1)\rho_J) + \frac{\sigma_H^2}{n_H^2}(n_H + n_H(n_H-1)\rho_H) - \frac{2}{n_J n_H}(n_J n_H \gamma)$$

$$= \frac{\xi_{J,0}}{n_J} + \frac{\xi_{H,0}}{n_H} - 2\gamma.$$

The trouble is that we have no estimators of $\xi_{J,0}$, $\xi_{H,0}$ or γ. It may be reasonable to suppose that $\gamma = 0$ if the different breeds are on different farms, but it is less reasonable to suppose that $\rho_J = \rho_H = 0$ if the experiment used whole herds. So we cannot test whether there is a difference between the two breeds. However, it is not the purpose of this experiment to compare Jerseys with Holsteins.

What we *can* do is compare the effect of added monensin on Jerseys with its effect on Holsteins. If treatment 1 is 'no monensin' then we estimate this effect as

$$\frac{\text{SUM}_{B=J,T=2}}{r_J} - \frac{\text{SUM}_{B=J,T=1}}{r_J} - \frac{\text{SUM}_{B=H,T=2}}{r_H} + \frac{\text{SUM}_{B=H,T=1}}{r_H},$$

whose variance is

$$\frac{\sigma_J^2}{r_J^2}\left(2r_J + 2r_J(r_J-1)\rho_J - 2r_J^2\rho_J\right) + \frac{\sigma_H^2}{r_H^2}\left(2r_H + 2r_H(r_H-1)\rho_H - 2r_H^2\rho_H\right) = \frac{2\xi_{J,1}}{r_J} + \frac{2\xi_{H,1}}{r_H},$$

irrespective of the size of γ. We can estimate this variance, so we can indeed test whether monensin has the same effect on the two breeds.

Thus in that comparatively simple example, we can recover valid estimators and tests for everything except the difference between breeds. It may not be possible to do this when there are more complicated plot structures involved.

14.3 Writing a protocol

We can now revisit the topics of Chapter 1. The protocol for an experiment is written by the scientist and statistician together. It should include at least the following headings, many of which are interrelated.

14.3.1 What is the purpose of the experiment?

This section comes primarily from the scientist, but the statistician's questions should help him to refine it.

If the answer is something vague like 'to investigate new varieties of sunflower' or 'to find out about the nutritional effects of margarine', then there may be a perfectly worthwhile exploratory experiment to be done but a statistician is unlikely to be able to help.

Usually it is better to state a specific question that is to be answered, such as:

- to estimate how much better drug A is than drug B in reducing inflammation (of course, the two drugs would need to be named more precisely);

- to test the hypothesis that a new type of preservative for building stone is as effective as the one currently used;

- to fit a model for how much sucrose is absorbed at different distances from the surface of the liver.

14.3.2 What are the treatments?

Here the scientist needs to give a precise description of the treatments that he intends to apply to the experimental units. Give complete technical details, such as '5 mg of ciprofloxacin 4 hours after contact'. It is usually helpful to state how many treatments there are as well as what they are.

Sometimes treatments are simple; sometimes they are combinations. In the ciprofloxacin example, there may be four different doses combined with two times of administration: this would give eight treatments. Any such factorial structure should be spelt out explicitly. If all doses are administered at 4 hours then the information about '4 hours' belongs in the Methods section rather than here. Likewise, if all doses are the same and the purpose of the experiment is to find the best time to administer the drug, then the treatments are just times of administration and all details about dose and drug go into the Methods section.

In Example 14.1, we should ask the professor if he just wants to compare the two new pills with each other or whether he wants to compare them both with the effect of doing nothing. If the latter then there is a third treatment, 'do nothing', which we usually call control. Does the experiment need a control, or is this a waste of resources? If there is a control, say so explicitly.

In experiments on people, 'do nothing' should often be replaced by a placebo, so that everyone involved thinks that something is being done (Chapter 7).

For a straightforward treatment structure, give each treatment a simple code like A, B, C ... or 1, 2, 3, ... If the treatments are factorial, give each treatment factor a short name and give codes to the levels of each factor.

14.3. Writing a protocol

Explain the treatment structure as fully as possible: see Chapters 3, 5 and 10. It may be helpful to draw the Hasse diagram.

14.3.3 Methods

This section is for the scientist to write. This is where she describes exactly how the treatments will be applied to the experimental units, and what will be done from then on until all measurements have been taken. There should be sufficient detail for other scientists to replicate the work.

14.3.4 What are the experimental units?

Exactly what are the experimental units? How many are there? How are they structured before treatments are applied? Describe then carefully, and give details of any relevant blocking factors.

14.3.5 What are the observational units?

If the observational units are the same as the experimental units, then simply say so. If there are several observational units per experimental unit, then say how many there are and how they are defined.

If the relationship between the experimental units and the observational units is any more complicated than this, think again.

Explain the plot structure as fully as possible: see Chapters 4, 6, 8 and 10. I usually draw the Hasse diagram.

Of course, the scientist needs to know what he is going to measure before he can say what objects he is going to measure, so details under the next heading should be filled at the same time as this one.

14.3.6 What measurements are to be recorded?

Write down everything to be recorded, for example 'weight in kg at 15 days old' or 'proportion of diseased plants in each of 10 samples of 100 plants from each plot'. (In the second case, the samples are the observational units.)

It is a good idea to prepare a data sheet in advance, either on paper or a spreadsheet, with one row for each observational unit and one column for each measurement. See Section 1.1.3.

14.3.7 What is the design?

Here you should explain the systematic, or combinatorial, design used. In a simple experiment this can be very straightforward: for example 'four feeds are each allocated to two pens'. A completely randomized design (Chapter 2) or a complete-block design (Chapter 4) can probably be described adequately in words. So can a standard split-plot design (Section 8.3), so long as you are careful to state which factors are applied to small blocks and which to plots. For anything else, the systematic design needs to be written out in full: for example, a row–column design (Chapters 6 and 9), an incomplete-block design (Chapters 11–12) or a fractional factorial design with specified aliasing (Chapter 13).

14.3.8 Justification for the design

You need to justify the amount of replication. If there is too much replication then the experiment may waste time and money. If animals are to be sacrificed, it is unethical to use too many.

On the other hand, if there is too little replication then any genuine differences between treatments may be masked by the differences among the experimental units. An experiment which is too small to give any conclusions is also a waste of resources. It is also an unethical use of animals or people.

Watch out for false replication (Section 8.1).

Comment on any blocking (see also Section 14.3.4). Are the blocks inherent and discrete, or chosen to split up a continuous trend, or used for management (Section 4.1)? How were block sizes chosen?

Are there any constraints on applying or changing treatments? For example, can some be applied only to large areas, or changed only infrequently? Does everyone involved accept that there may be low power and lack of precision for such factors (Chapter 8)?

If an incomplete-block design is used, justify the choice of which one. If it is not balanced, do the treatment contrasts of most interest have higher efficiency factors (Chapter 11)? If it is a confounded factorial design (Chapter 12), justify the choice of what to confound. Is it assumed that any interactions are zero? If so, is Principle 12.1 satisfied? If it is a fractional factorial design (Chapter 13) then justify your choice of aliasing: list what you assume to be zero and what you intend to estimate, and make sure that each alias set contains at most one effect to be estimated, with the remainder assumed zero.

14.3.9 Randomization used

Explain the method that you used: such as 'I randomized blocks, then randomized plots within each block independently'. Keep a record of the random numbers that you used, or of the seed given to your randomization software. If you are replaced by another statistician, she should be able to trace your work and obtain the same plan.

14.3.10 Plan

This gives the exact details of which (coded) treatment is allocated to which (explicitly named) experimental unit. It is the outcome of the randomization. Some information from it may be put into the data-recording sheet.

Remember that sometimes some details of the plan must be hidden from those involved with management of the trial and actual measurement.

14.3.11 Proposed statistical analysis

Write down guidelines for the statistical analysis that you propose to do once the data are collected. Do this before any data are collected.

What will be the assumed expectation model for the responses? What will be the assumed covariance structure? Are these consistent with the method of randomization used?

For a simple experiment with two treatments and no blocks, you would probably write 'two-sample t-test'. For three treatments and no blocks, 'analysis of variance with three treatments and no blocks' suffices.

However, in general, you should give the skeleton analysis of variance. This enables you to check which stratum contains which treatment subspace. Are the more important treatment subspaces in strata which are likely to have smaller stratum variances? Are there enough residual degrees of freedom in each stratum containing treatment subspaces?

If you plan to transform the data before analysis, say so.

Do not wait till you have the data before verifying that you can do the necessary statistical calculations, either with a calculator or using appropriate software. Unless you have recently analysed an experiment of exactly the same type, it is a good idea to invent some dummy data (or copy some from an old experiment) and check that you know how to get your software to do the correct analysis.

By the time the data are collected, a different statistician may be involved. This part of the protocol should be detailed enough for him to know exactly what you intended to do.

Of course, you may end up doing a different analysis from the one you proposed. This can happen if something goes wrong during the experiment, or if further relevant information comes to light.

14.4 The eight stages

In the process of design and analysis of an experiment, the concepts of *model*, *randomization*, *design* and *analysis* are interwoven. Their inter-dependence can be summarized in eight stages.

(i) An additive **model** for expectation is assumed, like the one in Equation (14.1). As discussed in Section 5.2, this involves choice of an appropriate scale of measurement.

(ii) The method of **randomization** is specified. This is usually based on the structure of the experimental units.

(iii) The method of randomization in Sections 10.11 and 14.1.2 leads to a simplification of the **model**. If there are factors like those in Section 14.2, the model may contain rather more parameters than we would like.

(iv) We choose a combinatorial **design**.

(v) The design and the model give the skeleton **analysis** of variance.

(vi) The previous two steps are iterated until the proposed analysis seems satisfactory. This gives us the **design** we will actually use.

(vii) This design is **randomized** by the prescribed method. This gives the experimental layout, so the experiment can proceed.

(viii) Data are collected, and the real **analysis** is performed.

14.5 A story

Of course, the ideas do not always come in the order listed in the draft protocol given in Section 14.3. The following story demonstrates how the different considerations interplay. At the same time, it provides a review of most of the book.

Example 14.10 (Example 4.3 continued: Road signs) A road traffic researcher tells you that he is planning an experiment to compare two types of bilingual road sign: the languages are English and Welsh, and either may be on top. On each of four afternoons he will ask twelve volunteers to drive around the special test track at the traffic research station. Different signs will be set up, and a researcher will sit in the passenger seat and ask the driver questions during the drive, to assess his level of distraction.

At this point you jot down '2 treatments, 4 afternoons, 12 people, so the experimental unit is a person-afternoon and there are 48'.

You ask him about the treatments. Does he need a control as well as the two treatments he has specified? What question is he trying to answer (Chapter 1)? Should he be thinking about another factor at the same time (Chapter 5)?

There are two likely answers here. The first is that some local government department has issued a fiat that road signs shall be bilingual, so he wants to find out whether the amount of distraction depends on which language is on top. In this case he should stick to his two proposed treatments, and replicate them as equally as possible (Chapter 2).

The other is that some well-meaning person has suggested that bilingual road signs would be generally helpful. Now the researcher probably wants to know whether the bilingual signs distract drivers more than the current monolingual signs. So he should include the current signs as a control treatment. The relevant factors are as follows.

	Treatment		
	1	2	3
number of languages	2	2	1
language on top	Welsh	English	n/a

Now you can discuss replication. The two bilingual treatments should have the same replication, say r, but there is no need for the monolingual treatment to have the same replication as these two. If the comparison of one language with two is of over-riding importance then the levels of number of languages should be equally replicated: hence the monolingual signs should have replication $2r$. If the researcher wants to compare *each* type of bilingual sign with the monolingual then the monolingual should have replication approximately $1.4r$ (so you might choose $1.5r$) (Chapter 3). If he prefers to compare the two bilingual types with each other, then equal replication of all three treatments may be best.

For simplicity, the remainder of this example assumes that the experimenter has decided to have just the two bilingual treatments.

You know that people are variable and so it is better to use each person more than once if this is feasible (Chapter 7). So you ask him how many afternoons each volunteer will be asked to attend. He does not seem to understand you, so you sketch out the modified crossover design (Chapter 6) in Figure 14.4. (Of course, you do know the Hasse diagrams and skeleton analysis of variance that go with this: Chapter 10.) His face clears. 'Oh, no', he says,

14.5. A story

Afternoon	Person			
	1–6	7–12	13–18	19–24
1	English	Welsh		
2	Welsh	English		
3			English	Welsh
4			Welsh	English

Fig. 14.4. Modified cross-over design suggested in Example 14.10

'we can't do that. For one thing, the people can't do the test drive twice, because they would learn the track. For another, it takes us quite a long time to set up the signs, so we can't do more than one sort in an afternoon.' The second part of his reply is more important, in this context, than the first: he has just told you that *the experimental units are the afternoons, not the test drives*.

Now you have to convince him that he has much less true replication than he thought (Chapter 8) and has only two degrees of freedom for residual, which will give him low power. Cajole him into using more afternoons. This will be hard. Let us suppose that you persuade him to use eight afternoons instead of four.

You wonder whether the afternoons should be blocked (Chapter 4). He may not understand the word 'block' in this context, so you ask him if the afternoons are all similar. For example, is he planning to use four afternoons in January and four in March? You establish that he is going to use eight weekday afternoons in quick succession, so there is no need to block.

You ask him the analogous question about people. Rather sheepishly, he admits that his volunteers are of two types: retired people and university students. Trying to be helpful, he suggests a simple scheme: use retired people on four of the afternoons and students on the other four. If we do this and do not block the afternoons by type of people then we increase the variability between afternoons, and hence the variance of the estimator of the treatment difference. If we do block the afternoons by type of people then we lose one of our precious residual degrees of freedom. Reducing, say, 20 residual degrees of freedom to 19 hardly matters, but reducing six to five does.

You ask him if it would be possible to test six retired people and six students each afternoon. He says that this will be no problem. After discussing whether he needs to invite, say, eight of each type each afternoon to allow for no-shows, you wonder if there is any other source of potential systematic difference that you have overlooked. He has got the hang of this by now, and replies that the shadows are longer later in the afternoon. You therefore decide to divide each afternoon into the *early* part (first six test drives) and the *late* part. This is just a nuisance factor, like the division into types of people, so it will be sensible to alias these two factors (Chapter 10) by getting the retired people to do the test drives in the early part of the afternoon.

You go away and draw up the randomized plan in Figure 14.5 and the skeleton analysis of variance in Table 14.5. You randomly allocate treatments to afternoons by writing down a systematic list and then applying a random permutation of the eight afternoons. Within each part of each afternoon separately, you randomize the volunteers (Section 14.1.6). Here I am assuming that you cannot randomly allocate volunteers to afternoons, because they may not

		Afternoon							
		1	2	3	4	5	6	7	8
language on top		English	Welsh	Welsh	English	Welsh	Welsh	English	English
Early	2.00	R1	R10	R15	R19	R29	R36	R37	R47
	2.20	R4	R12	R18	R23	R26	R33	R42	R43
	2.40	R6	R11	R14	R24	R28	R32	R40	R46
	3.00	R2	R7	R13	R20	R30	R35	R38	R44
	3.20	R5	R9	R16	R22	R27	R34	R41	R48
	3.40	R3	R8	R17	R21	R25	R31	R39	R45
Late	4.00	S1	S12	S14	S21	S27	S31	S42	S48
	4.20	S4	S7	S17	S22	S29	S33	S41	S47
	4.40	S3	S8	S16	S20	S25	S35	S38	S44
	5.00	S5	S10	S18	S23	S30	S34	S39	S46
	5.20	S2	S9	S13	S19	S28	S32	S40	S43
	5.40	S6	S11	S15	S24	S26	S36	S37	S45

Fig. 14.5. *One possible plan for the experiment in Example 14.10*

be available every day, but that you can randomize the order of the six people allocated to each part of each afternoon. Thus the retired volunteers who attend on the first afternoon are labelled R1–R6.

It is perfectly possible that the researcher will accept this plan and skeleton analysis. This is a good design for the problem originally posed.

However, the researcher's reaction to your mentioning of nuisance factors may be to object and say that he would be quite interested to find out whether retired people and students are affected differently by the new road signs. He wonders whether we could even find out whether the difference between the two languages is the same for students as it is for retired people; he is talking about 'interaction' (Chapter 5) but is not familiar with the word. Thus he wants to change the treatment structure from two unstructured treatments to a 2×2 factorial, with person-type as a treatment factor rather than a plot factor (Section 14.2). This will be all right so long as we are clear that an observational unit is now a test drive, not a person-afternoon.

Now aliasing the type of person with part of afternoon will be disastrous. Instead, you

Table 14.5. *Skeleton analysis of variance for the plan in Figure 14.5*

Stratum	Source	Degrees of freedom
mean	mean	1
afternoon	language on top	1
	residual	6
	total	7
part	part	1
afternoon ∧ part	afternoon ∧ part	7
test drive	test drive	80
Total		96

14.5. A story

		Afternoon							
		1	2	3	4	5	6	7	8
language on top		Welsh	English	English	Welsh	Welsh	English	English	Welsh
Early	2.00	R2	S8	R14	S19	S27	R33	R37	S43
	2.20	S2	S9	S15	S21	S25	S31	S37	S45
	2.40	R1	R7	S14	R20	S26	S33	S39	S44
	3.00	S1	R9	S13	S20	R26	S32	R38	R43
	3.20	S3	R8	R15	R21	R27	R32	S38	R45
	3.40	R3	S7	R13	R19	R25	R31	R39	R44
Late	4.00	R4	R12	S16	R22	R30	S35	R40	S48
	4.20	R6	R10	R17	S22	R28	R35	S40	R46
	4.40	S4	S12	R18	R24	R29	R34	R42	R47
	5.00	R5	R11	S18	R23	S30	S34	S41	S46
	5.20	S5	S10	S17	S23	S28	S36	R41	R48
	5.40	S6	S11	R16	S24	S29	R36	S42	S47

Fig. 14.6. *Another possible plan for the experiment in Example 14.10*

allocate three retired people and three students to each part of each afternoon, and randomize the order of the six people in each part of each afternoon. The people are now distinguishable instances of the levels of a treatment factor, so they must be randomized within parts of afternoons (Section 14.1.5), but there is no harm in allocating the first three of each type to volunteer to the early part of the afternoon, as this merely aliases the nuisance factors part and time of volunteering. This gives a plan such as the one in Figure 14.6 and the skeleton analysis in Table 14.6.

Perhaps the researcher now grumbles that he does not like the complication of having different random orders in the different afternoons. You explain to him why the treatment factor person-type should not be confounded with the potential plot factor time-slot. You offer to write out the data sheet for him if he gives you the list of names for each afternoon. He agrees to this, and you produce a sheet (Chapter 1) like the one in Figure 14.7.

Alternatively, now that you have alerted him to time-slot, he may object to the plan in

Table 14.6. *Skeleton analysis of variance for the plan in Figure 14.6*

Stratum	Source	Degrees of freedom
mean	mean	1
afternoon	language on top	1
	residual	6
	total	7
part	part	1
afternoon ∧ part	afternoon ∧ part	7
test-drive	person-type	1
	language ∧ type	1
	residual	78
	total	80
Total		96

Day	Language on top	Time	Person	Type	First bend	Long straight	...
14 April	Welsh	2.00	Lloyd George	R			...
14 April	Welsh	2.20	Dylan Thomas	S			
14 April	Welsh	2.40	Owain Glyndwr	R			
⋮	⋮	⋮	⋮	⋮			

Fig. 14.7. Possible data sheet for the plan in Figure 14.6

Figure 14.6 on the grounds that all the people who have their test-drive at 2.20 are students. You discuss with him whether it is really necessary to take account of such a fine division of afternoons as the one into time-slots, but the numbers fit nicely and you can make a design using twenty-four 2×2 Latin squares (Chapter 6): the time-slots are the rows and the afternoons are the columns. The new design is fairly comparable to the second one, and there is no harm in letting the researcher choose between them.

Notice that none of the three designs recommended in this story is exactly the same as any design given earlier in the book. This demonstrates the construction of a design fit for purpose as opposed to choosing one from a list.

Questions for discussion

14.1 Find out how your usual statistical software analyses block designs with unequal block sizes. What model does it assume?

14.2 Give an argument like the one in Example 14.3 to justify the model in Section 4.6.

14.3 What is the statistical advantage in choosing a 6×6 Latin square at random from among all 6×6 Latin squares, rather than simply starting with one 6×6 Latin square and then randomizing rows and columns?

14.4 In Example 14.1, what are the advantages and disadvantages, if any, of randomizing the order of the eleven chosen students?

14.5 Plot the four means in Table 14.2 on a suitable graph to demonstrate the interaction in Example 14.9.

14.6 Construct the final design in Example 14.10. Then randomize it. Write out the skeleton analysis-of-variance table.

14.7 Retell the story in Example 14.10 on the assumption that the monolingual treatment is needed as well.

14.8 What other matters might be pertinent to the experiment in Example 14.10?

Exercises

Here is a collection of exercises which are not tied to the individual chapters.

Exercise 1 A professional cat-breeder has written to you. She is proposing to do an experiment next spring. Her letter asks for an appointment to discuss details of the experiment with you. Part of her letter reads as follows.

> In the past I have always fed newly-weaned kittens on a diet which I cook and prepare myself. However, there are some good commercial cat-foods on the market now, and I wonder if I could save myself the trouble of preparing a special diet.
>
> Three of the new cat-foods—Purrfect, Qualicat and Rumpuss—seem particularly attractive. I should like to do an experiment to compare these three, to see which produces the healthiest kittens.
>
> My mature cats always produce four or five kittens in a litter. If I use the kittens from three mother cats I shall therefore have at least twelve kittens available for the experiment. Will this be enough?

Make clear notes on the most important points to discuss with her at your meeting. Explain the statistical motives for discussing these points.

Exercise 2 A horticulturist wants to compare three varieties of tomato, two types of compost in which to grow them, and two watering regimes. He has three glasshouses, each of which contains twelve chambers. In each chamber he will grow six plants, all of the same variety, in the same type of compost and with the same watering regime. He will record the total weight of tomatoes from each chamber.

(a) Identify the experimental units, the observational units and the treatments in this experiment, and state how many there are of each.

(b) Draw the Hasse diagram for the factors on the observational units (ignoring the treatment factors) and the Hasse diagram for the factors on the treatments.

(c) Describe how to construct and randomize the design for the trial.

(d) Write down the skeleton analysis of variance, showing stratum, source and degrees of freedom.

Table E.1. *Data for Exercise 3*

Board divided?	Coloured pens?	Number of pairs	Number of trials	Total y	Total y^2
No	No	6	216	997	6548
No	Yes	6	216	933	6110
Yes	No	6	216	819	6999
Yes	Yes	6	216	1259	10785
Total		24	864	4008	30442

Exercise 3 A computer scientist has done an experiment to investigate how people communicate with each other when drawing on a shared whiteboard. He used 24 pairs of people. Each pair of people undertook 36 trials. Two treatment factors were involved in the experiment: (a) sometimes the board was divided into two halves and each person was allowed to write only in their own half; in the other trials each person could write wherever they liked in the board (b) sometimes the people had a collection of coloured pens to draw with; in the other trials they had only black pens. For each pair of people, these two factors were set at the same combination of levels for all 36 trials. The whiteboard was attached to a computer. After each trial, the total amount of ink (in thousands of computer screen pixels) used in that trial was recorded as the response variable y. Thus 864 numbers were recorded.

The computer scientist gives you the summary of the data in Table E.1. He also tells you that the sum of the squares of the 24 totals of the y values for the pairs is 707467.

(a) Calculate the analysis-of-variance table, with an appropriate decomposition of the treatment sum of squares.

(b) Briefly interpret the analysis. Then calculate appropriate table(s) of means for the treatments.

(c) Briefly state how you would advise the computer scientist to design the next experiment of this type.

Exercise 4 (a) Explain what it means for an incomplete-block design to be (i) *balanced* (ii) *resolved*. State why each of these properties is desirable.

(b) A tropical agronomist wishes to compare 13 varieties of banana tree. He intends to plant four trees of each variety. Thirteen farmers have agreed to let him use their land, so his experiment will involve planting four trees on each of 13 different farms.

(i) Construct a suitable design for the agronomist.

(ii) The agronomist's supervisor suggests that it would be simpler to use four whole villages, rather than individual farms, so that 13 experimental trees—one of each new variety—could be planted in each village. However, the plots-within-villages variability is likely to be greater than the plots-within-farms variability. A good estimate is that the plots stratum variance within the farms is only 3/4 of the plots stratum variance within the villages. Calculate which of the two proposed designs is more efficient.

Exercises

Exercise 5 A sports scientist wants to find an alternative form of training for athletes who are temporarily unable to run on a normal hard surface because of minor injury. She plans to ask nine fit athletes to run 400 metres under each of three different conditions and then measure their heartbeats. The three different conditions are:

n: a normal running track;
t: a treadmill;
w: a flat-bottomed pool with 30 cm of water in it.

She meets this group of athletes every Wednesday afternoon. She will ask each of these athletes to do one of these 400-metre runs each Wednesday, for a total of six Wednesdays. By the end of the six weeks each athlete will have run 400 metres twice under each condition.

(a) Identify the experimental units, observational units and any suitable blocks in this experiment. Briefly explain your decisions.

(b) Construct the design and randomize it.

(c) Present the plan in a form suitable for the sports scientist.

(d) After the experiment, you will analyse the data with your usual statistical computing package. What factors must be declared? After the data and factor values have been entered, what commands should be given, or what should you enter in the relevant dialogue boxes?

Exercise 6 Mountain bags are often carried by walkers in the Scottish Highlands. If a walker is trapped by injury or bad weather, he or she gets into the mountain bag, which is supposed to stop them getting too cold.

Three new makes of mountain bag are going to be assessed at a Scottish university. Thirty-six students will take part in the experiment. Twelve mountain bags of each make will be bought and allocated to students. The thirty-six students will go into a special temperature-controlled room and get into their mountain bags. The temperature in the room will be reduced. After one hour, the skin temperature of each student will be measured on the chest, arm, thigh and calf.

(a) Identify the experimental units, the observational units and the treatments in this experiment, and state how many there are of each.

(b) Draw the Hasse diagram for the factors on the observational units (ignoring the treatment factors) and the Hasse diagram for the factors on the treatments.

(c) Describe how to construct and randomize the design for the trial.

(d) Write down the skeleton analysis of variance, showing stratum, source and degrees of freedom.

(e) Someone suggests that it would be better to let each student try all three makes of mountain bag. In this case only twelve students would be needed, but the temperature-controlled room would need to be used on three different days. Explain two important ways in which this new design is different from the old.

Exercise 7 A small company is experimenting with new workstations, to see which type to buy for the whole workforce. They want to compare five types of keyboard and also five types of screen. They believe that there is no interaction between type of keyboard and type of screen. The experiment will use five workers for the five days of one working week. Each day each worker will be given one combination of type of keyboard with type of screen, and given a programme of work to do. At the end of each day one of the managers will score each worker out of 100 according to the amount of work completed.

(a) Construct a design for this experiment in such a way that both main effects can be estimated.

(b) Write down the skeleton analysis-of-variance table, showing stratum, source and degrees of freedom.

(c) Explain how the design should be randomized.

(d) The first manager proposes to give every worker the same programme of work to do every day. The second manager thinks that this is a bad idea, because the workers will get better day by day. He thinks that there should be five different programmes of work, and that every worker should do each one. How should the second manager design his experiment? Which design is better?

Exercise 8 An experiment was conducted on tomatoes growing in a greenhouse. The purpose of the experiment was to compare five spray treatments using a chemical growth regulator. The five treatments were as follows.

A: 75 parts per million of chemical, sprayed early
B: 150 parts per million of chemical, sprayed early
C: 75 parts per million of chemical, sprayed late
D: 150 parts per million of chemical, sprayed late
E: no spray

The tomatoes were grown in thirty separate chambers in the greenhouse, to prevent drift of spray. The chambers were arranged in six blocks of five, with each block in the East–West direction (so that one block was the furthest North and was relatively shaded, while another was the furthest South and so received the most sunshine).

The yield of tomatoes (in units not specified by the experimenter) in each of the thirty chambers is shown in Table E.2. Here the data are *not* in greenhouse order, but have been rearranged to facilitate hand calculation.

The sum of the squares of all the yields is 616.3632.

(a) Calculate the analysis-of-variance table, with an appropriate decomposition of the sum of squares for treatments.

(b) Briefly interpret the analysis. Then calculate appropriate table(s) of means for the treatments.

Table E.2. *Data for Exercise 8*

	Treatment					Treatment
	A	B	C	D	E	total
Block I	3.38	3.33	1.79	3.19	2.98	14.67
Block II	3.35	3.02	3.16	2.67	1.83	14.03
Block III	4.30	3.02	2.11	1.96	1.54	12.93
Block IV	6.15	5.53	5.39	4.32	3.64	25.03
Block V	7.28	6.06	5.17	3.83	3.19	25.53
Block VI	8.22	7.35	5.91	5.68	5.75	32.91
Treatment total	32.68	28.31	23.53	21.65	18.93	125.10

Exercise 9 Eight high-quality club cyclists will take part in a study to compare the performance of a new racing bicycle with that of a standard racing bicycle and also to see whether the relative performance depends on the cyclist's workload. Two sets of cycling exercises have been devised, one involving a higher workload than the other. The study will take place over four afternoons, with each cyclist performing each set of exercises on each type of bicycle, one combination per afternoon. During each set of exercises, each cyclist's oxygen consumption will be measured.

(a) Identify the experimental units, the observational units and the treatments in this experiment, and state how many there are of each.

(b) Draw the Hasse diagram for the factors on the observational units (ignoring the treatment factors) and the Hasse diagram for the factors on the treatments.

(c) Describe how to construct and randomize the design for the trial.

(d) Write down the skeleton analysis of variance, showing stratum, source and degrees of freedom.

Exercise 10 Five two-level treatment factors A, B, C, D, E are to be tested for their effect on an industrial process. It is believed that the only nonzero interactions are the two-factor interactions.

(a) Construct a single-replicate design for these five two-level treatment factors in four blocks of eight plots each, in such a way that all main effects and all two-factor interactions can be estimated.

(b) Write down the skeleton analysis-of-variance table, showing stratum, source and degrees of freedom.

(c) Now the industrialist changes his mind, and tells you that he thinks that the B-by-D-by-E interaction may also be nonzero. Can this interaction be estimated from your design? Give reasons for your answer.

Table E.3. *Primary schools, and marks in arithmetic, for Exercise 11*

School	Arithmetic mark
St. Anne's	93
Broadhaven	66
Cuthbert Ellis	88
St. David's	63
East Wormly	71
Green Lanes	77
High Lodge	78
King Edward's	95
Leagrove	81

Exercise 11 A research worker in a local education authority (L.E.A.) wishes to investigate the effect that different methods of practice have on pupils' ability in mental arithmetic. There are three methods:

- a : the teacher holds a session with the whole class for 40 minutes every day, calling out questions and asking children to put up their hands if they know the answer;
- b : the teacher divides the class into four small groups, and holds a similar 10-minute session with each group every day;
- c : the teacher tests each child individually and privately, for 10 minutes per week.

She plans to test these three methods on the Year 3 pupils at nine primary schools in her L.E.A. The schools are shown in Table E.3, together with the average mark on a standard arithmetic test obtained by Year 3 pupils at that school during the past four years.

The experiment will last for the whole autumn term. The research worker does not wish any of the pupils to know that they are part of an experiment, and so all Year 3 classes in the same school will use the same method of practice at any one time.

(a) Identify the experimental units, observational units and any suitable blocks in this experiment. Briefly explain your decisions.

(b) Construct the design and randomize it.

(c) Present the plan in a form suitable for the research worker.

(d) After the experiment, you will analyse the data with your usual statistical computing package. What factors must be declared? After the data and factor values have been entered, what commands should be given, or what should you enter in the relevant dialogue boxes?

Exercise 12 (a) Construct a pair of mutually orthogonal 5×5 Latin squares.

(b) Give two experimental situations (including details of both plot structure and treatment structure) for which the above pair of Latin squares could provide a suitable design. In each case, explain how to construct the design from the Latin squares, state the replication, and explain how the design should be randomized.

Exercise 13 A professional vegetable-grower has written to you, making an appointment to discuss an experiment which he is proposing to conduct in the coming growing season. Part of his letter reads:

> One problem with growing broad beans is that they are often infested with blackfly. For the last five years I have been controlling the blackfly by spraying the bean plants with my own secret mixture. However, a new chemical called Black Death has recently come onto the market, and it is claimed to be very effective. I should like to compare it with my secret mixture.
>
> The area where I intend to grow broad beans next year is naturally divided into three parts, separated by paths. I propose spraying the bean plants on one part with my secret mixture; spraying those on the second part with Black Death; and leaving those on the third part with no spray against blackfly. I shall then see which part does best.
>
> Looking ahead, perhaps you could also advise me on the experiment I plan for the following year. I saw a television programme which suggested that blackfly colonies are maintained by ants as a source of food, and that therefore destroying the ants will also get rid of the blackfly. Once I know which chemical to use on the blackfly themselves, I can then do another experiment to see if using ant-killer improves my broad beans.

Make clear notes on the most important points to discuss with him—and why!—at your meeting.

Exercise 14 (a) Show that $\{0, 1, 2, 4\}$ is a difference set modulo 7.

(b) A car manufacturer wishes to compare seven modifications to the engine which are supposed to cut down the amount of fuel used when the car is driven in town conditions. He intends to make four standard cars with each type of engine. Seven medium-sized towns will be involved in the experiment. Four experimental cars will be sent to each town, where each car will be allocated to an employee of the company. Each such employee will drive the car for a week to get used to it. In the subsequent week the employee will drive the test car only over specified routes within the town, at specified times of day. In each town, the total length of the specified routes is 200 miles. The amount of fuel consumed by each test car in the second week will be measured.

  (i) Construct a suitable design for the car manufacturer.

  (ii) Explain how the design should be randomized.

  (iii) Someone from the head office of the company suggests that it would be simpler to use four large towns, so that seven employees could be used in each large town, and every type of engine could be tested in each large town. However, the variance of driving times is likely to be greater in large towns than in medium-sized towns. A good estimate is that the variance in medium-sized towns is only $3/4$ of the variance in large towns. Calculate which of the two proposed designs is more efficient.

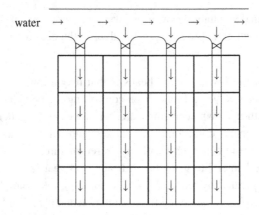

Fig. E.1. Irrigation in one block of the garlic experiment in Exercise 15

Exercise 15 Vegetables are grown in irrigated land on the East side of the Andes. A garlic grower wants to study the effects of four different amounts of water (in the irrigation) and four different sizes of the garlic clove which is planted. He uses three rectangular blocks of sixteen plots each, irrigated as shown in Figure E.1. Taps (shown by ⋈ in the figure) control the flow of water into each column. In each plot, all garlic cloves planted have the same size.

(a) Identify treatments, stating how many there are.

(b) Draw the Hasse diagram for the factors on the treatments.

(c) Draw the Hasse diagram for the factors on the plots (ignoring the treatment factors).

(d) Describe how to construct the design for the trial.

(e) Describe how to randomize the design.

(f) Write down the skeleton analysis of variance, showing stratum, source and degrees of freedom.

Exercise 16 A breakfast-cereal company wishes to improve the flavour of the dried fruit in its muesli. It has been suggested that the addition of very small amounts of certain chemicals to the fruit will preserve (and so improve) the flavour. There are three chemicals which the company would like to test: it proposes to test each chemical in quantities of 0, 5 and 10 mg per tonne of dried fruit. Company policy precludes the use of more than one of these chemicals at a time in the dried fruit.

The mueslis made with the different chemical additives will be tested by 28 volunteer families during a seven-week period. Each week each family will be given a supply of one of the mueslis, which they will eat for breakfast and assess for flavour. At the end of the week the family will give the muesli a single rating on a scale 1–15 provided and explained by the company.

(a) Identify plots and treatments, stating how many of each there are.

(b) Draw the Hasse diagram for the factors on the plots (ignoring the treatment factors).

Exercises

(c) Draw the Hasse diagram for the factors on the treatments.

(d) Describe how to construct the design for the trial.

(e) Describe how to randomize the design.

(f) Write down the skeleton analysis of variance, showing stratum, source and degrees of freedom.

Exercise 17 A commercial bakery wishes to compare 16 possible recipes for crusty white bread. The bakery bakes four batches of loaves per day. Each batch consists of several trays full of loaves. The experiment will be conducted during the normal commercial operation of the bakery. At each batch, space will be made in the ovens for four trays of experimental loaves. On each of these trays all the loaves must be made from a single recipe. The experiment will last for four days.

(a) Explain why a resolved incomplete-block design is needed.

(b) Construct a suitable design.

(c) Explain how the design should be randomized.

(d) Choose two treatments that never occur in the same block. Suppose that these treatments are i and j. Show that the simple contrast for estimating the difference between i and j is the vector $\mathbf{x}$ with coordinates

$$x_\alpha = \begin{cases} \dfrac{1}{4} & \text{if } T(\alpha) = i \\ -\dfrac{1}{4} & \text{if } T(\alpha) = j \\ -\dfrac{1}{12} & \text{if } T(\alpha) \neq i \text{ but } i \text{ occurs in block } B(\alpha) \\ \dfrac{1}{12} & \text{if } T(\alpha) \neq j \text{ but } j \text{ occurs in block } B(\alpha) \\ 0 & \text{if neither } i \text{ nor } j \text{ occurs in block } B(\alpha). \end{cases}$$

Hence show that the efficiency factor for this simple contrast is equal to $3/4$.

Exercise 18 (a) The raw ingredients for an industrial process are three substances L, M and N, each of which can be provided in five different quantities. There are thus 5^3 possible combinations of ingredients. It is possible to test five such combinations in any one week. Five weeks are available for finding the best combination. It is believed that there are no interactions between the three substances.

Construct a suitable design.

(b) Suppose that a balanced incomplete-block design is required for ten treatments in blocks of size eight. Show that there must be at least 45 blocks. Explain how to construct a balanced incomplete-block design with exactly 45 blocks.

Exercise 19 A biochemist is investigating the effect of two treatment factors on the quantity of a certain enzyme in the blood of mice. He uses 96 mice, from the standard breed of laboratory mice, housed in eight cages of twelve mice each. One treatment factor is the type of food, of which there are two. Food rations are put into the cages for mice to eat ad lib. The other treatment factor is the presence or absence of an injection of a certain chemical. Each of the four combinations

| Food A | Food A | Food B | Food B |
| Chemical | No chemical | Chemical | No chemical |

is applied to two whole cages. The mice are individually injected with chemical (or not) then kept in the cages for two weeks eating one of the two types of food. At the end of the two weeks a sample of blood is drawn from each mouse, and the biochemist measures the quantity of the enzyme present in a standard volume of the blood.

(a) Fill in the blank spaces in the following incomplete analysis-of-variance table.

Stratum	Source	SS	df	MS	VR
mean	mean	2875	1		
cages	food	32			
	chemical	60			
	food ∧ chemical	3			
	residual				
	total	112			
mice	mice	163			
Total		3150	96		

(b) What conclusions can you draw about the presence or absence of treatment effects? Give reasons.

(c) Estimate the variance of the estimator of the difference between the two types of food. Also estimate $\sigma^2(\rho_1 - \rho_2)$, where σ^2 is the variance of each observation, ρ_1 is the correlation between observations on different mice from the same cage and ρ_2 is the correlation between observations on mice from different cages. You may use the fact that

$$\xi_{cages} = \sigma^2(1-\rho_1) + k\sigma^2(\rho_1 - \rho_2)$$
$$\xi_{mice} = \sigma^2(1-\rho_1),$$

where ξ_{cages} is the stratum variance for cages, ξ_{mice} is the stratum variance for mice, and k is the number of mice per cage.

(d) Now the biochemist is planning to do another similar experiment. Comment briefly on the advantages and disadvantages of each of the following suggested modifications to the previous design.

(i) Use 15 mice in each of the eight cages.

Exercises

(ii) Use twelve cages with only eight mice each.

(iii) Use twelve mice in each of the eight cages, apply the types of food to whole cages but inject six randomly chosen mice per cage.

Exercise 20 There is an air pollution research unit in the centre of London. On the roof of the research unit are five machines for measuring the amount of smoke in the air. These machines are labelled M1–M5. Each machine sucks in air and passes it through a filter paper, where the particles of carbon are trapped. At the end of each hour the filter paper is removed and placed under a bright lamp in the laboratory. The amount of light reflected is recorded: the more carbon there is then the darker is the paper and the less light is reflected.

The director of the research unit wants to compare four new sorts of filter paper ('fine', 'grainy', 'hard' and 'smooth') with the sort that is being used now. In order not to spoil the ongoing records, he will continue to use the current sort of filter paper in each of the five machines for most of the time. However, he will use the hour 1400–1500 on each of the five days Monday–Friday in one working week to experiment with the filter papers.

The research unit has five technicians who know how to put the filter papers into the machines, start and stop the machines, remove the filter papers and record the amount of reflected light: Alan, Brian, Chris, Diana and Elizabeth. On each day of the experiment, each technician can work only one of the machines, because all the machines must be worked at exactly the same time, to ensure that sudden atmospheric changes do not perturb the experiment.

(a) Identify the plots and any suitable blocks in this experiment.

(b) Construct the design and randomize it.

(c) Present the plan in a form suitable for the director of the research unit.

(d) During the experiment, the amounts of reflected light will be recorded in a single column in a spreadsheet. What other columns should you provide in that spreadsheet before the experiment starts? After the experiment, how will you analyse the data with your usual statistical computing package?

Exercise 21 A psycholinguistics researcher wants to find out whether people's reaction times to words depend on whether those words are nouns, verbs or adjectives. She has made a list which contains ten nouns, ten verbs, ten adjectives and 40 other non-words that sound similar to real words. She plans to read the list of 70 items aloud to about 50 people, one person at a time. Each person will be asked to press a buzzer each time that they think that the item is a real word. For each real word, the time taken to press the buzzer will be noted. She has written to you as follows.

> I could sit with the person and read out the list of items. Then I could change the order if I feel like it and also have visual feedback about their reactions. Or would it be better for me to record the items on tape and play the tape to each person?

You have made an appointment to see her in order to discuss this.

Make clear notes on the most important points to discuss with her at your meeting. Explain the statistical motives for discussing these points.

Village 1	1	2	3	4	5	6	7	8	9	10
	i	*d*	*h*	*b*	*c*	*g*	*e*	*j*	*a*	*f*
Village 2	11	12	13	14	15	16	17	18	19	20
	c	*a*	*j*	*b*	*g*	*d*	*h*	*e*	*f*	*i*
Village 3	21	22	23	24	25	26	27	28	29	30
	e	*c*	*f*	*d*	*i*	*b*	*g*	*h*	*j*	*a*
Village 4	31	32	33	34	35	36	37	38	39	40
	j	*h*	*d*	*a*	*e*	*b*	*i*	*c*	*g*	*f*
Village 5	41	42	43	44	45	46	47	48	49	50
	c	*g*	*d*	*a*	*j*	*e*	*f*	*i*	*b*	*h*

Fig. E.2. Plan for last year's experiment in Exercise 22

Exercise 22 Farmers in five neighbouring Nigerian villages have agreed to let a scientist at the local Tropical Agriculture Research Institute use some of their oil-palm trees for his experiments. Ten trees have been made available in each village. The trees are labelled 1–50 for identification.

Last year the scientist conducted an experiment on the oil-palm trees. There were ten treatments $a, b, \ldots, j$, and each village formed a complete block. The plan is shown in Figure E.2.

(a) This year the scientist wishes to conduct a further experiment on these trees, using five new treatments A, B, C, D, E. Construct a design for the experiment, randomize it, and present the plan in a form suitable for the scientist.

(b) After the experiment, the quantities of palm-oil obtained from trees 1–50 will be typed into a text file, in a single column. You will analyse the data with your usual statistical computing package. How will you read in the data? What factors must be declared? In order to analyse the data, what commands should you give, or what should you enter in the relevant dialogue boxes?

Exercise 23 At a horticultural research station in New Zealand, scientists conduct experiments on kiwi-fruit, which grow on vines. The main stem of each vine is trained into a horizontal position, as shown in Figure E.3. The branches, which are called *canes*, are trained back down to the earth and tied down. The canes are thinned, only the best six being kept on each vine. Fifteen to twenty shoots grow on each cane, with two kiwi-fruit on each shoot.

The experiment is to compare two different methods of training the vines: either along a pergola or along a T-bar. These methods are to be investigated in combination with three different chemical sprays which keep pests off the fruit. To apply a spray, a plastic tube is placed around a whole shoot and the spray is squirted into the tube, wetting both fruit.

Six vines will be used for the experiment. Only the twelve highest shoots on each cane will be used. At harvest, the weight of each fruit will be recorded.

(a) Identify observational units, experimental units and treatments, stating how many of each there are.

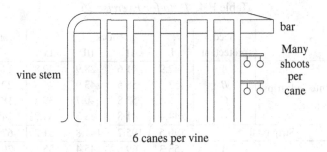

Fig. E.3. Kiwi-fruit vine in Exercise 23

(b) Draw the Hasse diagram for the factors on the observational units (ignoring the treatment factors).

(c) Draw the Hasse diagram for the factors on the treatments.

(d) Describe how to construct the design for the trial.

(e) Describe how to randomize the design.

(f) Write down the skeleton analysis of variance, showing stratum, source and degrees of freedom.

Exercise 24 (a) Name the three most important experimental designs for comparing two drugs (A and B) on animals. For each design, briefly state how to construct and randomize it, and state one advantage and one disadvantage of the design.

(b) A vet has made an appointment to see you to discuss a proposed trial of two new drugs A and B on horses, for a disease which you have never heard of. Make clear notes on TWO of the important points to discuss with her at your meeting. Include reasons for discussing these points.

Exercise 25 There are three treatment factors, A, B and C, each with three levels. It is assumed that the three-factor interaction is zero. There are 27 plots, divided into three blocks of size nine.

(a) Construct a single-replicate design in such a way that all main effects and two-factor interactions are orthogonal to blocks.

(b) Write down the skeleton analysis-of-variance table, showing stratum, source and degrees of freedom.

(c) After the experiment has started, 27 more plots become available, also in three blocks of size nine. Briefly explain how to construct the second replicate in such a way that all main effects and two-factor interactions are orthogonal to blocks in the second replicate also, but that no treatment effect is confounded with blocks in both replicates.

Table E.4. *Data for Exercise 26*

	Seed protectant	Block				Total
		I	II	III	IV	
Infected strips	A	42.9	41.6	28.9	30.8	144.2
	B	53.8	58.5	43.9	46.3	202.5
	C	49.5	53.8	40.7	39.4	183.4
	D	44.4	41.8	28.3	34.7	149.2
Strip total		190.6	195.7	141.8	151.2	679.3
Uninfected strips	A	53.3	69.6	45.4	35.1	203.4
	B	57.6	69.6	42.4	51.9	221.5
	C	59.8	65.8	41.4	45.4	212.4
	D	64.1	57.4	44.1	51.6	217.2
Strip total		234.8	262.4	173.3	184.0	854.5
Block total		425.4	458.1	315.1	335.2	1533.8

Exercise 26 An experiment was conducted to compare the effectiveness of four sorts of seed protectant in protecting oats against a particular pest. The four sorts of protectant were coded A, B, C and D. The farmers wanted to know the effect of these protectants not only on a crop infested with the pest, but also on a pest-free crop.

A field was divided into four large blocks. Each block was divided into two strips. One strip in each block was chosen at random and infected with the pest. Each strip was divided into four plots, to which the four seed protectants were randomly allocated.

The yield of oats (in bushels per acre) on each of the 32 plots is shown in Table E.4. The data are *not* in field order, having been rearranged to facilitate hand calculation.

The totals for each of the four seed protectants are as follows.

$$\begin{array}{cccc} A & B & C & D \\ 347.6 & 424.0 & 395.8 & 366.4 \end{array}$$

The sum of the squares of all the yields is 77342.78.

(a) Calculate the analysis-of-variance table.

(b) Briefly interpret the analysis.

(c) State briefly what else should be given for a complete analysis of these data.

Exercise 27 A sports coach wants to test the effectiveness of various exercise programmes as preparation for running. There are two types of exercise (A and B): he wants each to be tried for 15 minutes before running and for 30 minutes before running. He also wants to compare these with no preparatory exercises at all.

He will conduct his experiment during the five days of one working week, using 30 student volunteers who run regularly. Each day at 3.30 the volunteers will assemble in the gym. Each volunteer will be assigned an exercise programme for the day: those with 30 minutes of exercise start straight away; those with 15 minutes start at 3.45; those with no exercise simply rest. At 4 p.m. all the volunteers will run a standard 5-mile course. The sports coach will record how long each runner takes to complete the course.

(a) Identify plots and treatments, stating how many of each there are.

(b) Draw the Hasse diagram for the factors on the plots (ignoring the treatment factors).

(c) Draw the Hasse diagram for the factors on the treatments.

(d) Describe how to construct the design for the trial.

(e) Describe how to randomize the design.

(f) Write down the skeleton analysis of variance, showing stratum, source and degrees of freedom.

Exercise 28 A small bakery has a row of six ovens. Each oven has three shelves, each of which has room for three trays of loaves of bread.

The head baker wants to compare three different temperature settings for the ovens. He also wants to compare three different methods of kneading the dough.

The head baker asks:

> Should I use the three ovens at one end of the row for comparing the temperatures and the other three for comparing loaves kneaded by different methods, or would it be better to use alternate ovens for the temperature experiment and the remainder for the kneading experiment?

(a) How do you reply?

(b) Identify the plots and any suitable blocks in this experiment.

(c) Construct the design for the whole experiment and randomize it.

(d) Present the plan in a form suitable for the head baker.

Exercise 29 (a) Explain what it means for an incomplete-block design to be *balanced*. State one advantage and one disadvantage of balance.

(b) Show that $\{1,2,3,5\}$ is a difference set modulo 7.

(c) A cheese maker wants to compare seven different recipes for cheddar cheese.

   (i) Seven members of the public have volunteered to taste cheeses for him and to give the tasted cheeses a mark out of ten for strength of cheddar flavour. The cheese maker thinks that each volunteer can taste four cheeses without becoming ineffective, and that the order of tasting does not matter. Construct a suitable design for the cheese maker.

   (ii) Alternatively, the cheese maker could employ professional tasters. He believes that such people can taste all seven types of cheese without becoming ineffective, and that order of tasting still does not matter. However, he can afford to employ only three professional tasters. What design should he now use? On the assumption that the variance per response is the same in each case, calculate which of the two proposed designs is more efficient.

Exercise 30 Consider a disease which is not very serious but does cause its victims to stay at home for a month, unable to work. At the moment there is no known cure: patients are simply advised to stay at home, keep warm and rest until the disease passes. Now a pharmaceutical company has produced a drug which, it claims, will cure the disease in only two weeks. A clinical trial is being designed to test the drug.

(a) In this context, explain

   (i) what a *placebo* is and why one should be used;

   (ii) what *matching* means and why it is desirable;

   (iii) what *double-blind* means and why the trial should be double-blind.

(b) It is decided to conduct the first trial of this drug on 24 patients, giving twelve the drug and twelve a placebo. The 24 patients will all come from the same health centre. The next 24 patients who come to this health centre with the disease will be entered into the trial, except that patients with certain other medical conditions will be excluded, in case the new drug affects them adversely. The doctors in the health centre estimate that it will take about a year to recruit 24 suitable patients into the trial.

The following three methods have been proposed for randomly allocating drug or placebo to the 24 patients. Explain what is wrong with each of them.

   (i) For each patient, toss a fair coin. If it falls heads, give the patient the drug; if it falls tails, give the patient the placebo.

   (ii) For the first patient, randomly choose between drug and placebo. Then allocate drug and placebo alternately to succeeding patients.

   (iii) Write the word *drug* on twelve pieces of paper, write the word *placebo* on twelve other pieces of paper, and thoroughly shuffle the 24 pieces of paper. Copy the words from the pieces of paper onto a single list according to the shuffled order. For example, the list might begin

   1 Placebo
   2 Drug
   3 Drug
      ...

   Give this list to the doctor in charge of the trial and tell him to allocate each suitable patient to the next treatment on the list.

Exercise 31 (a) Construct a half-replicate fractional factorial design for six two-level treatment factors A, B, C, D, E, F in such a way that all main effects and two-factor interactions can be estimated if it is assumed that all interactions among three or more factors are zero.

(b) Write down the skeleton analysis-of-variance table for this design.

(c) Explain how the data analysis can be enhanced to investigate whether any three-factor interactions are nonzero.

Table E.5. *Data for Exercise 32*

Nitrogen	Phosphate	Block I		Block II		Block III		Block IV		Treatment total
N0	P0	186	260	268	239	275	242	259	267	1996
N1	P0	365	281	343	366	368	401	413	409	2946
N2	P0	373	293	498	522	475	542	396	400	3499
N0	P1	284	313	297	250	344	277	244	228	2237
N1	P1	431	336	471	412	359	429	453	389	3280
N2	P1	480	542	536	443	395	464	512	504	3876
	Block total	4144		4645		4571		4474		17834

Exercise 32 In 1927 an experiment was conducted to compare the effectiveness of six sorts of fertilizer in promoting growth of barley. The six sorts were all combinations of

$N0$: no nitrogen
$N1$: single nitrogen   with $P0$: no phosphate
$N2$: double nitrogen $P1$: phosphate.

The experiment was conducted in four blocks, each of which consisted of twelve plots with the same area. Each fertilizer was used on two plots in each block. The yield of barley grain (in units of 2 ounces) on each of the 48 plots is shown in Table E.5. The data are *not* in field order, but have been rearranged to facilitate hand calculation.

The sum of the squares of all the yields is 7074040.

(a) Calculate the analysis-of-variance table, with an appropriate decomposition of the treatment sum of squares.

(b) Briefly interpret the analysis. Then calculate appropriate tables of means for the treatments.

(c) State briefly what else should be given for a complete analysis of these data.

Exercise 33 A cookery-book writer wants to compare 16 recipes for fruit cake. She asks twelve housewives to help her. Four of the housewives have gas ovens, four have electric ovens and four have solid-fuel ovens. Each housewife will be given four of the recipes and asked to bake four fruit cakes, using her own oven and following each recipe exactly.

(a) Explain why a resolved incomplete-block design is needed.

(b) Construct a suitable design.

(c) Explain how the design should be randomized.

(d) Choose two recipes which are used by the same housewife. Calculate the standard error of the estimator of the difference between these two recipes, in terms of the housewives-stratum variance ξ.

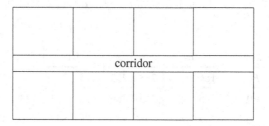

Fig. E.4. Diagram of the glasshouse in Exercise 34

Exercise 34 At an institute for horticultural research, a glasshouse is to be used to study the effects of two factors on the growth of tomatoes. The factors to be studied are fertilizer and ladybirds. There are four different types of fertilizer. There are four different quantities of ladybirds to be released in the neighbourhood of the plants.

The glasshouse is divided by interior glass walls into eight compartments, as shown in Figure E.4, with a central corridor running East–West giving access to all of the compartments. Each compartment contains four gro-bags. Each gro-bag has a rail above it. All tomato plants in a single gro-bag will have their stems wound around the same rail.

(a) Identify plots and treatments, stating how many there are.

(b) Draw the Hasse diagram for the factors on the treatments.

(c) Draw the Hasse diagram for the factors on the plots (ignoring the treatment factors).

(d) Describe how to construct the design for the trial.

(e) Describe how to randomize the design.

(f) Write down the skeleton analysis of variance, showing stratum, source and degrees of freedom.

Exercise 35 A professional grower of soft fruit has written to you to make an appointment to discuss an experiment which she is proposing to conduct next summer. Part of her letter reads:

> One problem with growing strawberries is that they are often attacked by slugs. For the last six years I have been controlling the slugs by spreading homemade slug pellets on the ground around the strawberries. However, a new type of commercial pellet called Slay-the-Slugs has recently come onto the market, and it is claimed to be very effective. I should like to compare it with my homemade pellets.
>
> The piece of land where I intend to grow strawberries next summer is naturally divided into three patches, separated by paths. I propose spreading my homemade pellets on one patch; spreading Slay-the-Slugs pellets on the second patch; and putting no pellets at all on the third patch. I shall then see which patch does best.

While I am talking to you, perhaps you could also advise me on the experiment I plan for the following summer. Some people say that slugs like beer and so can be trapped and drowned in cups of beer which have their tops level with the ground. Once I know which type of pellet to use on the ground, I can then do another experiment to see if using beer-traps improves my strawberries.

Make clear notes on the most important points to discuss with her at your meeting. Explain the purpose of raising each of these points.

Exercise 36 Three seven-level treatment factors F, G, H are to be tested for their effect on an industrial process. It is believed that there are no interactions among these factors.

(a) Construct a design for these three seven-level treatment factors in seven blocks of seven plots each, in such a way that all main effects can be estimated.

(b) Write down the skeleton analysis-of-variance table, showing stratum, source and degrees of freedom.

(c) Explain how the design should be randomized.

Exercise 37 The director of a chain of small supermarkets wants to find out what type of music should be played in her shops to encourage customers to spend money. She wants to compare the following three types of music.

r: rock music
c: classical music
e: 'easy listening' music

She decides to use nine of her supermarkets for her experiment, from the following different parts of London.

Acton   Brixton   Camberwell
Deptford   Eltham   Forest Hill
Greenford   Highgate   Islington

Each of these supermarkets is open from 0800 to midnight Monday to Saturday inclusive. The experiment will last for one week. At the end of each day the manager of each supermarket will add up how much money has been spent in that shop that day, and will report this to the director.

(a) Identify the plots and any suitable blocks in this experiment.

(b) Construct the design and randomize it.

(c) Present the plan in a form suitable for the director.

(d) After the experiment, you will analyse the data with your usual statistical computing package. What data should you use? What factors do you need? After the data and factor values have been entered, what must you type, either as commands or as entries in dialogue boxes, to analyse the data?

					Row total
F2 36.0	F0 21.3	F4 63.9	F1 34.1	F3 41.7	197.0
F3 41.0	F1 21.0	F2 24.5	F4 62.1	F0 11.4	160.0
F1 28.3	F2 26.4	F0 11.8	F3 59.8	F4 69.0	195.3
F4 70.7	F3 45.4	F1 25.1	F0 11.0	F2 29.0	181.2
F0 15.6	F4 59.3	F3 49.5	F2 26.8	F1 31.2	182.4
Column total 191.6	173.4	174.8	193.8	182.3	915.9

Fig. E.5. Layout and data for Exercise 38

Exercise 38 An experiment was conducted to compare the effectiveness of five sorts of fungicide in controlling blight in potatoes. The five sorts were:

$F0$: no fungicide
$F1$: substance A, applied early
$F2$: substance A, applied late
$F3$: substance B, applied early
$F4$: substance B, applied late.

The experiment was conducted in the Latin square shown in Figure E.5. All plots had the same area. The yield of potatoes in pounds per plot is shown in the square in Figure E.5, together with row and column totals.

The treatment totals are as follows.

$F0$	$F1$	$F2$	$F3$	$F4$
71.1	139.7	142.7	237.4	325.0

The sum of the squares of all the yields is 41970.79.

(a) Calculate the analysis-of-variance table, with an appropriate decomposition of the sum of squares for treatments.

(b) Briefly interpret the analysis. Then calculate appropriate tables of means for the treatments.

(c) State briefly what else should be given for a complete analysis of these data.

Exercise 39 An experiment is to be conducted to compare six nutrients for tobacco plants. Each plant will be grown in an individual pot in the laboratory. There are eighteen pots altogether, spread over three benches, with six pots per bench. Each nutrient will be applied to the compost in one pot per bench. After a certain time, the six lowest leaves on each plant will be picked and subjected to a chemical analysis.

(a) Identify the experimental units, the observational units and the treatments, and state how many there are of each.

(b) Draw the Hasse diagram for the plot factors (ignoring the treatment factors).

(c) Construct the design and randomize it, showing your working.

(d) Write down the skeleton analysis of variance, showing stratum, source and degrees of freedom.

Exercise 40 (a) Construct a pair of mutually orthogonal 5×5 Latin squares.

(b) A catering company looks after the managers' dining room in four large firms. It wants to compare 25 types of cheese for popularity among the managers. So that the managers are not confused by too much choice, the catering company will put out five types of cheese at lunch each day, and record how much of each is taken. In the first week, it will do this at the first firm, on each of the five working days. In the second week it will carry out a similar programme at the second firm. The third and fourth firms will be covered in the third and fourth weeks respectively.

   (i) Explain why a resolved incomplete-block design is needed.
   (ii) Construct a suitable design for the catering company.
   (iii) Explain how the design should be randomized.
   (iv) If your design is used, the difference in popularity between a pair of cheeses will be estimated with different variances, depending which two cheeses are compared. State a pair of cheeses for which the variance will be smallest.

Exercise 41 A consumer organization wishes to test all combinations of six new washing machines and four new biological detergents. Six families have volunteered to help with the tests. Only one sample of each new washing machine is available, but there are unlimited quantities of the new detergents.

The trial will last for six months. Each month each sample machine will be installed in a different family's house. During that month the family will try out the new detergents in that washing machine. For each of four weeks during the month, the major family wash will be tested, and scores for cleanliness, whiteness, degree of creasing, etc. recorded.

(a) Identify plots and treatments, stating how many of each there are.

(b) Draw the Hasse diagram for the factors on the plots (ignoring the treatment factors).

(c) Describe how to construct the design for the trial.

(d) Describe how to randomize the design.

(e) Write down the skeleton analysis of variance, showing stratum, source and degrees of freedom.

Exercise 42 (a) Define what it means for two $n \times n$ Latin squares to be *orthogonal* to each other.

(b) Construct a pair of mutually orthogonal 8×8 Latin squares.

(c) Describe an experimental situation where the above pair of mutually orthogonal Latin squares could be used to construct a suitable design. Explain the plot structure and the treatment structure; explain how to use the pair of Latin squares to construct the design; and explain how to randomize the design.

Exercise 43 An agronomist is planning a fertilizer experiment to compare all combinations of the following three factors:

superphosphate	two levels (with or without)
source of nitrogen	four levels (sulphate of ammonia, chloride of ammonia, cyanamide, urea)
quantity of nitrogen	three levels (1, 2, 3)

(a) Draw the Hasse diagram for the factors on the treatments, showing numbers of levels and degrees of freedom.

(b) Now the agronomist tells you that the three levels of the factor quantity of nitrogen are

      1   no nitrogen
      2   the standard amount of nitrogen
      3   double the standard amount.

Redraw the Hasse diagram for the factors on the treatments, showing numbers of levels and degrees of freedom.

Sources of examples, questions and exercises

Almost all of the examples in this book are real. On the other hand, almost none of them is the whole truth. Sometimes I have merely simplified the details, so that I can concentrate on a particular issue. At other times I have adapted the original: for example, by using a subset of the data, omitting levels of one or more factors or omitting a factor completely. If data are given in such cases, I created them from the original data by using the fitted values and a random selection of the original residuals. Thus the data should be fairly true to the original context.

I am fairly certain that my memory of the sources of the experimental stories is accurate. However, when I first started teaching I was by no means as careful as I should have been about recording the sources of data used in classes or assignments. I have therefore had to reverse engineer the derivation of most data sets which I have been using for more than ten years. I apologize if the result is that I have omitted to credit any source that I did use.

I spent 1981–1990 in the Statistics Department at Rothamsted Experimental Station, which was at that time an institute of the Agriculture and Food Research Council. Examples 1.5 (rain at harvest), 1.8 (wheat varieties), 1.17 (oilseed rape), 4.4 (laboratory measurement of samples), 4.5 (field trial), 5.4 (modern cereals) and 10.15 (bean weevils) are from Rothamsted, some coming to me directly, some reported to me by my statistical colleagues. There are more details of Example 10.15 in [106]. Question 9.5 is also from Rothamsted. Example 5.13 (Park Grass) is based partly on the pamphlet *Rothamsted Experimental Station: Guide to the classical field experiments*, published by the Lawes Agricultural Trust, Harpenden, in 1984 (ISBN 0-7084-0302-6), and partly on my own observation while at Rothamsted.

One of my duties at Rothamsted was to provide statistical advice to the Agricultural Development and Advisory Service (ADAS). Examples 1.2 (calf feeding), 1.3 (leafstripe), 1.7 (rye-grass), 1.14 (simple fungicide), 1.15 (fungicide factorial), 1.16 (fungicide factorial plus control), 3.5 (reducing feed for chickens), 4.7 (piglets), 5.9 (counts of bacteria) and 14.4 (factorial rotation experiment), as well as Questions 4.1 and 5.5, are taken from my ADAS work.

The Ministry of Agriculture, Fisheries and Foods consulted me about the experiment in Example 1.1 (ladybirds).

The story in Example 1.4 (kiwi fruit) was told to me by statisticians in the Ruakura Statistics Group at AgResearch in Hamilton, New Zealand, during a visit in December 2004.

Examples 1.9 (asthma) and 10.13 (cross-over with blocks), as well as Exercise 20, are based on my experience working in the Medical Research Council's Air Pollution Research Unit in London in 1965.

Example 1.10 (mental arithmetic) is a synthesis of later concerns about primary education with the experiment reported in [22, pages 93–94], which is itself based on [116, page 263]. As described in those two books, the experiment had false replication.

Example 1.11 (detergents), Question 9.7, Example 10.2 (car tyres) and Exercise 41 are based on my contact with Rosamund Weatherall, statistician at the Consumers' Association in the 1970s.

Example 1.12 (tomatoes) and Exercise 2 are based on my visits to the West of Scotland Agricultural College at Auchincruive, near Ayr, in about 1977 and the Glasshouse Crops Research Institute in Littlehampton in about 1980. Exercise 34 comes from a visit to Rodney Edmondson at Horticulture Research International, Wellesbourne, Warwick, after this had absorbed the work of G. C. R. I. in the 1990s. He also told me about Example 4.9 (mushrooms) in May 2004.

Example 1.13 (pullets) is a reworking of Example 8.6 on pages 92–96 of [67], which is itself using data from [29]. The original experiment appears to suffer from false replication.

Question 1.2 is based on a report in *The Pulse* section of *The Weekend Australian* on 28 August 2004.

Question 1.3 and Example 11.5 (lithium carbonate) are taken from the worked example in [115].

The experiment described in Question 1.4 and Example 8.6 (concrete) was undertaken by engineering students at the University of Queensland in 2004. Heng Mok Kwee provided the information.

The data in Example 2.2 (milk production) are taken from the paper [59].

The idea in Example 2.3 (limited availability) is based on discussions with Paul Darius at the Katholieke Universiteit Leuven in April 1999.

Question 2.2 is based on an experiment carried out by a PhD student in Engineering at Queen Mary, University of London, in the 2000s.

Question 2.5 uses data presented on page 97 of Cochran and Cox [28].

Example 3.2 (fungicide on potatoes) is adapted from [35].

Example 3.3 (rubber trees) uses the information in Example 9.6 of [67, pages 104–109], which in turn is a simplified version of the experiment described in [48, page 121].

Example 3.4 (drugs at different stages of development) is a simplified version of an experiment described by Peter Colman in a talk presented at a meeting of the British Region of the International Biometric Society held at Pfizer, Sandwich, in June 2001.

Questions 3.5 and 11.10 describe an experiment reported in Section 20.1 of [103] and in [104].

Example 4.1 (insect repellent) is taken from [37].

Example 4.2 (irrigated rice) is based on my visit to the Asian Vegetable Research and Development Center in Tainan, Taiwan in June 1990.

The story in Example 4.3 (road signs) was told to me by Brian Ripley in March 2004.

I visited the Escola Superior de Agricultura "Luiz de Quieros", University of São Paulo, in Piracicaba, Brazil, from March to July 2005. During my stay, Paulo Justiniano Ribeiro explained the details in Example 4.6 (citrus orchards) in a seminar talk; ESALQ geneticists told me about Example 8.2 (insemination of cows); I was shown the biotechnology experiment in Question 8.5, which was being carried out by a PhD student at ESALQ; and Clarice Demétrio described the experiment on eucalyptus trees in Example 10.6 (soil fungicide) and Question 10.11.

The advice in Example 4.8 (weed control) is taken from *Field Manual for Weed Control Research* by L. C. Burrill, J. Cardenas and E. Locatelli, published by the International Plant Protection Center, Oregon State University, Corvallis, Oregon, U. S. A. in 1976. Example 5.2 (herbicides) is from the same source. I am grateful to Keotshepile Kashe for showing me this booklet while we were both at the University of Queensland in August–September 2004.

Example 4.11 (wine tasting) comes from Chris Brien during his time at Roseworthy Agricultural College in South Australia.

The data in Example 4.13 (metal cords) are taken from [34].

Example 4.14 (hay fever) is adapted from Section 20.2.5 of [103].

Example 4.15 (pasture grass) was reported to me by David Baird and David Saville of the Biometrics Unit of the Ministry of Agriculture and Fisheries at Lincoln, New Zealand, in January 1987.

Question 4.3 and Example 10.5 (nematodes) are based on an experiment reported on page 176 of the 1935 Rothamsted Experimental Station Annual Report. Ken Ryder tracked down the original field-plan, and Donald Preece gave me a copy. The data also appear on page 46 of [28].

The experiment in Question 4.5 and Example 14.7 (glass jars) is adapted from one performed by Julia Reiss, a post-doctoral researcher in Biology at Queen Mary, University of London, in 2007.

Example 5.3 (cow-peas) comes from pages 117–119 of Pearce, Clarke, Dyke and Kempson [87], who took it from Rayner [94], pages 439–440.

Example 5.5 (enzyme in blood) is based on Example R on pages 135–138 of Cox and Snell [33]. Their data have been adapted for use in Exercise 19. See also [30, Section 7.4].

Example 5.6 (saplings) is based on an experiment described by I. D. Mobbs at Alice Holt Research Station of the Forestry Commission during a visit there in September 1991 by the conference on The Optimal Design of Forest Experiments and Surveys held at the University of Greenwich under the auspices of the International Union of Forest Research Organizations. I thank Hervé Monod for help in recording and recalling the details.

In some years Phil Woodward of Pfizer comes to Queen Mary, University of London, to give a talk to undergraduates in mathematics and statistics about the uses of statistics, especially the design of experiments, in the pharmaceutical industry. Examples 5.7 (tablet manufacture) and 12.8 (pill manufacture) are based on two different examples from his talk in March 2004; Example 13.6 (chromatograph) on one from the talk in March 2002.

Example 5.10 (vetch and oats) is in Section 9c of Frank Yates's wonderful *The Design and Analysis of Factorial Experiments*, published as 'Technical Communication 35' by the Imperial Bureau of Soil Science, Harpenden, in 1937. Statisticians who followed Yates at Rothamsted regularly referred to this publication, affectionately calling it just *TC35*. In this example, Yates does not appear to have noticed the linear component of the interaction. Also taken from *TC35* are Example 12.6 (sugar beet) (Section 10b) and Example 12.9 (field beans) (Section 7). Questions 10.8 and 10.9 are discussed in Sections 15a and 15b of *TC35*.

Example 5.12 (protein in feed for chickens) is adapted from Example K on pages 103–106 of Cox and Snell [33]. They credit Section 3.6 of John and Quenouille [63], who say that the experiment was performed by 'J. Duckworth and K. Carpenter'. It appears to be the second experiment described in [21].

I have been using Question 5.1 since 1989 but cannot remember where I took the data from. They are a subset of the data given on page 430 of Rice [95], who took them from [20].

Question 5.3 is from pages 88–89 of Lane, Galwey and Alvey [72], who took it from [38].

The data in Question 5.4 were reported in a poster presented at the twenty-first International Biometric Conference, which was held in Freiburg in July 2002. See 'Infostat for data analysis' by J. Di Rienzo, C. W. Robledo, M. Balzarini, F. Casanoves, I. Gonzales, M. Tablada and W. Guzmán, in *The XXIst International Biometric Conference, Proceedings—Abstracts of Special and Contributed Paper Presentations*, ISSN 1606-8653, page 173.

Example 6.3 (straw) is taken from page 175 of the 1935 Annual Report from Rothamsted Experimental Station.

Question 6.4 is adapted from Example 4.7 of [26, page 56].

I have not been involved personally in clinical trials, so I am indebted to Deborah Ashby, Anthony Atkinson, Sheila Bird, Michael Healy, John Matthews, Stuart Pocock and Stephen Senn for conversations helping me to clarify the material in Chapter 7.

Example 7.1 (regenerating bone) is based on an experiment performed by a PhD student in Biomaterials at Queen Mary, University of London, in 2006.

Example 7.4 (the Lanarkshire milk experiment) was reported by W. S. Gosset, writing as "Student", in [111].

Example 7.6 (AIDS tablets) was reported in the *New Scientist* on 6 July 2002, page 13, citing [79, 99, 88].

Example 7.7 (incomplete factorial) is based on the papers [56, 57].

Example 7.8 (the British doctors' study) begins with the paper by Doll and Hill in [42] and concludes with the paper by Doll, Peto and others in [43].

Example 7.9 (frogs) is taken from the *New Scientist* of 4 May 2002, citing [58].

Examples 7.10 (educating general practitioners) and 7.11 (maternal dietary supplements in Gambia) are both taken from Sandra Eldridge's PhD thesis [47].

Questions 7.4 and 7.5 are taken from a talk by A. K. Smilde on 'Megavariate analysis-of-variance' at the First Channel Network Conference of the International Biometric Society, which was held at Rolduc in the Netherlands in May 2007.

The quotations in Question 7.6 are taken from the *New Scientist* of 13 January 2007. The quotations refer to the papers [36], [16] and [73] respectively.

Michael Healy, retired from the London School of Hygiene and Tropical Medicine, told me about the randomization in Question 7.7, in July 2002.

Example 8.3 (cider apples) is taken from page 21 of the 1941 Annual Report for Long Ashton Research Station.

Example 8.5 (animal breeding) is based on conversations with Robin Thompson when he was at the Animal Breeding Research Organisation in the 1980s.

Example 8.7 (insecticides on grasshoppers) and Questions 8.6 and 10.6 come from the papers [74], [55] and [70] in a single issue of *Environmental Toxicology and Chemistry* in 1996. I am grateful to Tim Sparks, of the Institute of Terrestrial Ecology, and David Elston, of

Biomathematics and Statistics Scotland, for drawing my attention to the large amount of false replication in experiments reported in the environmental science literature around that time.

The email message in Question 8.2 is genuine.

The stories in Question 8.3 and Exercise 15 were passed on to me by Violeta Sonvico and other statisticians in the Departamento de Estadistica in the Instituto Nacional de Tecnologia Agropecuria in Buenos Aires during my visit there in April 1992.

Question 8.4 is from page 135 of Pearce, Clarke, Dyke and Kempson [87], who in turn took it from page 370 of Snedecor and Cochran [107].

Example 9.4 (oceanography) is adapted from information which Wojtek Kranowski gave me in early 2005. He had helped to design the trial reported in [118].

Question 9.6 is based on my visit to East Malling Research Station in the late 1980s, and also on many conversations with Donald Preece.

Example 10.8 (rats) comes from page 393 of the 1987 Genstat manual [85], and before that from page 305 of Snedecor and Cochran [108].

Example 10.12 (soap pads) is a simplified version of Example N on pages 116–120 of Cox and Snell [33]. Question 10.14 is a development of this.

Examples 10.16 (rugby) and 10.17 (carbon dating) are adapted from student projects which I read as an external examiner in the Statistics Department at the University of Glasgow in 1998–2002. So are Exercises 5, 6, 9 and 21.

Example 10.19 (unwrinkled washing) is adapted from the paper [80].

Example 10.20 (molybdenum) is taken from a poster presented to the twentieth International Biometric Conference, which was held in Berkeley in July 2000. See 'Analysis of experiments with leguminous with non traditional designs' by Ivani P. Otsuk, M. B. Gláucia, Schammass Ambrosano, A. Eliana, Edmilson J. Ambrosano and José Eduardo Corrente, in *The XXth International Biometric Conference, Volume I, Proceedings*, ISSN-1606-8653, pages 115–116. Versions of Figure 10.35 and Table 10.18 are given on page 198 of [10].

Question 10.5 is loosely based on an experiment described to me in 1980 by Tony Grassia of the Commonwealth Scientific and Industrial Research Organisation in Perth, Western Australia. The original version is discussed in [15].

The large-scale experiment in Question 10.10 was described by Peter Rothery, Suzanne Clark and Joe Perry in a talk at the twenty-first International Biometric Conference, which was held in Freiburg in July 2002. The corresponding paper is 'Design and analysis of farmscale evaluations of genetically modified herbicide-tolerant crops' by Peter Rothery, Suzanne J. Clark and Joe N. Perry, pages 351–364 of *The XXIst International Biometric Conference, Proceedings—Manuscripts of Invited Paper Presentations*, ISSN 1606-8653. This work was described in more detail in [25, 51].

Question 10.12 is based on 'The *Horizon* homeopathic dilution experiment' by Martin Bland, published in *Significance*, Volume 2, Issue 3, September 2005, pages 106–109. I am grateful to Martin Bland for further explanation of the details.

The problem in Question 10.16 is based on one posed to me by Dr. S. Raman of the Gujurat Agricultural University in 1996.

The designs for microarray experiments in Example 11.9 are given by Kerr in [69].

The cross-over design in Example 11.10 (phase I cross-over trial with a placebo) was described by Steven Julious of Glaxo-Smith-Kline at a meeting of PSI (Statisticians in the Pharmaceutical Industry) at Chester in May 2001.

Question 11.8 is based on discussions with statisticians at Horticulture Research International, Wellesbourne, Warwick, in the early 2000s.

Question 11.9 is adapted from an example described by Jan Engel in a talk given to the 6th Workshop on Quality Improvement Methods at the Universiätskolleg Bommerholz of Universität Dortmund in May 2007. Further details of this type of experiment are in [93].

Example 12.3 (watering chicory) is a simplified version of an experiment carried out within the Institut National de la Recherche Agronomique in France, which was reported to me by André Kobilinsky in the early 1990s.

The statistician involved in the experiment in Example 13.5 (glass insulators) was Fang Kai-Tai. He told this story at the 2006 International Conference on Design of Experiments and its Applications at Tianjin, China in July 2006.

Example 14.8 (silicon wafers) is taken from [78].

Example 14.9 (Jerseys and Holsteins) is based on a short report 'The effect of monensin on milk production of Jersey and Holstein cows' on page 21 of the 1999 issue of *World Jersey Research News*. The report cites the paper [114].

Exercise 3 is based on an experiment conducted by Pat Healey in the Department of Computer Science at Queen Mary, University of London, in 2003.

Exercise 8 is adapted from Section 4.1 of Mead [76].

Exercise 23 is based on a talk contributed by H. N. De Silva at the XVIIIth International Biometric Conference, which was held in Amsterdam in July 1996. See 'Analysis of within-vine variation for fruit size with applications for sampling' by H. N. Da Silva, R. D. Ball and J. H. Maindonald, in *The XVIIIth International Biometric Conference, Contributed Papers*, page 37.

Exercise 26 is taken from Exercise 7B of Pearce, Clarke, Dyke and Kempson [87], who in turn took the data from page 384 of Steel and Torrie [109].

Exercise 29 is loosely based on conversations with Tony Hunter, who is now retired from Biomathematics and Statistics Scotland, where for many years he was involved in the design and analysis of cheese-tasting experiments.

Exercise 32 is a simplified form of the experiment given in Section 52 of [54]. Exercise 43 is based on the same source.

Exercise 38 seems to be adapted from Exercise 6B of Pearce, Clarke, Dyke and Kempson [87], which is in turn taken from page 466 of [94].

Some of my stories of bad practice come from informants who have asked to remain anonymous so that the actual experiment cannot be identified. Examples 1.6 (eucalypts), 7.3 (educational psychology), 7.5 (doctor knows best) and 14.5 (acacia) are of this type. I fear that none of these is an isolated instance. For example, I have recently come across an experiment on reproduction in flour beetles which used the method of randomization in Example 7.3.

Further reading

I have found nothing to equal [27] or [30] for a clear introduction to the important general considerations in designing experiments. The books [26], [32] and [76] are in a similar vein but include some more advanced topics. The classic [54] is still worth reading.

The practical aspects of designing experiments in the agricultural and forestry contexts are well explained by [86] and [119] respectively. Experiments in the chemical and engineering industries are covered by [18]; those in the pharmaceutical industry by [103]. For more information about experiments in medical research, see [1], [60], [75] and [89].

The book [119] contains a very good chapter on recording data, while [96] is excellent on the interface between the statistician and the scientist. For field experiments, [45] has good advice on everything from marking out the plots to how many decimal places should be retained in recorded data.

For more details about the method of orthogonal projection introduced in Chapter 2, see [24], [100] or [101].

The classic book of designs is [28], which tabulates many designs and comments on their properties but gives no information on how to construct new designs yourself. The best books on constructions are [92] and [110], but they do assume familiarity with finite groups and finite fields.

Chapter 7 deliberately stops short of cross-over trials for the case when carry-over effects can be expected. In [102], Senn argues that cross-over designs should not be used in such circumstances. Cox and Reid discuss the question carefully in [32, Section 4.3]. For those who disagree with Senn, some suitable designs are given in [50, Section 7.4], [66], [98, Section 11.4] and [105, Chapter 6].

The methods of sequential randomization discussed briefly in Section 7.7 are still being debated in the literature. See [2], [12], [32, Section 8.2], [46], [90], [97] and [112].

For an alternative introduction to the use of Hasse diagrams to show the relationships between factors, see [113], but be aware that several of the conventions used there are different from those used in Chapter 10. Orthogonal plot structures (Section 10.10) are described in more detail in [10, Chapter 6], where they are called *orthogonal block structures*. The special subclass called *simple orthogonal block structures* was introduced by Nelder in [82]. Between these two is the class of *poset block structures*, described in [10, Chapter 9].

It is these poset block structures for which the randomization in Section 10.11 gives the covariance matrix in Section 10.10. The proof of this is the 'general argument' mentioned at the end of Section 14.1.2. It is very technical, and can be found in [13] and [8].

For designs with the sort of neighbour balance described in Example 10.15, see [3] and

[110, Chapter 14]. Example 10.21 mentions a type of design called a *semi-Latin square*: [9] gives more information about such designs, including a better non-orthogonal design for this example.

To find out more about incomplete-block designs than is covered in Chapter 11, including the proof of Fisher's Inequality, read one of [10], [64], [65], [92] and [110]. The first two of these also cover the type of row–column design discussed in Section 11.10. Bose's Inequality is proved in [65]. A full treatment of optimality, for both incomplete-block designs and row–column designs, is given in [105]. Theorem 11.7 was proved by Kshirsagar in [71], Theorem 11.8 by Cheng and Bailey in [23]. Hall's Marriage Theorem (Section 11.10) and the corresponding algorithm are given in Chapter 6 of [19].

The approach to factorial design in Chapters 12–13, using Abelian groups, is essentially the one given for block designs by Fisher in [52], [53] and Chapter 7 of [54], and extended to fractions by Finney [49] and Kempthorne [68]. This approach is given at greater length in [40, Chapters 13–15] and [120, Chapters 3–6]. Although similar in spirit, [18, Chapters 10–13] is restricted to treatment factors with two levels each: it gives more information about resolution. A more discriminatory property of fractions, called *minimum aberration*, is discussed in [120, Chapter 4].

The type of fraction constructed in Chapter 13 is now called a *regular* fraction. Non-regular fractions are given by [41], [65, Chapters 8–9] and [120, Chapter 7].

In [17], Bose recast the theory of confounding in factorial designs in terms of affine geometry. This is used in [92, Chapter 13]. However, this limits the theory to treatment factors whose levels are all powers of the same prime number. The approach using Abelian groups is extended to other numbers of levels in [4, 6, 11]. It is applied to designs with more complicated plot structures, such as row–column designs and split-plot designs, in [4, 6, 84].

The Ordering Principle (Principle 12.1) is called the *Marginality Principle* by Nelder [83] and the *Effect Heredity Principle* by Wu and Hamada [120, page 111].

The book [39] considers factorial designs for a large number of treatment factors where the Sum Principle (Principle 5.2) is broken.

The approach to randomization summarized in Section 14.1.2, and given in more detail in [5] and [8], is that randomization justifies the assumed model, which in turn dictates the way in which the data should be analysed. By Section 14.1.7, the point of view has shifted: we know how we intend to analyse the data and we want the randomization to be in some sense compatible with this, while avoiding certain layouts. This is explained further in [14] and [81]. Restricted randomization depends very much on the positions of the plots in space and time, so typically has to be worked out afresh for each new situation. Some solutions for a line of plots are given in [12, 44, 117, 121]; some for a rectangular array in [7, 61].

For other views on the problem of factors like age and sex, discussed in Section 14.2, see [31] and [91].

A longer version of Section 14.4 was published as [5].

References

[1] ALTMAN, D. G.: *Practical Statistics for Medical Research*, Chapman and Hall/CRC, Boca Raton, (1991).

[2] ATKINSON, A. C.: The comparison of designs for sequential clinical trials with covariate information, *Journal of the Royal Statistical Society, Series A*, **165**, (2002), pp. 349–373.

[3] AZAÏS, J.-M., BAILEY, R. A. & MONOD, H.: A catalogue of efficient neighbour-designs with border plots, *Biometrics*, **49**, (1993), pp. 1252–1261.

[4] BAILEY, R. A.: Patterns of confounding in factorial designs, *Biometrika*, **64**, (1977), pp. 597–603.

[5] BAILEY, R. A.: A unified approach to design of experiments, *Journal of the Royal Statistical Society, Series A*, **144**, (1981), pp. 214–223.

[6] BAILEY, R. A.: Factorial design and Abelian groups, *Linear Algebra and its Applications*, **70**, (1985), pp. 349–368.

[7] BAILEY, R. A.: One-way blocks in two-way layouts, *Biometrika*, **74**, (1987), pp. 27–32.

[8] BAILEY, R. A.: Strata for randomized experiments (with discussion), *Journal of the Royal Statistical Society, Series B*, **53**, (1991), pp. 27–78.

[9] BAILEY, R. A.: Efficient semi-Latin squares, *Statistica Sinica*, **2**, (1992), pp. 413–437.

[10] BAILEY, R. A.: *Association Schemes: Designed Experiments, Algebra and Combinatorics*, Cambridge Studies in Advanced Mathematics, 84, Cambridge University Press, Cambridge, (2004).

[11] BAILEY, R. A., GILCHRIST, F. H. L. & PATTERSON, H. D.: Identification of effects and confounding patterns in factorial designs, *Biometrika*, **64**, (1977), pp. 347–354.

[12] BAILEY, R. A. & NELSON, P. R.: Hadamard randomization: a valid restriction of random permuted blocks, *Biometrical Journal*, **45**, (2003), pp. 554–560.

[13] BAILEY, R. A., PRAEGER, C. E., ROWLEY, C. A. & SPEED, T. P.: Generalized wreath products of permutation groups, *Proceedings of the London Mathematical Society*, **47**, (1983), pp. 69–82.

[14] BAILEY, R. A. & ROWLEY, C. A.: Valid randomization, *Proceedings of the Royal Society, Series A*, **410**, (1987), pp. 105–124.

[15] BAILEY, R. A. & SPEED, T. P.: Rectangular lattice designs: efficiency factors and analysis, *Annals of Statistics*, **14**, (1986), pp. 874–895.

[16] BINSWANGER, I. A., STERN, M. F., DEYO, R. A., HEAGERTY, P. J., CHEADLE, A., ELMORE, J. G. & KOEPSELL, T. D.: Release from prison — A high risk of death for former inmates, *The New England Journal of Medicine*, **356**, (2007), pp. 157–165.

[17] BOSE, R. C.: Mathematical theory of the symmetrical factorial design, *Sankhyā*, **8**, (1947), pp. 107–166.

[18] BOX, G. E. P., HUNTER, W. G. & HUNTER, J. S.: *Statistics for Experimenters*, Wiley, New York, (1978).

[19] CAMERON, P. J.: *Combinatorics: Topics, Techniques, Algorithms*, Cambridge University Press, Cambridge, (1994).

[20] CAMPBELL, J. A. & PELLETIER, O.: Determination of niacin (niacinamide) in cereal products, *Journal of the Association of Official Analytical Chemists*, **45**, (1962), pp. 449–453.

[21] CARPENTER, K. J. & DUCKWORTH, J.: Economies in the use of animal by-products in poultry rations. I. Vitamin and amino-acid provision for starting and growing chicks, *Journal of Agricultural Science*, **41**, (1941), pp. 297–308.

[22] CHATFIELD, C.: *Problem Solving: A Statistician's Guide*, Chapman and Hall, London, (1988).

[23] CHENG, C.-S. & BAILEY, R. A.: Optimality of some two-associate-class partially balanced incomplete-block designs, *Annals of Statistics*, **19**, (1991), pp. 1667–1671.

[24] CHRISTENSEN, R.: *Plane Answers to Complex Questions: The Theory of Linear Models*, Springer-Verlag, New York, (1987).

[25] CLARK, S. J., ROTHERY, P. & PERRY, J. N.: Farm Scale Evaluations of spring-sown genetically modified herbicide-tolerant crops: a statistical assessment, *Proceedings of the Royal Society, Series B*, **273**, (2006), pp. 237–243.

[26] CLARKE, G. M. & KEMPSON, R. E.: *Introduction to the Design and Analysis of Experiments*, Arnold, London, (1997).

[27] COBB, G. W.: *Design and Analysis of Experiments*, Springer-Verlag, New York, (1998).

[28] COCHRAN, W. G. & COX, G. M.: *Experimental Designs*, 2nd edition, Wiley, New York, (1957).

[29] COMMON, R. H.: Observations on the mineral metabolism of pullets. III., *Journal of Agricultural Science*, **26**, (1938), pp. 85–100.

[30] COX, D. R.: *Planning of Experiments*, Wiley, New York, (1958).

[31] COX, D. R.: Interaction, *International Statistical Review*, **52**, (1984), pp. 1–31.

[32] COX, D. R. & REID, N.: *The Theory of the Design of Experiments*, Chapman and Hall/CRC, Boca Raton, (2000).

[33] COX, D. R. & SNELL, E. J.: *Applied Statistics: Principles and Examples*, Chapman and Hall, London, (1981).

[34] CROWDER, M. & KIMBER, A.: A score test for the multivariate Burr and other Weibull mixture distributions, *Scandinavian Journal of Statistics*, **24**, (1997), pp. 419–432.

[35] DAGNELIE, P.: La planification des expériences et l'analyse de la variance: une introduction, In: *Plans d'Expériences: Applications à l'Enterprise* (eds. J.-J. Droesbeke, J. Fine & G. Saporta), Éditions Technip, Paris, (1997), pp. 13–67.

[36] DAR-NIMROD, I. & HEINE, S. J.: Exposure to scientific theories affects women's math performance, *Science*, **314**, (2006), p. 435.

[37] DAVID, O. & KEMPTON, R. A.: Designs for interference, *Biometrics*, **52**, (1996), pp. 597–606.

[38] DAVIES, O. L.: *Statistical Methods in Research and Production*, Oliver and Boyd, London, (1947).

References

[39] DEAN, A. & LEWIS, S. (EDITORS): *Screening: Methods for Experimentation in Industry, Drug Discovery and Genetics*, Springer Science, New York, (2006).

[40] DEAN, A. & VOSS, D.: *Design and Analysis of Experiments*, Springer-Verlag, New York, (1999).

[41] DEY, A. & MUKERJEE, R.: *Fractional Factorial Plans*, Wiley, New York, (1999).

[42] DOLL, R. & HILL, A. B.: The mortality of doctors in relation to their smoking habits. A preliminary report, *British Medical Journal*, **228**, (1954), pp. 1451–1455.

[43] DOLL, R., PETO, R., BOREHAM, J. & SUTHERLAND, I.: Mortality in relation to smoking: 50 years' observations on male British doctors, *British Medical Journal*, **328**, (2004), pp. 1–10.

[44] DYKE, G. V.: Restricted randomization for blocks of sixteen plots, *Journal of Agricultural Science*, **62**, (1964), pp. 215–217.

[45] DYKE, G. V.: *Comparative Experiments with Field Crops*, 2nd edition, Griffin, London, (1988).

[46] EFRON, B.: Forcing a sequential experiment to be balanced, *Biometrika*, **58**, (1971), pp. 403–417.

[47] ELDRIDGE, S.: *Assessing, understanding and improving the efficiency of cluster randomised trials in primary care*, Ph. D. thesis, University of London, 2005.

[48] FEDERER, W. T.: *Experimental Design: Theory and Application*, Macmillan, New York, (1955).

[49] FINNEY, D. J.: The fractional replication of factorial experiments, *Annals of Eugenics*, **12**, (1945), pp. 291–301.

[50] FINNEY, D. J.: *An Introduction to the Theory of Experimental Design*, University of Chicago Press, Chicago, (1960).

[51] FIRBANK, L. G., HEARD, M. S., WOIWOD, I. P., HAWES, C., HAUGHTON, A. J., CHAMPION, G. T., SCOTT, R. J., HILL, M. O., DEWAR, A. M., SQUIRE, G. R., MAY, M. J., BROOKS, D. R., BOHAN, D. A., DANIELS, R. E., OSBORNE, J. L., ROY, D. B., BLACK, H. I. J., ROTHERY, P. & PERRY, J. N: An introduction to the Farm-Scale Evaluations of genetically modified herbicide-tolerant crops, *Journal of Applied Ecology*, **40**, (2003), pp. 2–16.

[52] FISHER, R. A.: The theory of confounding in factorial experiments in relation to the theory of groups, *Annals of Eugenics*, **11**, (1942), pp. 341–353.

[53] FISHER, R. A.: A system of confounding for factors with more than two alternatives, giving completely orthogonal cubes and higher powers, *Annals of Eugenics*, **12**, (1945), pp. 282–290.

[54] FISHER, R. A.: *Design of Experiments*, 8th edition, Oliver and Boyd, Edinburgh, (1966).

[55] GARDNER, S. C. & GRUE, C. E.: Effects of Rodeo® and Garlon® 3A on nontarget wetland species in central Washington, *Environmental Toxicology and Chemistry*, **15**, (1996), pp. 441–451.

[56] GERAMI, A. & LEWIS, S. M.: Comparing dual with single treatments in block designs, *Biometrika*, **79**, (1992), pp. 603–610.

[57] GERAMI, A. & LEWIS, S. M.: Completely randomized designs for comparing dual with single treatments, *Journal of the Royal Statistical Society, Series B*, **56**, (1994), pp. 161–165.

[58] GILBERTSON, M.-K., HAFFNER, G. D., DROUILLARD, K. G., ALBERT, A. & DIXON, B.: Immunosuppression in the Northern leopard frog (*Rana pipiens*) induced by pesticide exposure, *Environmental Toxicology and Chemistry*, **22**, (2003), pp. 101–110.

[59] GOAD, C. L. & JOHNSON, D. E.: Crossover experiments: A comparison of ANOVA tests and alternative analyses, *Journal of Agricultural, Biological and Environmental Statistics*, **5**, (2000), pp. 69–87.

[60] GORE, S. M. & ALTMAN, D. G: *Statistics in Practice*, The British Medical Association, London, (1982).

[61] GRUNDY, P. M. & HEALY, M. J. R.: Restricted randomization and quasi-Latin squares, *Journal of the Royal Statistical Society, Series B*, **12**, (1950), pp. 286–291.

[62] HU, J., COOMBES, K. R., MORRIS, J. S. & BAGGERLY, K. A.: The importance of experimental design in proteomic mass spectrometry experiments: some cautionary tales, *Briefings in Functional Genomics and Proteomics*, **3**, (2005), pp. 322–331.

[63] JOHN, J. A. & QUENOUILLE, M. H.: *Experiments: Design and Analysis*, 2nd edition, Griffin, London, (1977).

[64] JOHN, J. A. & WILLIAMS, E. R.: *Cyclic and Computer-Generated Designs*, Chapman and Hall, London, (1995).

[65] JOHN, P. W. M.: *Statistical Design and Analysis of Experiments*, MacMillan, New York, (1971).

[66] JONES, B. & KENWARD, M. G.: *Design and Analysis of Cross-over Trials*, Chapman and Hall, London, (1989).

[67] JOSHI, D. D.: *Linear Estimation and Design of Experiments*, Wiley Eastern, New Delhi, (1987).

[68] KEMPTHORNE, O.: A simple approach to confounding and fractional replication in factorial experiments, *Biometrika*, **34**, (1947), pp. 255–272.

[69] KERR, M. K.: Design considerations for efficient and effective microarray studies, *Biometrics*, **59**, (2003), pp. 822–828.

[70] KRUGH, B. W. & MILES, D.: Monitoring the effects of five "nonherbicidal" pesticide chemicals on terrestrial plants using chlorophyll fluorescence, *Environmental Toxicology and Chemistry*, **15**, (1996), pp. 495–500.

[71] KSHIRSAGAR, A. M.: A note on incomplete block designs, *Annals of Mathematical Statistics*, **29**, (1958), pp. 907–910.

[72] LANE, P., GALWEY, N. & ALVEY, N.: *Genstat 5: An Introduction*, Clarendon Press, Oxford, (1987).

[73] LORENZ, M., JOCHMANN, N., VON KROSIGK, A., MARTUS, P., BAUMANN, G., STANGL, K. & STANGL, V.: Addition of milk prevents vascular protective effects of tea, *European Heart Journal*, **28**, (2007), pp. 219–223.

[74] MARTIN, P. A., JOHNSON, D. L. & FORSYTH, D. J.: Effects of grasshopper-control insecticides on survival and brain acetylcholinresterase of pheasant (*Phasianus Colchicus*) chicks, *Environmental Toxicology and Chemistry*, **15**, (1996), pp. 518–524.

[75] MATTHEWS, J. N. S.: *An Introduction to Randomized Controlled Clinical Trials*, Arnold, London, (2000).

[76] MEAD, R.: *The Design of Experiments: Statistical Principles for Practical Application*, Cambridge University Press, Cambridge, (1988).

[77] MEAD, R.: The non-orthogonal design of experiments, *Journal of the Royal Statistical Society, Series A*, **153**, (1990), pp. 151–201.

[78] MEE, R. W. & BATES, R. L.: Split-lot designs: experiments for multi-stage batch processes, *Technometrics*, **40**, (1998), pp. 127–140.

References

[79] METADILOGKUL, O., JIRATHITIKAL, V. & BOURINBAIAR, A. S.: Survival of end-stage AIDS patients receiving V-1 Immunitor, *HIV Clinical Trials*, **3**, (2002), pp. 258–259.

[80] MILLER, A.: Strip-plot configurations of fractional factorials, *Technometrics*, **39**, (1997), pp. 153–161.

[81] MONOD, H., AZAÏS, J.-M. & BAILEY, R. A.: Valid randomisation for the first difference analysis, *Australian Journal of Statistics*, **33**, (1996), pp. 91–106.

[82] NELDER, J. A.: The analysis of randomized experiments with orthogonal block structure. I. Block structure and the null analysis of variance, *Proceedings of the Royal Society of London, Series A*, **283**, (1965), pp. 147–162.

[83] NELDER, J. A.: A reformulation of linear models, *Journal of the Royal Statistical Society, Series A*, **140**, (1977), pp. 48–77.

[84] PATTERSON, H. D. & BAILEY, R. A.: Design keys for factorial experiments, *Applied Statistics*, **27**, (1978), pp. 335–343.

[85] PAYNE, R. W., LANE, P. W., AINSLEY, A. E., BICKNELL, K. E., DIGBY, P. G. N., HARDING, S. A., LEECH, P. K., SIMPSON, H. R., TODD, A. D., VERRIER, P. J., WHITE, R. P., GOWER, J. C., TUNNICLIFFE WILSON, G. AND PATERSON, L. J.: *Genstat 5 Reference Manual*, Clarendon Press, Oxford, (1987).

[86] PEARCE, S. C.: *The Agricultural Field Experiment: A Statistical Examination of Theory and Practice*, Wiley, Chichester, (1983).

[87] PEARCE, S. C., CLARKE, G. M., DYKE, G. V. & KEMPSON, R. E.: *A Manual of Crop Experimentation*, Griffin, London, (1988).

[88] PHANUPHAK, P., TAN-UD, P., PANITCHPAKDI, P., TIEN-UDOM, N., NAGAPIEW, S. & CAWTHORNE, P.: V-1 Immunitor, *HIV Clinical Trials*, **3**, (2002), pp. 260–261.

[89] POCOCK, S. J.: *Clinical Trials: A Practical Approach*, Wiley, Chichester, (1983).

[90] POCOCK, S. J. & SIMON, R.: Sequential treatment assignment with balancing for prognostic factors in the controlled clinical trial, *Biometrics*, **31**, (1975), pp. 103–115.

[91] PREECE, D. A.: Types of factor in experiments, *Journal of Statistical Planning and Inference*, **95**, (2001), pp. 269–282.

[92] RAGHAVARAO, D.: *Constructions and Combinatorial Problems in Design of Experiments*, John Wiley and Sons, New York, (1971).

[93] RAJAE-JOORDENS, R. & ENGEL, J.: Paired comparisons in visual perception studies using small sample sizes, *Displays*, **26**, (2005), pp. 1–7.

[94] RAYNER, A. A.: *A First Course in Biometry for Agriculture Students*, University of Natal Press, Pietermaritzburg, (1969).

[95] RICE, J. A.: *Mathematical Statistics and Data Analysis*, Wadsworth, Pacific Grove, California, (1988).

[96] ROBINSON, G. K.: *Practical Strategies for Experimenting*, Wiley, New York, (2000).

[97] ROSENBERGER, W. F. & LACHIN, J. L.: *Randomization in Clinical Trials: Theory and Practice*, Wiley, New York, (2002).

[98] RYAN, T. P.: *Modern Experimental Design*, Wiley, Hoboken, (2007).

[99] SABIN, C.: Editorial comments, *HIV Clinical Trials*, **3**, (2002), pp. 259–260.

[100] SAVILLE, D. J. & WOOD, G. R.: *Statistical Methods: The Geometric Approach*, Springer-Verlag, New York, (1991).

[101] SAVILLE, D. J. & WOOD, G. R.: *Statistical Methods: A Geometric Primer*, Springer-Verlag, New York, (1996).

[102] SENN, S. J.: *Cross-over Trials in Clinical Research*, Wiley, Chichester, (1993).

[103] SENN, S. J.: *Statistical Issues in Drug Development*, Wiley, Chichester, (1997).

[104] SENN, S. J., LILLIENTHAL, J., PATALANO, F. & TILL, D.: An incomplete blocks cross-over in asthma: a case study in collaboration, In: *Cross-over Clinical Trials* (eds. J. Vollmar & L. Hothorn), Fischer, Stuttgart, (1997), pp. 3–26.

[105] SHAH, K. R. & SINHA, B. K.: *Theory of Optimal Designs*, Springer-Verlag, New York, (1989).

[106] SMART, L. E., BLIGHT, M. M., PICKETT, J. A. & PYE, B. J.: Development of field strategies incorporating semiochemicals for the control of the pea and bean weevil, *Sitona lineatus* L., *Crop Protection*, **13**, (1994), pp. 127–135.

[107] SNEDECOR, G. W. & COCHRAN, W. G.: *Statistical Methods*, 6th edition, Iowa State University Press, Ames, Iowa, (1967).

[108] SNEDECOR, G. W. & COCHRAN, W. G.: *Statistical Methods*, 7th edition, Iowa State University Press, Ames, Iowa, (1980).

[109] STEEL, R. G. D. & TORRIE, J. H.: *Principles and Procedures of Statistics: A Biometrical Approach*, 2nd edition, McGraw-Hill, New York, (1978).

[110] STREET, A. P. & STREET, D. J.: *Combinatorics of Experimental Design*, Oxford University Press, Oxford, (1987).

[111] "STUDENT": The Lanarkshire milk experiment, *Biometrika*, **23**, (1931), pp. 398–406.

[112] TAVES, D. R.: Minimization: a new method of assigning patients to treatment and control groups, *Clinical Pharmacology and Therapeutics*, **15**, (1974), pp. 443–453.

[113] TJUR, T.: Analysis of variance models in orthogonal designs, *International Statistical Review*, **52**, (1984), pp. 33–81.

[114] VAN DER WERF, J. H. J., JONKER, L. J. & OLDENBROEK, J. K.: Effect of monensin on milk production by Holstein and Jersey cows, *Journal of Dairy Science*, **81**, (1998), pp. 427–433.

[115] WESTLAKE, W. J.: The use of balanced incomplete block designs in comparative bioavailability trials, *Biometrics*, **30**, (1974), pp. 319–327.

[116] WETHERILL, G. B.: *Elementary Statistical Methods*, 3rd edition, Chapman and Hall, London, (1982).

[117] WHITE, L. V. & WELCH, W. J.: A method for constructing valid restricted randomization schemes using the theory of D-optimal design of experiments, *Journal of the Royal Statistical Society, Series B*, **43**, (1981), pp. 167–172.

[118] WIDDICOMBE, S. & AUSTEN, M. C.: The interaction between physical disturbance and organic enrichment: An important element in structuring benthic communities, *Limnology and Oceanography*, **46**, (2001), pp. 1720–1733.

[119] WILLIAMS, E. R., MATHESON, A. C. & HARWOOD, C. E.: *Experimental Design and Analysis for Tree Improvement*, 2nd edition, CSIRO Publishing, Collingwood, (2002).

[120] WU, C. F. J. & HAMADA, M.: *Experiments: Planning, Analysis, Parameter Design and Optimization*, Wiley, New York, (2000).

[121] YOUDEN, W. J.: Randomization and experimentation, *Technometrics*, **14**, (1972), pp. 13–22.

Index

abelian group, 246, 249
acacia, 275, 318
accidental bias, 20
additive model, 77, 82, 87, 90
age, 279
AIDS, 124, 316
alias set, 260
aliasing, 170, 243, 259, 284
analysis, 226, 230, 253, 266, 285
analysis of variance, 33, 40, 46, 64, 67, 87, 90, 112, 113, 145, 200
    full, 134, 149, 201
    null, 133, 140, 149
    skeleton, 134, 138, 140, 149, 200, 201, 285
animal breeding, 146
antagonism, 78
apples, 16, 168
archaeology, 206
asthma, 9, 52, 118, 313

bacteria, 82, 313
balanced incomplete-block design, 219, 223, 226, 233, 257
baseline measurement, 117
bean weevils, 203, 313
best linear unbiased estimator, 24, 25, 28, 39, 62, 143, 226
bias, 68, 122, 124
    accidental, 20
    assessment, 122
    selection, 20, 124
    systematic, 20
biotechnology, 155, 315
blind, 122
block, 6, 13, 53, 119, 121
    contrast, 62
    design, 159, 166
        incomplete, 219–238, 241–258
        orthogonal, 57, 59, 99
    different sizes, 272
    factor, 57
    large, 146
    principal, 246
    small, 146
    subspace, 57
blocking, 6, 53–71, 105, 146, 283, 284

blood, 80, 315
Bose's Inequality, 221
bottom stratum, 200
breed, 279

calculations ignoring treatments, 133, 140, 145, 149, 200
calf feeding, 2, 9, 131, 137, 138, 144, 313
car tyres, 168, 170, 314
carbon dating, 206, 317
carry-over effect, 119
case-control study, 125
catalyst, 81
cattle breeder, 154
cereals, 79, 313
chain, 176, 177, 181, 197
character, 242
chemicals, 168
chicken feeding, 50, 88, 91, 313, 316
chicory, 241, 242, 245, 257, 318
chromatograph, 267, 315
cider apples, 145, 147, 160, 161, 188, 258, 316
citrus orchard, 56, 314
class, 93, 169
clinical trial, 237
coarser, 171
cohort study, 125
column, 105
    contrast, 111
    factor, 111
combinatorial design, 8
common cold, 127, 157
comparative experiments, 40
complete-block design, 58
completely randomized design, 19, 120
concrete, 18, 147, 170, 314
concurrence, 219
confounded, 241
    with blocks, 230
confounding, 241–258
consultation, 1
control treatment, 7, 11, 12, 43–47, 52, 72, 99, 124, 234, 237, 282
correction for the mean, 31
coset, 246, 260
cost, 6

covariance, 24, 38, 59, 60, 111, 112, 194, 284
cow-peas, 78, 97, 315
criss-cross design, 208
cross-over trial, 118, 194, 197, 237, 313, 317
crossover interaction, 78
crude sum of squares, 93, 178
    for treatments, 27

data
    collection, 2, 283
    copying, 2
    scrutiny, 3
defining contrast, 259, 260
degrees of freedom, 26, 133, 134, 184, 185, 200
design, 13, 283, 285
detergents, 9, 59, 224, 234, 257, 314
difference set, 220
difference table, 220
direct sum
    internal, 23
disordinal interaction, 80
distribution
    exponential, 254
    F, 34, 40, 144, 146, 268
    normal, 29, 34, 35, 40, 253
    t, 29, 40, 144
    χ^2, 34, 40, 253
double-blind, 122
drugs, 48, 86, 118, 119, 122, 185, 191

educating general practitioners, 126
eelworms, 72, 315
effect, 84, 93, 184, 187
    fixed, 60–68
    random, 60, 67–69
efficiency, 229–234
efficiency factor, 229–234, 242, 257
eigenspace, 113
eigenvalue, 39, 67, 113
eigenvector, 39, 67
enzyme, 80, 315
equality factor, 171
equivalent factors, 170
estimation, 24, 84, 86, 113
estimator, 5, 202, 253
ethical issues, 124
eucalypts, 4, 8, 182, 315, 318
excipient, 17, 249
expectation, 45, 48, 59, 60, 111, 112
expectation model, 50, 77, 84–87, 95, 190, 284
expected mean square, 27, 39, 64, 90, 133, 135, 201
experimental unit, 6–8, 10, 12, 117, 131, 275, 283
exponential distribution, 254

F-distribution, 34, 40, 144, 146, 268
F-probability, 64

factor, 169–214
factorial
    design, 11, 241–270
    experiment, 97
    treatments, 11, 75–99, 124, 248, 316
false replication, 135, 170, 284
field beans, 250, 315
field trial, 55, 56, 313
finer, 171
Fisher's Inequality, 221
fit, 178
fitted value, 23
fixed effects, 15, 60–68, 110, 132, 226, 253, 279
fractional replicate, 161, 166, 167, 259–270
frogs, 125
full analysis of variance, 134, 149, 201
fungicide, 11, 47, 182, 186, 313, 315

genomics, 236
glass jars, 276, 315
Graeco-Latin square, 162–167
    finite field, 163
    prime order, 163
    product method, 165
grafting skin, 154
grasshoppers, 152, 316
grazing, 275

half replicate, 260
Hall's Marriage Theorem, 235
Hasse diagram, 175–177, 185–189, 195, 196, 198–202, 243, 283
hay fever, 68, 315
herbicides, 78, 315
historical controls, 118, 125
hypothesis test, 5, 33–35, 44, 70, 84, 113, 135, 140, 141, 202, 253, 266

incomplete-block design, 219–238, 241–258
    balanced, 219, 223, 226, 233, 257
    complement, 221
    cyclic, 220, 236
    from Latin squares, 221
    lattice, 221
    optimal, 233
    resolved, 221, 223, 230
    unreduced, 220
infimum, 171
informed consent, 126
insect repellent, 53, 121, 314
insulators, 266, 318
intention to treat, 126
interaction, 77–84, 88, 90, 93, 95, 97, 122, 140, 141, 144, 146, 152, 159, 161, 166, 167, 202, 214, 237, 242, 244, 252, 259, 265, 279
    crossover, 78

Index

decomposing, 242
disordinal, 80
three-factor, 95
threshold, 80
trigger, 81
two-factor, 95
Intersection Principle, 84
irrigation, 54, 55, 211, 314

kiwi fruit, 4, 313

laboratory measurement, 55, 313
ladybirds, 2, 170
Latin square, 106, 157–167, 222, 242
    cyclic, 106
    group construction, 106
    product construction, 107
lattice design, 221–223, 230, 234, 238
layout, 13
leafstripe, 3, 313
linear model, 14, 23, 24, 59–61
lithium carbonate, 17, 226, 314
litters, 56
local control, 6, 13
lucerne, 155

main plot, 146
main effect, 84, 88, 90, 140, 141, 144, 159, 242
main-effects-only design, 159, 161, 166, 167, 192, 214, 241, 259
management, 55, 56, 68, 114, 221, 275
marine engineering, 41
matched pairs design, 119
maternal dietary supplements, 126
matrix, 26, 195
maximal model, 85, 86
mean, 5
mean square, 26, 44, 64, 86
    for residual, 28
    for treatments, 32
mental arithmetic, 9, 202, 314
metal cords, 64, 68, 69, 315
microarrays, 236, 317
milk, 25, 28, 29, 32, 34, 123, 280, 316, 318
minimization, 122
model, 45, 48, 132, 148, 253, 285
molybdenum, 211, 317
mouthwash, 126
multiplicative model, 82
mushrooms, 57, 314

nematodes, 175, 181, 192
normal distribution, 29, 34, 35, 40, 253
null analysis of variance, 15, 133, 140, 149
null model, 30, 88

oats, 83, 315

observational study, 125
observational unit, 8, 10, 12, 117, 131, 276, 283
oceanography, 166, 317
oilseed rape, 11, 13, 44, 313
optimal incomplete-block design, 233
Ordering Principle, 252
orthogonal
    basis, 23
    complement, 23
    decomposition, 182
    design, 197–214
    factors, 178
    Latin squares, 162, 222
    plot structure, 193
    projection, 23
    to blocks, 230
    treatment structure, 189
    vectors, 23
Orthogonality Principle, 85
overall mean, 30

Park Grass, 98, 208
pasture grass, 70, 315
piglets, 56, 313
pill manufacture, 249, 315
placebo, 122, 237, 282, 317
plan, 13, 284
planned analysis, 4, 284
plot, 10
    factor, 170, 193, 196, 197, 276
    structure, 5, 12, 193, 198, 272, 279, 283
potatoes, 42, 47, 52, 168, 314
power, 5, 6, 15, 35–38, 69, 71, 120, 121, 126, 144, 152, 230, 284
principal block, 246
prospective study, 125
protocol, 10, 12, 282–285
pseudo-replication, 135
pseudofactor, 197, 251, 262
psychology, 41, 122, 318
pullets, 11, 314
purpose of the experiment, 5, 282

quantile, 254
quantile plot, 254, 269
quantitative factor, 12, 79, 92, 266
quarter replicate, 161, 264

rain, 4, 313
random effects, 15, 60, 67–69, 112, 132, 148, 193, 228, 253, 279
random sample, 5, 271
randomization, 8, 13, 15, 19–21, 38, 41, 59, 98, 108, 118, 120, 122, 123, 132, 138, 147, 159, 161, 196, 223, 271–279, 284, 285
    sequential, 121

randomized controlled trial, 118
rats, 185, 192, 317
repeated measurements, 147
repetition, 131
replication, 5, 19, 30, 35–38, 41, 43–44, 52, 70, 85, 87, 97–99, 121, 126, 131, 181, 197, 284
residual, 27, 35, 85, 134, 140, 226
    degrees of freedom, 202, 285
    mean square, 28, 45, 84, 86
    sum of squares, 27, 201, 202
residual effect, 119
resolution, 265
resolved incomplete-block design, 221, 223, 230
retrospective study, 125
rice, 54
road signs, 54, 286, 314
rotation experiment, 274, 313
row, 105
    contrast, 111
    factor, 111
row–column design, 105–114, 118, 157, 158, 235–238, 272
rubber, 48, 314
rugby, 205, 317
rye-grass, 7, 9, 75, 97, 146, 170, 173, 174, 191, 200, 241, 313

saplings, 80, 315
scalar product, 22
selection bias, 20
semi-Latin square, 211
sex, 279
significance level, 35, 70, 121, 144
silicon wafers, 276, 318
single replicate, 159, 166, 241, 250, 253
skeleton analysis of variance, 15, 134, 138, 140, 149, 200, 201, 285
smoking, 125, 316
soap pads, 193, 317
source, 33
split-plot design, 146–153
squared length, 22
standard error, 29, 202
    of a difference, 29, 86, 90, 95, 140, 141, 143, 230
    of a mean, 29
strata, 67
stratum, 40, 120, 132, 133, 138, 148, 149, 196, 198, 241, 285
stratum variance, 196, 226
straw, 114, 316
strictly orthogonal factors, 241
strip-plot design, 209
subgroup, 246, 249, 259
subplot, 146
subspace, 178, 182
sugar beet, 247, 253, 254, 257, 315

sum of squares, 22, 23, 26, 39, 52, 86, 94, 133, 134, 145, 184, 187, 200, 253
    crude, 27, 178
    for the mean, 31
    for treatments, 32, 49, 51, 201
    total, 28
Sum Principle, 84
superimposed design, 166
supplemented balance, 234
supremum, 172
synergism, 78
systematic bias, 20

t-distribution, 29, 35, 40, 70, 144
t-test, 35
table of means, 86, 90, 95, 178
tablet manufacture, 81, 249, 315
three-factor interaction, 95
threshold interaction, 80
time, 147, 279
time-course experiments, 147
tomatoes, 9, 314
total sum of squares, 28
tractor, 56
treatment, 7, 8, 11, 282
    contrast, 22, 62, 111
    effect, 32
    factor, 21, 75, 170, 189, 197, 242
    structure, 5, 12, 43–51, 189, 198, 279, 283
    subspace, 21, 149, 198, 202, 241, 285
    vector, 22
trees, 53
trigger interaction, 81
two-factor interaction, 95

uniform factor, 175
universal factor, 171
unstructured
    plots, 12, 19
    treatments, 12

variance, 5, 6, 15, 24, 28, 30, 37, 39, 40, 43, 62, 69, 121, 137, 141, 194, 229–234, 253
variance ratio, 34, 40, 64, 68, 135, 141, 202
vetch, 83, 315

wash-out period, 119
washing, 208, 317
weeds, 56, 315
weight, 262
wheat, 8, 313
whole plot, 146
wine, 16
wine tasting, 59, 105, 315

χ^2-distribution, 34, 40, 253

Youden square, 236

Printed in the United States
By Bookmasters